W0091349

**SAGE** was founded in 1965 by Sara Miller McCune to support the dissemination of usable knowledge by publishing innovative and high-quality research and teaching content. Today, we publish more than 750 journals, including those of more than 300 learned societies, more than 800 new books per year, and a growing range of library products including archives, data, case studies, reports, conference highlights, and video. SAGE remains majority-owned by our founder, and after Sara's lifetime will become owned by a charitable trust that secures our continued independence.

Los Angeles | London | Washington DC | New Delhi | Singapore | Boston

# Multi-level Forest Governance in Asia

Thank you for choosing a SAGE product!
If you have any comment, observation or feedback,
I would like to personally hear from you.
*Please write to me at* **contactceo@sagepub.in**

**Vivek Mehra,** Managing Director and CEO,
SAGE Publications India Pvt Ltd, New Delhi

## Bulk Sales

SAGE India offers special discounts
for purchase of books in bulk.
We also make available special imprints
and excerpts from our books on demand.

*For orders and enquiries, write to us at*

Marketing Department
SAGE Publications India Pvt Ltd
B1/I-1, Mohan Cooperative Industrial Area
Mathura Road, Post Bag 7
New Delhi 110044, India

*E-mail us at* **marketing@sagepub.in**

## Get to know more about SAGE

Be invited to SAGE events, get on our mailing list.
*Write today to* **marketing@sagepub.in**

This book is also available as an e-book.

# Multi-level Forest Governance in Asia

Concepts, Challenges and the Way Forward

Edited by
Makoto Inoue
Ganesh P. Shivakoti

THE UNIVERSITY OF TOKYO

ASNET Network for Education and Research on Asia

**SAGE**   www.sagepublications.com

Los Angeles • London • New Delhi • Singapore • Washington DC • Boston

*Copyright © Makoto Inoue and Ganesh P. Shivakoti, 2015*

All rights reserved. No part of this book may be reproduced or utilised in any form or by any means, electronic or mechanical, including photocopying, recording, or by any information storage or retrieval system, without permission in writing from the publisher.

*First published in 2015 by*

 **SAGE Publications India Pvt Ltd**
B1/I-1 Mohan Cooperative Industrial Area
Mathura Road, New Delhi 110 044, India
*www.sagepub.in*

**SAGE Publications Inc**
2455 Teller Road
Thousand Oaks, California 91320, USA

**SAGE Publications Ltd**
1 Oliver's Yard, 55 City Road
London EC1Y 1SP, United Kingdom

**SAGE Publications Asia-Pacific Pte Ltd**
3 Church Street
#10-04 Samsung Hub
Singapore 049483

Published by Vivek Mehra for SAGE Publications India Pvt. Ltd, typeset in 10/12pt Times New Roman by Diligent Typesetter, Delhi and printed at Chaman Enterprises, New Delhi

**Library of Congress Cataloging-in-Publication Data**

Multi-level forest governance in Asia : concepts, challenges, and the way forward / edited by Makoto Inoue and Ganesh P. Shivakoti.
    pages cm
Includes bibliographical references and index.
    1. Forest policy—Asia. 2. Forest management—Asia. 3. Decentralization in management—Asia. I. Inoue, M. (Makoto), 1960– II. Shivakoti, Ganesh.
    SD561.M85        333.75095—dc23        2015        2015013705

**ISBN:** 978-93-515-0259-3 (HB)

**The SAGE Team:** Shambhu Sahu, Alekha Chandra Jena, Nand Kumar Jha and Ritu Chopra

# Contents

# List of Tables

# List of Figures

# Foreword

The Network for Education and Research on Asia was established in 2010 as an organisation of the University of Tokyo with the aim of enabling researchers whose research is related to Asia and Japan to collaborate and share information and explore new possibilities in education and research. It is known as ASNET, an abbreviation of its old name 'Asian Studies Network'. Some of its activities began in 1999. This network is designed to function as a virtual network for linking researchers across their departmental and disciplinary affiliations and responding flexibly to advances and new needs in research and social circumstances. Whether humanities or natural sciences, 'Asian Studies' includes not only research related to Asia excluding Japan—the conventional concept of Asian Studies in Japan—but also research related to Japan as an integral part of Asia. By providing a forum for researchers in various departments to participate and collaborate, this network aims to support the studies of individual researchers and cultivate new collaborations and exchanges in research and education. Furthermore, we hope that a variety of people from Japan, Asia and the rest of the world will participate and enjoy this network.

To promote the international and disciplinary exchange of faculty members, we have been hosting interdisciplinary and international symposiums on the broad theme of Asia. This book is a result of such a workshop organised by Professor Inoue, who is one of the core members of ASNET. He has been contributing greatly to the activities of ASNET. The workshop was held in October 2012 at the University of Tokyo, which was financially supported by ASNET and Laboratory of Global Forest Environmental Studies (GFES) of the University of Tokyo, Japan. Most of the authors of this book participated in this workshop. The topic of this workshop and, therefore, of this book, namely degradation of forest resources, has been a serious concern in Asia for decades. We hope that the ideas expressed in this book will contribute to govern and manage forest resources in rural areas of the Asian region more effectively and sustainably.

**Yukio Ikemoto**
Vice Director, Network for Education and Research
on Asia (ASNET)
The University of Tokyo

# SECTION I

# Introduction

# 1

# Multi-level Forest Governance in Asia: An Introduction

*Makoto Inoue, Ganesh P. Shivakoti and Hemant R. Ojha*

## Background

Throughout Asia, forest degradation continues to be a primary environmental concern. Increasing cases of calamities associated with climate change have drawn increasing attention to the issue of deforestation. A significant portion of forest conversion is undertaken through rural livelihood activities, where resource degradation is often overexploited by users who make resource-use decisions under insecure tenure regimes and distorted market signals. Governments, NGOs and academics have been searching for appropriate policy options to reverse the trend of rural resource degradation. While the management of forest at the community level has at times helped to curb degradation, it is still unclear how such efforts are integrated with policy and multi-level institutional processes in forest governance. There is a clear need to think beyond communities and beyond government (Berkes, 2008; Ojha, 2014) to explore how multiple stakeholders can engage in the process of decision making at multiple levels of governance. This view highlights the need for promoting an effective policy and building the capacity of key stakeholders with the presumption that sustainable rural development can be promoted by integrating both the top-down and bottom-up perspectives. This view also emphasises the need for capacity building in the field of natural resource management and poverty alleviation. This book explores how such a multi-scalar process of forest governance can be unfolded in Asia and built on the prior works of the editors, who have suggested several policy alternatives (Inoue & Isozaki, 2003; Webb & Shivakoti, 2008).

The importance of multi-level thinking in policy guidance for the sustainable governance and management of common pool resources (CPRs) in general and forestry resources in particular is clearly demonstrated on conflicting and competing demands for and uses of these resources in the changing economic context in Asia (Balooni & Inoue, 2007; Nath, Inoue, & Chakma, 2005; Pulhin, Inoue, & Enters, 2007; Shivakoti & Ostrom, 2008; Viswanathan & Shivakoti, 2008). Forestry resources are unique in a sense that the management of these resources is a collective effort of the state and local community, while the benefits of these resources accrue at the individual and private levels in the form of forest products. In the larger virtual environmental context, however, the benefits and costs of forest resource management have global implications (such as through carbon sequestration for combating climate change). There are several modes of governance and management that can be arranged on a private–public continuum, linking local-level practices of forest management with global discourses on forest conservation and development. In view of the increasingly multi-level reality of forest governance, several issues need to be addressed in order to improve governance and management, including the ongoing policy efforts of decentralisation and poverty reduction in Asia.

The complexity of forest governance is also demonstrated in the relationships between forest and other sectors of environment and development. While a significant number of studies have been carried out to develop forestry policy options either from the point of view of national development or from the community perspective at the local level, hardly any studies have addressed the inter-relationships of forest with other resources such as water and land as mediated by prevailing institutional arrangements. As a result of the lack of such integrated analysis, the full potential of forest to contribute to development and environmental conservation remains unlocked. In our previous research, we identified several such anomalies involving the disconnect between forest and other sectors, and then tried to explore better management regimes for the forestry and related resources of several Asian countries (Dorji, Webb, & Shivakoti, 2006; Dung & Webb, 2008; Gautam, Shivakoti, & Webb, 2004; Kitjewachakul, Shivakoti, & Webb, 2004; Mahdi, Shivakoti, & Schmidt-Vogt, 2009; Shivakoti, Varughese, Ostrom, Shukla, & Thapa, 1997; Yonariza & Shivakoti, 2008). However, several issues remain in relation to the policy and management guidelines for integrated as well as multi-level management of forest resources. Specifically, the following issues are pertinent:

- How can the sustainability of efforts to improve the production capacity of forestry CPR systems be assessed in the context of the current debate on the effects of climate change and the implementation of new programmes such as REDD+?
- How do changing economic contexts affect partnerships between the private sector and CPR management groups and influence the dynamics of use and conditions of forestry CPR systems in particular and natural resource systems in general?
- How can multiple methods of information-gathering and analysis be used, meaning both the triangulation of methods to obtain accurate information and the combination of socio-economic methods with the biological sciences, including the combination of micro–macro analytic methods such as remotely sensed data over time verified by ground-truthing and additional GPS sample point verification? How can the application of multiple methods to community forestry CPRs be integrated into national forestry policy guidelines, and how can those results be used by local managers and users of CPRs, government agencies and scholars?
- What are the effective polycentric policy approaches for the governance and management of environmentally sustainable and gender-balanced forest CPRs?

## Multi-level Governance of Forest: Rationale and Theory

For centuries, societal attempts to tackle environmental management challenge have relied on human's capacity to self-organise at communities. Scholars claim that community-based solutions rest on a number of expected socio-environmental benefits—ensuring effective environmental stewardship, and forging collective action to address social and environmental issues together (Ostrom, 1990). The community optimism has gone further in the wake of recent economic crisis resulting from the markets, which has triggered more community-oriented and participatory approach to governance. However, it has also become clear that local community practice is not isolated from the rest of the world. Multi-level dynamics are an essential part of community, private and public activities in resource management. As 'glocalisation' is taking place rapidly with both localisation and globalisations occurring simultaneously (Robertson, 1995), local-level resource management activities are becoming part of the multi-level world. In such a situation, it is hard for local communities to remain limited to a local level (Berkes, 2007). Indeed, as some argue, to ignore non-local forces means to strategically play the politics of scale from outside (Blaikie, 2006). While the urge to explore cross-scale dynamics is a welcome innovation, there is still a limited exploration—both conceptual and empirical—of how such politics unfolds in specific social, ecological and political circumstances.

The starting point to explore multi-level resource governance is to recognise the diversity of stakeholders at each level of the society, across both the public and private sectors as well as other arenas of practice. Understanding stakes and stakeholders is an important step towards developing policy options to enhance policy and strengthen stakeholder capacity. These stakeholders and the potential challenges they face are listed in Table 1.1.

As Table 1.1 summarises, a variety of stakes are involved in forest governance, and these stakes are pursued by an increasing number of stakeholders. The next question for us is how do these stakeholders interact across scales and how do we conceptualise these processes. Berkes (2002) uses the term 'cross-scale interactions' to refer to linking institutions both horizontally (across space) and vertically (across organisational levels), and has showed the importance of the effects of higher-level institutions on local-level practices. In the study conducted by Ojha (2014) on Nepal's community-based forest resource management regimes, he found that politics happening at the national scale is driving local-level institutions and practices of forest governance in Nepal. Current theoretical research indicates that these levels are highly interactive and overlapping; therefore, it is justifiable to undertake coordinated activities that lead to better information-sharing and capacity building at the national, district and local levels (Cash et al., 2006). The impact of earlier intervention efforts (various policies in general and decentralisation in particular) on effective outcomes has been limited due to the unwillingness of higher-level forest administrative officials to give up their authority, the lack of trust and confidence of forest officials in the ability of local communities to manage the forest, the capture of the benefits of decentralisation by local elites, and the more frequent occurrence of conflicts among multiple stakeholders at the local

**Table 1.1:**
*Multiple levels of stakeholders*

| Social level | Stakeholders | Potential challenges |
|---|---|---|
| National | National government politicians; Minister of the Environment; Directors of national-level environmental departments | What are the state-of-the-art development models and how do they relate to each particular country; how are national-level policies enacted at the local level; how does land tenure affect resource security; how have changes in the property right structures in the current economic context altered the land use, and what is the role of each sector in shaping new forms of CPR regimes? |
| Province/ district | Provincial/district-level politicians; Conservators of Forests, Heads of regional environmental divisions; Provincial/ district Forest Officers; Educational institutions; New Private Plantation Frontiers, REDD+ and other initiatives | How are national/provincial/district-level programmes implemented at the local level; how are communities using the forest, and what are the main factors leading to the behavioural patterns seen in that region; what are the main options for forest conservation and management in that region; what changes have occurred in the roles of private stakeholders in mediating intermediate-level policies and programmes? |
| Local | Rural communities; rural households; local governments; community user groups, private non-timber forest product collectors, and timber and other forest product contractors and dealers | What are the specific problems facing each community in terms of sustainable use of the forest resource; is it appropriate for the community to act collectively to manage and protect the forest; is assistance available to learn the foundations of a successful community-based organisational structure; how public–private partnerships developing in the context of the changing economy and local climate change initiatives can be established? |

*Source:* Authors.

level (IGES, 2007; Ojha, Cameron, & Kumar, 2009; Ribot, Agrawal, & Larson, 2006).

Most of the discussion on cross-scale interactions, cross-scale linkages and interrelations among multiple levels of stakeholders are based on the notion that vertical levels of organisation are stable as shown in Figure 1.1. In accordance with this notion, Mwangi and Wardell (2012) stated that multi-level governance (MLG) of forest resources involves complex interactions among the state, the private sector and civil society actors at various levels and among institutions linking higher levels of social and political organisation. Poteete (2012) also provides a comprehensive discussion of these issues, especially the difference between the terminologies of 'multi-level institutions' in policy-oriented research and 'multi-scale linkages' in ecological and socio-ecological analysis.

Here, however, we try to think about the issue of cross-scale interactions and MLG from a different point of view, removing the vertical barrier by granting unfettered access to trustworthy outsiders as shown in Figure 1.2. This figure shows that foreign outsiders (outsider X), domestic outsiders who live in other provinces/districts (outsider Y) and

**Figure 1.1:**
*Cross-scale interactions (horizontal and vertical)*

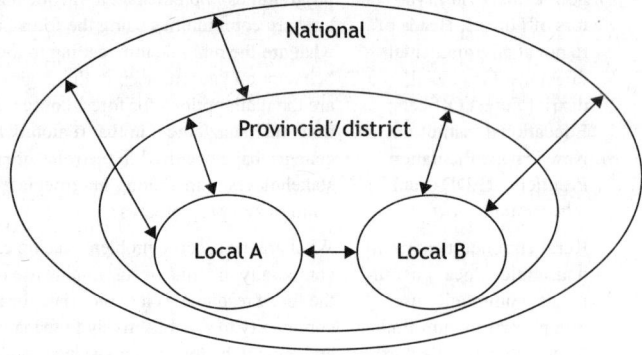

*Source:* Authors.

**Figure 1.2:**
*Unfettered outsiders to demolish vertical barrier*

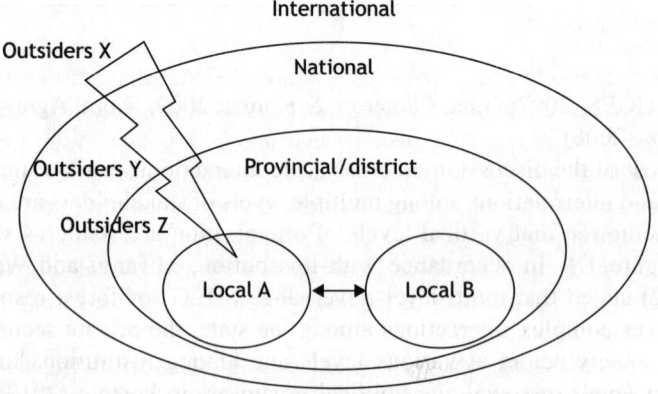

*Source:* Authors.

local outsiders who live in other villages (outsider Z) can be involved in the activities of forest use and management in a local village A. This involvement and the activities of these individuals are expected to lead to breakthroughs in the local problems.

As the collaborative governance literature proliferates, several policy options have also been identified, with regard to how forestry stakeholders at different scales can work together towards consensus-based decision making and action (Colfer, 2005; Ojha, Hall, & Rasheed Sulaiman, 2013). The first author previously suggested that collaborative forest governance should be organised through collaborations among various stakeholders who have a range of interest in local forest use and management through a consensual 'principle of involvement' (Inoue, 2003). This principle recognises the rights of stakeholders to speak and make decisions in a capacity that corresponds to their degree of involvement in and commitment to forest use and management. Under this principle, the local people who often visit and take care of the forest might be expected to have more power over the decision-making process than 'selfish outsiders' who might express commitment but without actual contributions. This principle also stipulates that conscientious outsiders who devote their time or money to local forest management might be given more power corresponding to their level of contribution. If appropriate institutional framework is in place to facilitate negotiations under this principle, it is likely that various stakeholders agree on a fair and workable system of resource governance. The form of governance that is derived through this principle could also enhance legitimacy and accountability of the decisions made, while minimising the conflicts that usually characterise local–outsider relationships in forest and natural resource governance.

However, studies show that collaborative governance principles are extremely difficult to actually work in practice (Ojha & Hall, 2012; Wollenberg et al., 2007). Inoue (2011) showed that collaboration among concerned stakeholders, including outsiders, is indispensable to sustainable forest use and management in Japan and in tropical and sub-tropical Asian countries in the era of globalisation, in which economic and social activities extend across national borders. But the processes of collaboration are not straightforward, and, to a large extent, are determined by the strategies and attitudes of local communities towards the role of external stakeholders. The attitudes of the local people can take the form of three different strategies.

The first is a 'resistance strategy' or localisation strategy, in which people do not want to adapt to globalisation and mostly refuse involvement by outsiders in order to preserve their autonomy. Under this strategy, local communities emphasise the reconstruction of local systems as a locally defined and autonomous process involving reciprocity among the members within the communities. The use and management of the local resources and environment is seen as being embedded into the livelihood strategies of local people. The village community positions itself as the focal actor of local forest governance, and the community is given exclusive membership to the group governing the forest. This strategy does not fit with either 'liberalism' or 'social democracy', but might be promoted under 'conservative' politics.

The second is an 'adjustment strategy' or globalisation strategy, in which local individuals are eager to assimilate the benefits of globalisation. This strategy intends to design open systems characterised by public access to local resource governance. This strategy might value the local resources and environment as components of broader social welfare, separate from the context of the livelihoods of local people. The expected focal actors in local forest governance are associations such as NGOs and NPOs (non-profit organisations) that are formed in civil society, and the viewpoints of these organisations inherently conflict with those of local people. This strategy does not fit within either 'liberalism' or 'conservatism', but might be promoted under 'social democratic' politics.

The third is an 'eclectic strategy' or glocalisation strategy, which represents a compromise between the first two strategies. In this approach, closure and openness are held in balance as are inherent values and universal values, and a partial resistance strategy and a limited adjustment strategy are adopted. This strategy can help achieve 'collaborative governance' (*kyouchi* in Japanese) of natural resources. This type of governance is organised through collaboration among various stakeholders who have a range of interests in local forest use and management (Inoue, 2004).

In the field, however, neither easy coordination nor happy consensus among all stakeholders can be accomplished. Even though most may hope for equal participation by all stakeholders, government policies ultimately may not reflect the voices of the people residing in forest regions, who are usually minorities with low levels of political power (Borrini-Feyerabend, 1996). Typical examples can be seen in the establishment and management of national parks and other protected areas in the tropics.

The sphere of collaborative governance is not necessarily the administrative area and scale. This sphere may be formed within a local community, beyond communities and local governments, or even beyond national boundaries.

# Translating Multi-level Governance Principles into Actionable Framework: Prototype Design Guidelines

In this book, we have adopted a common framework to understand the possibility of collaborative governance, with the analytical focus on understanding how and when communities and outsiders manage to overcome conflicts and foster cooperation in resource governance dilemmas. To tackle the barriers to the 'eclectic strategy' or the glocalisation strategy, we have used what Inoue (2009) has proposed as the prototype design guidelines for the collaborative governance of forests. The prototype guidelines comprise: (1) a degree of local autonomy, (2) clearly defined resource boundaries, (3) graduated membership, (4) the commitment principle, (5) fair benefit distribution, (6) a two-storied monitoring system, (7) two-storied sanctions, (8) a nested conflict management mechanism and (9) building trust.

Those guidelines, or *kyouchi* principles, were derived from and evolved out of the design principles for CPRs (McKean, 1999; Ostrom, 1990, 2005; Stern, Dietz, Ostrom, & Stonich, 2002), which pointed out the importance of linkages with outside organisations and nested enterprises but did not further develop those themes. Among these guidelines, we will focus on the two vital design guidelines of 'graduated membership' and 'commitment principle', as these guidelines have the potential to make an original contribution to enriching the conditions for 'group characteristics' and 'institutional arrangements,' respectively, described by Agrawal (2002). We will also examine the 'fair benefit distribution' guideline, which states that while the benefit distribution may not necessarily be equal, it should be fairly distributed in accordance with the distribution of costs.

## Graduated Membership of the Executive Management Body

Collaborative governance, in which local people and outsiders successfully build a consensus, cannot be established if local people remain guided only by their cultural traditions without interacting with outsiders. Thus, 'open-minded localism' is required, in which local people consent to open their resources and environments to outsiders. This principle agrees well with the principle of subsidiarity, in which the larger-scale political and administrative unit supplements the smaller-scale unit or basic autonomous unit.

In 'open-minded localism', some of the local people act as core members (first-class members) and are assigned the highest authority, and these core members cooperate with other graduated members (second- and third-class members) with less authority. Clear and graduated membership boundaries imply the exclusion of non-members. As such, executive bodies should deal with the exclusion issue to ensure fairness and to acquire legitimacy from relevant stakeholders.

In line with the notion that participation by all local people is neither possible nor favourable (Edmonds & Wollenberg, 2001), we propose something like a forest management committee, in which representatives of the local people form a core of (first-class) members; the local government administration, NGOs, and academics/scientists make commitments as second-class members; and others support activities as third-class members. The second- and third-class members should be afforded definite legitimacy by core members.

## The Commitment Principle for Decision Making

To avoid the deterioration of local autonomy, it is essential for all stakeholders to consent to the 'principle of involvement' (Inoue, 2003) as described earlier. While the 'principle of involvement' is the concept that embraces the authority of a variety of individuals to provide input to the forum, the 'commitment principle' (Inoue, 2009) refers to the clear assignment of the authority to make decisions. Here, we define the 'commitment principle' as a principle of decision making in which the authority of stakeholders is recognised to an extent that corresponds to their degrees of commitment to relevant activities.

Under this principle, local people who often enter and care for the forest might be expected to have greater power over the decision-making process; outsiders who say a lot without doing much might be provided less power; and the conscientious outsiders who devote their time or money to local forest management might be given more power. In this way, various stakeholders are able to agree on the legitimacy of the opinions of outsiders as well as those of local people.

Decision making is not done on an equal basis or with one-person, one-vote ballots, but should be regarded as fair, equitable and just by the stakeholders. Whether the decision is perceived as fair depends on whether the decision-making process is considered legitimate. It is vital for all members involved in the process to reach a consensus on what degree of legitimacy they should grant to statements made by different individuals. Once this consensus is reached, the scale of the arena or the numbers of members involved in decision making should be limited appropriately, because all members should recognise the relative degrees of commitment to one another. A small-scale arena is an ideal trial setting for the commitment principle. Even though stakeholders are spread out over broad geographical areas, a small-scale arena can be organised under the guidelines of 'graduated membership'. When the organisation of a larger arena cannot be avoided, indicators must be established to evaluate the various degrees of commitment of different individuals, such as the contributions of labour and funding by individual members, and a weighted right to vote should be established despite the difficulty of this task.

The 'commitment principle', however, seems to contradict the principle of deliberative democracy (Cohen, 1997). All participants should have an equal say in a deliberative democracy. This principle is important to enable the participants to speak out freely regardless of their social status, and to be bound only by the results of the deliberation. However, we introduced the commitment principle to avoid the influence of social status, because otherwise it would seem to be impossible for the participants to speak out freely regardless of their social status in the real world.

## Contents of the Book

These questions are best addressed by senior policy researchers from various countries in Asia. While a good deal of information and research reports are available for several countries in Asia focussing on forest resources as main sources of local livelihood and contributors to the national economy, these complex issues have not been simultaneously addressed among different levels for different countries.

A workshop was held in October 2012, at the University of Tokyo to serve as a venue for scholars from Asian countries, with financial support from the Network for Education and Research on Asia (ASNET) and the Laboratory of Global Forest Environmental Studies (GFES) of the University of Tokyo. The scholars attending this conference presented papers addressing portions or the full span of governance issues addressed in the research questions described above, and received comments from their peers attending the conference. Thus, this edited volume can address the following two important concerns for different policy arenas:

(i) Provision of relevant empirical information that might be helpful in the review, revision or formulation of the design guidelines related to forest resource governance, particularly those guidelines associated with rural communities. This information can be used to develop policies for both legislative (national and district) and non-legislative (local) institutions.

(ii) Enhancement of the capacity of multiple societal levels to effectively govern and sustainably manage forest resources in rural areas of Asia. These efforts to enhance capacity include learning from diverse policy contexts based on precise, scale-relevant information (national, district and local), thereby bridging the multi-level outcomes.

Based on the presentations and discussion at the workshop, the editors requested the authors from each country to revise their chapters to cover information on 'policy background', 'case studies (more than one) including methodology' and 'discussion of the design guidelines and other important issues'. For the case studies, we requested the authors to select at least two representative cases from each country (see Figure 1.3).

**Figure 1.3:**
*Category of cases*

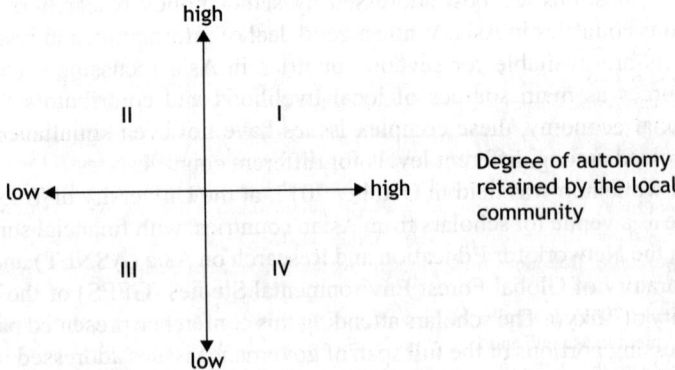

Dependency of local people on forests

*Source:* Authors.

Section II contains South Asian countries such as Bangladesh (Chapter 2), India (Chapter 3), Nepal (Chapter 4), Sri Lanka (Chapter 5) and Bhutan (Chapter 6); Section III includes Southeast Asian countries such as Indonesia, Malaysia (Chapter 10), the Philippines (Chapter 11), Thailand (Chapter 12), Vietnam (Chapter 13) and Laos (Chapter 14). We allocated three chapters (Chapters 7–9) for Indonesia, having the largest tropical forest area in Asia. Chapter 7 elaborates on policy issues; Chapters 8 and 9 examine the cases in East Kalimantan and West Sumatra, respectively. Section IV includes East Asian countries such as China (Chapter 15), Korea (Chapter 16) and Japan. We also allocated two chapters (Chapters 17 and 18) for Japan, because *iriai* forests or communal forests in Japan have become the focus of attention of scholars engaging in the issue of common-pool resource management. Chapter 17 refers to legalised two types of former *iriai* forests; Chapter 18 investigates a new mode of collaboration by various stakeholders for rural development. At last, all of the chapters are further analysed in Chapter 19.

This book reflects the remarkable diversity of policy implementation including governance and management challenges, structural variables and the outcomes of countries in South, Southeast and East Asia. These reports highlight the importance of context in defining flexible policies for policy makers, development practitioners and the academic communities of these countries.

# References

Agrawal, A. (2002). Common resources and institutional sustainability. In Ostrom et al. (eds), *The Drama of the Commons* (pp. 41–85). Washington, DC: National Academy Press.

Balooni, K.B. & Inoue, M. (2007). Decentralized forest management in South and Southeast Asia. *Journal of Forestry,*105(8), 414–420.

Berkes, F. (2002). Cross-scale institutional linkages: Perspectives from the bottom up. In E. Ostrom, T. Dietz, N. Dolsak, P. Stern, S. Stonich, & E. Weber (eds), *The Drama of the Commons*. Washington, DC: National Academy Press.

———— (2007). Community-based conservation in a globalized world. *Proceedings of the National Academy of Sciences*, 104(39), 15188–15193.

———— (2008). Commons in a multilevel world. *International Journal of the Commons,* 2(1), 1–6.

Blaikie, P. (2006). Is small really beautiful? Community-based natural resource management in Malawi and Botswana. *World Development*, 34(11), 1942–1957.

Borrini-Feyerabend, G. (1996). *Collaborative Management of Protected Areas: Tailoring the Approach to the Context*. Switzerland: IUCN-The World Conservation Union Gland.

Cash, D.W., Adger, W.N., Berkes, F., Garden, P., Lebel, L., Olsson, P., Pritchard, L., & Young, O. (2006). Scale and cross-scale dynamics: Governance and information in a multilevel world. *Ecology and Society*, 11(2), 8.

Cohen, J. (1997). Deliberation and democratic legitimacy. In J. Bohman, & Rehg, W. (eds.), *Deliberative Democracy: Essays on Reason and Politics* (pp. 67–91). Cambridge, MA: The MIT Press.

Colfer, C.J.P. (2005). *The Complex Forest: Communities, Uncertainty, and Adaptive Collaborative Management.* Washington, DC: Resources for the Future; Center for International Forestry Research.

Dorji, L., Webb, E., & Shivakoti, G.P. (2006). Forest property rights under nationalized forest management in Bhutan. *Environmental Conservation*, 33(2), 141–147.

Dung, N.T. & Webb, E. (2008). Incentives of the forest land allocation process: Implications for forest management in Nam Dong District, Central Vietnam. In E. Webb, & G.P. Shivakoti (eds). *Decentralization, Forests and Rural Communities: Policy outcome in South and South East Asia* (pp. 269–291). New Delhi: SAGE Publications.

Edmonds, D. & Wollenberg, E. (2001). A strategic approach to multistakeholder negotiations. *Development and Change*, 32, 231–253.

Gautam, A., Shivakoti, G.P., & Webb, E.L. (2004). A review of forest policies, institutions, and the resource condition in Nepal. *International Forestry Review*, 6(2), 136–148.

IGES. (2007). *Decentralization and State-sponsored Community Forestry in Asia.* Kanagawa: Institute for Global Environmental Studies.

Inoue, M. (2003). Diverse management of Indonesian forests: A new governor gives locals a greater say in their resources. *International Herald Tribune and the Asahi Shimbun*, April 4, 27.

—— (2004). *In Search of the Principle of Commons.* Tokyo, Japan: Iwanami Shoten (in Japanese).

—— (2009). Design guidelines for collaborative governance (*kyouchi*) of natural resources. In T. Murota (ed.), *Local Commons in Globalized Era* (pp. 3–25). Kyoto, Japan: Minerva Shobou (in Japanese).

—— (2011, January). *Prototype Design Guidelines for 'Collaborative Governance' of Natural Resource.* Presented at 13th Biennial Conference of the International Association for the Study of the Commons, Hyderabad, India.

Inoue, M. & Isozaki, H. (2003). *People and Forest-policy and Local Reality in Southeast Asia, the Russian Far East and Japan.* Netherlands: Kluwer Academic Publishers.

Kijtewachakul, N., Shivakoit, G., & Webb, E. (2004). Forest health, collective behaviors and management. *Environmental Management*, 33(5), 620–636.

Mahdi, M., Shivakoti, G.P., & Schmidt-Vogt, D. (2009). Livelihood change and livelihood sustainability in the uplands of Lembang Subwatershed, West Sumatra, Indonesia, in a changing natural resource management context. *Environmental Management*, 43, 84–99.

McKean, M.A. (1999). Common property: what it is, what it is good for, and what makes it work? In C. Gibson, M.A. Mckean, & E. Ostrom (eds), *Forest Resources and Institutions*. Rome: FAO.

Mwangi, E. & Wardell, A. (2012). Multi-level governance of forest resources. *International Journal of the Commons*, 6 (2), 79–103.

Nath, T.K., Inoue, M., & Chakma, S. (2005). Prevailing shifting cultivation in the Chittagong Hill Tracts, Bangladesh: Some thoughts on rural livelihood and policy issues. *International Forestry Review*, 7(5), 327–328.

Ojha, H. (2014). Beyond the 'local community': The evolution of multi-scale politics in Nepal's community forestry regimes. *International Forestry Review*, 16(3), 339–353.

Ojha, H., Hall, A., & Rasheed Sulaiman, V. (eds) (2013). *Adaptive collaborative approaches in natural resource governance: Rethinking participation, learning and innovation*. London: Routledge.

Ojha, H.R., Cameron, J., & Kumar, C. (2009). Deliberation or symbolic violence? The governance of community forestry in Nepal. *Forest Policy and Economics*, 11(5–6), 365–374. Retrieved from http://www.scopus.com/inward/record.url?eid=2-s2.0-72049092456&partnerID=40&md5=8f4f547bc9bbc51986f28e551660765b

Ojha, H.R. & Hall, A. (2012). Confronting challenges in applying adaptive collaborative approaches. In H. Ojha, A. Hall, & V. Rasheed Sulaiman (eds), *Adaptive Collaborative Approaches in Natural Resource Governance: Rethinking Participation, Learning and Innovation* (pp. 287–311). London and New York: Routledge.

Ostrom, E. (1990). *Governing the Commons*. Cambridge, UK: Cambridge University Press.

——— (2005). *Understanding Institutional Diversity*. Princenton, NJ: Princeton University Press.

Poteete, A. (2012). Levels, scales, linkages, and other 'multiples' affecting natural resources. *International Journal of the Commons*, 6(2), 134–150.

Pulhin, J.M., Inoue, M., & Enters, T. (2007).Three decades of community-based forest management in the Philippines: Emerging lessons for sustainable and equitable forest management. *International Forestry Review*, 9(4), 865–883.

Ribot, J.C., Agrawal, A., & Larson, A.M. (2006). Recentralizing while decentralizing: How national governments reappropriate forest resources. *World Development*, 34(11), 1864–1886.

Robertson, R. (1995). Glocalization: Time-space and homogeneity-heterogeneity. In M. Featherstone, C. Lash, & R. Robertsen (eds), *Global Modernities* (pp. 25–44). Thounsand Oaks: SAGE.

Shivakoti, G. & Ostrom, E. (2008). Facilitating decentralized policies for sustainable governance and management of forest resources in Asia. In E. Webb & G.P. Shivakoti (eds), *Decentralization, Forests and Rural Communities: Policy Outcomes in South and Southeast Asia* (pp. 292–310). New Delhi: SAGE Publications.

Shivakoti, G., Varughese, G., Ostrom, E., Shukla, A., & Thapa, G. (1997). *People and Participation in Sustainable Development: Understanding the Dynamics of Natural Resource System*. Proceedings of an International Conference held at Institute of Agriculture and Animal Science, Rampur, Chitwan, Nepal. 17–21 March, 1996, Bloomington, Indiana and Rampur, Chitwan.

Stern, P., Dietz, T., Ostrom, E., & Stonich, S. (2002). Knowledge and questions after 15 years of research. In E. Ostrom, T. Dietz, N. Dolsak, P. Stern, S. Stonich, & E. Weber (eds), *The Drama of the Commons*. Washington, DC: National Academy Press.

Viswanathan, P.K. & Shivakoti, G. (2008). Adoption of rubber integrated farm livelihood systems: Contrasting empirical evidences from Indian context. *Journal of Forest Research*, 13(1), 1–14.

Webb, E. & Shivakoti, G.P. (eds) (2008). *Decentralization, Forests and Rural Communities: Policy Outcomes in South and Southeast Asia*. New Delhi: SAGE Publications.

Wollenberg, E., Iwan, R., Limberg, G., Moeliono, M., Rhee, S., & Sudana, M. (2007). Facilitating cooperation during times of chaos: Spontaneous orders and muddling through in Malinau district, Indonesia. *Ecology and Society*, 12(1), 3. Retrieved from http://www.ecologyandsociety.org/vol12/iss11/art13/

Yonariza, Y. & Shivakoti, G. (2008). Decentralization and co-management of protected areas in Indonesia. *Journal of Legal Pluralism*, 57, 141–165.

# SECTION II

# South Asia

# 2

# Bangladesh: Do Changes in Policy Ensure Good Forest Governance?

*Tapan K. Nath, Mohammed Jashimuddin and Makoto Inoue*

## Introduction: Forest and Forest Policy in Bangladesh

Bangladesh has only 2.52 million hectares (Mha) of land designated as forests, which accounts for 17% of its total land mass (BFD, 2011); however, recent estimates show a total of 1.44 Mha (11%) of forest areas with 0.009 ha per capita forest, which is very low compared to the world per capita forest value of 0.597 ha (FAO, 2011). The distribution of forests in Bangladesh is highly skewed, with 29 out of 64 districts having no official forest area and only 12 districts having sizeable forest areas of 10% or more of their land mass (Jashimuddin, 2011). It has been reported that within the public forest land in Bangladesh, only about 15% is closed forest (more than 40% crown density), 19% is open forest (10–40% crown density), 12% is plantation forest and the remaining 54% is used for non-forestry purposes (FAO, 2000). The growing stock of forests in Bangladesh is also low (48 m³ha⁻¹) compared to the average values of South and Southeast Asia (99 m³ha⁻¹) and the world (131 m³ha⁻¹) (FAO, 2010).

The history of forest management in Bangladesh can be considered as a classic example of continued deforestation and degradation. The forests were exploited to earn revenue and supply raw materials for the ship and railway industries of the British colonial rulers (1757–1947) and to generate revenue and cater raw materials for forest industries under Pakistani rule (1947–1971), which also continued into the current period of independent Bangladesh sovereignty (Iftekhar, 2006). The conventional central forest management system in Bangladesh has been deemed unsuitable for the resource base and socio-economic situation of the country. Because of the inability to prevent widespread overexploitation of forest resources, many state forest areas have been rapidly degraded under population pressure and increasing demands for forest products (Biswas & Choudhury, 2007). The major weakness of forest management in Bangladesh has been the inability to secure participation of villagers and the community at large, which has led to large scale encroachment and pilferage; this can only be stopped by obtaining public participation (FMP, 1992). The national forest resources and the authority over them, the Forest Department (FD), have been centralised under the government, superseding traditional rights and communal authority (Millat-e-Mustafa, 2002). In response to forest degradation, increasing emphasis has been placed over the last two decades on the formulation of a decentralised forest policy.

The forest policy of Bangladesh has undergone several modifications since British colonial rule in order to cope with the changing political and socio-economic situations. The British colonial government formulated forest policy in the undivided Indo-Pakistan subcontinent for the first time in 1894, and this policy was modified in 1955, 1962, 1979 and 1994. Throughout the colonial era (up to 1947), forest policy has been oriented towards the commercial considerations of revenue generation and maximum resource exploitation. Forest policy established under Pakistani rule (in 1955 and 1962) showed a high degree of continuity with its colonial heritage and maintained an emphasis on commercial and industrial interests that continued after the independence of Bangladesh in 1971. The first national forest policy of Bangladesh was enacted in 1979. It clearly described the participatory approach to be followed in government-owned forest land and plantations on marginal lands (Muhammed, Koike, Sajjaduzzaman, & Sophanarith, 2005) and also paved the way for social forestry (SF) in Bangladesh; however, it failed to effectively address the issue of broader participation in forest management (Millat-e-Mostafa, 2002). The negative social impacts of this excessive, government-sponsored commercialisation of forest interests include the systematic alienation of local communities from forests and forestry, a disregard for local livelihood needs and the progressive diminution of traditional rights. However, the current forest policy formulated in 1994 supports a significantly more people-oriented forestry and demonstrates the government's determination to protect and develop forest resources through popular participation. In an effort to better integrate community forestry (CF) into forest management practices, the government also formulated the 2004 Social Forestry Rules. These policy reforms have increased the sharing of experiences by local communities with the FD and the participation of these communities in forestry activities, thus, contributing to poverty reduction. However, additional reforms are urgently needed to further increase the efficiency of the FD and improve its governance (ADB, 2007).

## Community Forestry in Bangladesh

CF, commonly known as SF, participatory forestry or agroforestry (AF), has been practised in Bangladesh for more than three decades. It has become a highly attractive and acceptable programme to the rural people, especially the landless people and small farmers. The main components of the CF projects implemented in Bangladesh include afforestation; the establishment of woodlot plantations; AF; the development of strip plantations along roads, railways and canal embankments; and the rehabilitation of landless farmers and shifting cultivators. The basic principle of these projects is the integration of local people in afforestation activities with the objective of generating ecological, economic and social benefits (Ahmed & Akhtaruzzaman, 2010). Since the 1980s, some CF programmes have been successfully implemented by the FD and some new programmes are currently in progress (Table 2.1). CF experiences in Bangladesh can be classified into three important categories based on who initiated the resource management: the government, non-government organisations (NGOs) or the community itself. However, all these CF initiatives have their own success or failure stories depending on their management practices and local conditions. While traditional forest management resulted in a net loss of forest resource cover, CF, on the other hand, is playing a vital role in the expansion of forest cover and is benefiting thousands of poor people (Muhammed et al., 2005).

So far, a total of 30,666 ha of woodlot plantations, 8,778 ha of AF plantations and 48,420 km of strip plantations have been established by the FD under the CF programmes in Bangladesh since the mid-1980s (Table 2.2). Some of the mature plantations have been harvested and the

**Table 2.1:**
*Historical development of CF programmes in Bangladesh*

| Programmes | Period |
|---|---|
| 1.   Taungya System | 1871 |
| 2.   Forestry Extension Service Phase I | 1962–1963 |
| 3.   Betagi-Pomra Community Forestry Project | 1979–1980 |
| 4.   Jhumia Rehabilitation Programme in CHT Phase I | 1979–1989 |
| 5.   Development of Forestry Extension Service Phase II | 1980–1985 |
| 6.   Community Forestry Project | 1982–1987 |
| 7.   Thana Afforestation and Nursery Development Project | 1987–1995 |
| 8.   Jhumia Rehabilitation Programme in CHT Phase II | 1990–1995 |
| 9.   Participatory Social Afforestation | 1991–1998 |
| 10.  Forest Resources Management Project | 1992–2001 |
| 11.  Extended Social Forestry Project | 1995–1997 |
| 12.  Coastal Greenbelt Project | 1995–2000 |
| 13.  Forestry Sector Project | 1997–2004 |
| 14.  Sundarban Biodiversity Conservation Project | 1999–2006 |
| 15.  Nishorgo Support Project | 1999–2008 |
| 16.  Char Development and Settlement Project-III (2nd Phase) | 2005–2010 |
| 17.  Reed land Integrated Social Forestry Project | 2005–2010 |
| 18.  Afforestation in the Denuded Hill Areas of Chittagong North Forest Division (2nd Phase) | 2008–2012 |
| 19.  Biodiversity Conservation and Poverty Alleviation Through Afforestation in the Greater Rajshahi and Kushtia Districts. | 2008–2012 |
| 20.  Participatory Social and Extension Forestry in Chittagong Hill Tracts | 2008–2012 |
| 21.  Community Based Adaptation to Climate Change through Coastal Afforestation | 2009–2012 |
| 22.  Re-vegetation of Madhupur Forests through Rehabilitation of Forest Depended Local and Ethnic Communities | 2010–2012 |
| 23.  Poverty Alleviation through Social Forestry | 2010–2013 |

CHT, Chittagong Hill Tract.
**Source:** Jashimuddin and Inoue (2012a).

profits of the harvests have been distributed among different stakeholders. A total of 19,790 ha of woodlot and AF plantations and 8,566 km of
strip plantations have been felled. As a result, approximately US$18.91
million have been distributed among 85,900 beneficiaries, and a tree

**Table 2.2:**
*Achievement of CF projects since mid-1980s in Bangladesh*

| Components | Achievement |
|---|---|
| Strip plantations | 48,420 km |
| Woodlot plantation | 30,666 ha |
| Agroforestry plantation | 7,738 ha |
| Embankment plantation | 1,338 ha |
| Foreshore plantation | 645 ha |
| Village afforestation | 7,421 villages |
| Seedling for sale and distribution | 201 million |

*Source:* Muhammed et al. (2005).

farming fund of US$4.17 million has been established for future plantations (BFD, 2011). However, widespread corruption and poor governance in the forestry sector (Muhammed, Koike, & Haque, 2008) are hindering the desired achievement of CF.

# Forest Governance in Community Forestry: Two Case Studies

Two case studies were conducted on the government-initiated Upland Settlement Project (USP) and the community-managed village common forests (VCFs). The first case is category II, with a high dependency on forest resources and a low degree of autonomy retained by the local authority, while the second case is category I, with a high dependency on forest resources and a higher degree of autonomy retained by the local community. First, we discuss the governance situations in the USP.

## Governance in the Upland Settlement Project

The Chittagong Hill Tracts Development Board (CHTDB), a premier government agency that conducts most development projects in the Chittagong Hill Tracts (CHT), implemented the USP in two phases. The ADB funded the first USP phase, running from 1985 to 1993, by settling 2,000 ethnic families (hereafter called participants) and establishing

1,620 ha of homestead AF and 3,240 ha of rubber plantations in 39 project villages in the Khagrachari district. Considering the success of the first phase, the second phase of the USP was formulated with the following objectives (CHTDB, 2001; Khisa & Hossain, 2002):

- Settlement of landless shifting cultivators in suitable areas in the Khagrachari and Bandarban districts;
- Development of suitable upland areas, which are currently unused or used for shifting cultivation, for diversified cropping with horticulture, AF and rubber plantations; and
- Long-term socio-economic development of target groups.

The specific objective of the project was to settle 1,000 landless ethnic participants in 20 project villages in the Bandarban and Khagrachari districts, with 10 villages in each district and each village accommodating 50 participants. This phase was financed by a government grant under the Ministry of CHT Affairs. Even though the original project period was for the 1993–2000 fiscal years (FY), it was extended until FY 2006–2007 in order to complete the establishment of the rubber processing units. In this phase, 506 ha of homestead AF and 1,620 ha of rubber plantations were established; each participant was granted 2.1 ha of degraded forest land, of which 0.5 ha was intended for homestead AF and 1.6 ha for rubber plantation. Due to the remote location, lack of accommodations, poor communication services and civil unrest, this case study was conducted in two villages that were in the USP second phase in the Bandarban District.

The project implementing agency, the CHTDB, has two branch offices in the Bandarban and Khagrachari hill districts, with the head office in Rangamati. Even though some of the authority for decision making by the CHTDB has been devolved, the regional manager of the USP must get approval for major decisions from the central project manager based in Khagrachari.

Although the USP was able to achieve its stated objectives differentially in the two studied villages [village A and village B; Prokopy (2008) proposed that it is significant to protect source confidentiality and location confidentiality when publishing social research findings] with its decentralised style of management, we observed some local governance issues that demand the consideration of a programme that will generate a better outcome. In analysing governance issues, we will first shed light on the participant's organisation and their participation in project functions. Then, we will discuss the other project-related governance issues that we observed during this study, including equity, accountability and transparency, information flow and responsiveness. These governance principles provide normative guidance for good forest and natural resource management governance (FAO, 2011; Lockwood et al., 2009, 2010).

## Participation in Project Activities

In both the villages, there are two social organisations in addition to the project village committee. These organisations maintain linkages with other agencies, mostly with NGOs, and conduct social development projects, such as temple development and road maintenance, in the village. Four to five NGOs are operating in the studied villages and provide informal education, livestock husbandry, ginger cultivation and credit. To facilitate social development projects, participants call meetings at their villages; we found some differences in meeting satisfaction and deliberative quality between the two villages. As the participants indicated, these differences were due to domination by vocal-rich participants who made decisions that benefited themselves. In village B, some participants, who are now organised, spoke frequently and made decisions in support of themselves, resulting in significant differences in meeting satisfaction and deliberative quality.

Even though local meetings are held at convenient times, we found differences in the frequency of meetings between the two villages. On average, the frequency was 1.5 and 11.5 times a year in the village A and village B, respectively. We also found a difference regarding the frequency of meeting attendance. A small fraction of participants in village B attended meetings once or twice a year, and the meeting they attended mostly involved the discussion of religious festivals. On the other hand, major fractions of participants, who are involved with several NGOs, arranged meetings frequently, even two or three times a month in some cases.

Although most participants attended local meetings, few people had roles in the decision-making process. Leaders and some elite participants usually took part in making the decisions. While the participants do not have a say on decisions made at the meetings, they had differing views on the responsiveness and activities of their leaders. For instance, all participants in village A said that their leader is very responsive and that they are satisfied with his activities. On the contrary, 42% of participants in village B mentioned their dissatisfaction with the leader's activities and responsibilities. They told us:

> The leader does not come to our village. He does not involve us in any social development work except religious festivals. When work starts in the rubber plantation, he employs more labour from his own village. He employs rubber plantation guards from his village even though we live very near to the plantation. He does not inform us about committee funds. We don't know how much money is left in the bank account or what he is doing with this money.

Gray, Williams and Phillips (2005) reported that leadership is an important element in community organisation that assists in the development of the community capacity. A good leader, with characteristics of honesty and fairness, can enhance the level of social capital among villagers, thereby helping people procure rights and improve livelihoods (Dolom & Serrano, 2005).

The main task of the nine-member project village committees, which were formed by the USP authority in every project village, is to motivate participants to protect rubber plantations once grown and to be involved in the day-to-day project functions (USP meeting minutes in Bengali dated 17 February 1998). We observed that the committee virtually does not have any village-level activities. The committee is not actively involved with the project authority and project planning processes. Some members of the committee sometimes attended project meetings just to listen. Schouten and Moriarty (2003, cited in Prokopy, 2005) mentioned that for participation to lead to the expected sustainable outcomes, people need to be involved at higher levels of decision making. To succeed in dealing with farmers' vulnerability and environmental degradation, it is not only necessary to build active farmer organisations (Dendi, Shivakoti, Dale, & Ranamukhaarachchi, 2005; Hong, 2005), but also to delegate power to make decisions at the local level. The delegation of power helps local people make timely decisions for the good of their society. More importantly, the devolution of power through participatory decision making is an important process of governance (Murali, Murthy, & Ravindranath, 2006).

Not only project village committee members, but also participants, visited the project office headquarters and participated in project meetings. All participants visited the project head office frequently, four to five times every month, in the early stage of the project. The purposes of their visits were to sign papers for land tenure, collect their salary and obtain information on job opportunities at respective project villages. During three to four years of the project period, 17% and 53% of the participants of village A and village B attended project meetings held at temporary offices near the project villages 5.33 and 6.63 times, respectively. The participants only listened and could not provide any input at the meetings. These meetings were only intended to motivate the participants to live harmoniously together and plant trees at their homesteads. However, the participants need to be engaged in all project decision-making processes, because full community participation and management are increasingly recognised as being important for the long-term sustainability of the investment in these projects (Hettige, 2006). Pini and McKenzie (2006) also reported that sustainability of natural resource management is dependent upon community engagement at the local level.

All participants commented that by joining the project, they became aware of their social organisation and the importance of local meetings where they can make decisions on social development issues.

## Equity

We observed a lack of equity in the selection of participants, the distribution of settlement money, offers of training and employment opportunities. According to the project proposals, only landless and marginal farmers were supposed to be selected as participants by a specialised selection committee that had been formed by executives from the project authority, a sub-district officer, a settlement officer, a local government council chairman, a headman (leader of *mouza*) and a village leader. We found, however, that some better-off participants who possessed first-class agricultural land were also selected in both villages. One project staff member told us:

> The selection of participants was done hurriedly and hence real landless people could not be chosen. Many people were selected in such a manner that they really did not know how they had become participants. The headman and leader selected them because they [the participants] were known to them.

When asked about the speedy selection, he replied:

> Because the project started and the budget had been released, we had to complete the work in due time. If we had judged the actual criteria, it would have taken a long time to choose the participants. Ultimately this would hamper the completion of other project activities such as infrastructure development and the rubber plantation.

The consequences of this selection process are that when there were full-stream project activities, relatively better-off participants took the benefits, and then many of them left the village while still holding project land. If proper selection had taken place, some real landless farmers could have received land and project benefits.

At the time the participants joined, the project authority gave them some project money as settlement. All participants were supposed to receive the same amount of money, but we found disparities. On average, participants of villages A and B received Tk. 917 and Tk. 594, respectively. There were even some participants who did not receive any money. We also found that influential participants (e.g., leaders, wealthy and vocal participants) got more money than poor participants.

In accordance with the project plan, the project authority was supposed to provide training to all participants on seedling planting techniques. Though some participants (45%) of village A received two days of training, all participants of village B said that they had not yet participated in any training programme. Apart from this, the project authority trained 16 participants (both male and female) of village A in tapping for 15 days. The participants were paid during these training sessions. The project staff members said that due to time constraints, they could not organise training programmes for all participants. However, they affirmed that when tapping starts, the project will provide training in tapping techniques to participants in all project villages.

There was also inequality in the project employment opportunities. During this study, we found that 21% and 83% of the sampled participants in village B and village A, respectively, had current employment in project activities. The higher percentage of employment for the village A was due to the start of rubber production in that village.

Equity affects people's genuine participation in project activities. Kessler (2007) reports that equity plays a major role in developing people's self-confidence and capacity for collaborative thinking and working.

### Accountability and Transparency

Discussions with project staff members and participants explored the lack of accountability in handling project money. They commented:

> In most of the villages, if the budget for a job was Tk. 1000, the manager allocated 60 percent of that money to field staff. Field staff members then gave 40 percent of that 60 percent to the village leader to perform all

activities by employing participants. The leader, on the other hand, did not spend more than 20–30 percent of the allocated money.

Even though the situation was not the same in all villages, the above quotation implies that much of the project money had been mishandled and that the project could not benefit the targeted poor participants to the fullest extent. The leader of village B said that after the completion of the jobs, all remaining money should be deposited into the village common fund, but the participants felt it was very rare for that to happen.

The involvement of several stakeholders in addition to the CHTDB may help prevent the mishandling of project budgets and ensure transparency and accountability. Broderick (2005) claims that if natural resource management strategies are to be successful, then a much wider and more inclusive view of community is needed that captures the different stakeholder groups beyond just the farmers. In the Philippines, for example, Ernesto and Aguirre (2005) observed that even though projects are initiated by local government authorities, the subsequent involvement of NGOs and the formation of effective planter organisations that manage and protect the forests create an environment for the long-term sustainability of mangrove restoration projects.

*Information Flow*

There were gaps in the information flow. We found that even participants were not aware of the inception of the USP. This was evident from a comment by the leader of village A:

> I was working on my agricultural land and some neighbours informed me that the *unnyan board* (CHTDB) was clearing our land. I went there and asked the officers to stop clearing our land.... After a few days, officers again started cutting our forests and jungles. When I told them to stop, they informed me that the government had acquired the land, that the USP would be implemented there, and that 'you [the leader] will be selected as a participant and will get compensation for the land'.

This shows that even the leader of the project village was not consulted before implementing the USP. The participants told us that they did not know about future project functions at their villages unless they were asked to be involved in plantation activities. They were unsure who would manage the rubber plantation after the project period. The project manager as well as other staff members said that a central management unit would be formed to look after the rubber plantation and its production and overall management. They further stated that this management unit would manage the rubber until the rubber plantation was 40 years old and then hand over the land to the participants and that the benefits, after deducting all costs, would be distributed to the participants of the respective villages. Participants reported that they heard about the management unit, but that the project staff had not informed them of anything related to it.

*Responsiveness*

We found a lack of responsiveness in the project staff. In order to fulfil the short-term objectives, the participants who were shifting cultivators and were not skilled in horticulture were supposed to be encouraged to develop horticulture on their homesteads. However, they opined that the project staff members visited the rubber plantation frequently, but very seldom came to observe the AF at their homesteads.

The project authority acquired government *khas* land for the project. However, there was some private land[1] within the project territory for which the authority was supposed to pay. Eleven years into the project, the participants had not received any compensation, and they cautioned that it might create serious conflict with the project authority if no reasonable solution was reached soon.

## Governance in Community-initiated VCF

A common property regime can emerge as a way to secure control over a territory or a resource, to exclude outsiders or to regulate the individual use of a territory by members of the community (Arnold, 1998). As such, the birth of community-managed VCFs in the CHT is a direct result of resource constraints caused by deforestation and the prevention of entry into and use of the resources of the newly acquired reserved forests (Baten, Khan, Ahammad, & Misbahuzzaman, 2010; Halim & Roy, 2006). According to Nayak (2002), the negative impacts of forest degradation on the local agriculture and animal husbandry practices

---

[1] For example, in village A the project authority acquired 16 ha of private land owned by four participants (including the leader) of which 8 ha was allocated as project land. The project authority still had not paid compensation for this land, and participants had filed suit against the project authority.

completely traumatised the forest-based livelihood of many of the participants. As a result, people started travelling to far-off forest areas in need of fulfilment, which led to conflicts with other communities and harassment by the FD. On the other hand, the ecological effects of forest degradation that is, loss of soil fertility, erratic rainfall and the drying-up of streams, have also played a significant role in inducing forest protection by local communities. In such circumstances, perhaps as a last resort, many communities gradually turned to their adjacent degraded forests and initiated protection measures to restore the forests and local livelihoods.

The use of common land by the indigenous people is not new in this region. Since the British colonial period, the indigenous villagers who lost access to what was formerly the common land due to nationalisation of forests eventually moved to state-owned reserve forests. The result of this migration was innovation to retain forest cover for long-term use based upon their traditional resource management patterns. This gave birth to the VCFs of today, which are not allowed to be cultivated by any means on the strength of sanctions and religious taboos (Roy and Halim, 2002). These VCFs are also directly managed, protected and used by indigenous village communities (Baten et al., 2010; Halim & Roy, 2006; Rahman, 2005) under the leadership of the *mouza* headman or village *karbaris,* by educational or religious institutions, or by a committee formed by leaders from one or more villages (AF, 2010; Halim, Roy, Chakma, & Tanchangya, 2007; Islam et al., 2009; Roy, 2000; Saha, 2010; Tiwari, 2003).

The VCFs are mostly small, averaging 20–120 ha in size and consisting of naturally grown or regenerated vegetation. There is controversy about the total number of VCFs, but it may be around 700–800 in CHT (Saha, 2010). The VCF management in CHT has set a standard model for the protection of biodiversity and natural environments (Baten et al., 2010). These play important roles in conserving forest resources that are usually very rich in biodiversity. They harbour rare plant and animal species that are not usually found in the state-owned reserve forests or un-classed state forests due to continued deforestation and land degradation. In the VCF of the Bandarban Hill District, a recent study recorded a total of 162 plant species from 60 families, including bigger trees of valuable species that are not usually found in other forests (Jashimuddin & Inoue, 2012b). These VCFs are still the source of fuel wood, herbs, roots, bamboo shoots, wild fruits, vines and leaves for cooking or medicinal use, necessary to sustain the lives of the indigenous communities in the

CHT. Some VCFs consist predominantly of bamboo brakes; some VCFs contain a more heterogeneous stand of flora and fauna; many VCFs also contain herbaria for the village concerned, which the local *vaidya*s or *ojha*s (village shamans) use to prepare their traditional medicines; and other VCFs are regarded as sacred (Roy & Halim, 2002). The use and extraction of produce in VCFs is need based, with each person taking only what he/she requires in order to prevent the depletion of the natural resources of the forest, which existed for the benefit of the entire community. This system still continues today in some villages and in most cases, VCFs are the only remaining natural forests in the surrounding area and are considered as the depository of traditional knowledge.

If we look into the governance situation in the VCFs, we can see that indigenous people have a long-term vision for the conservation of forest resources. These VCFs meet the local energy demand, provide materials for house construction and serve as a storehouse of livelihood means. According to the rules of VCF communities, all villagers who belong to their respective VCF have equal access to resources and equally contribute to the protection and development of the VCF. If needed, they sometimes restrict the extraction of a forest product to allow it to regenerate. For example, at the Komolchari VCF in Khagrachari, the executive committee has banned the extraction of bamboo for the last three to four years because the majority of the bamboo plants had died due to flowering; that period was used to regenerate the bamboo naturally. The executive committee oversees the overall management of the VCF, but in the case of any punishment, they ask the village development committee to impose the sentence. The executive committee collects revenue (if any) earned from the VCF, but they give that money to the village development committee for use in social development purposes. This indicates that they maintain a hierarchy in governing not only the VCF management but also the overall social development activities.

## Implications of Design Guidelines

Although current forest policies have many shortcomings, decentralisation has created opportunities for building collaboration with local stakeholders to promote sustainable CF management in developing countries. Collaboration among concerned stakeholders, including governments, farmers, businesses, communities, NGOs, etc., is indispensable for sustainable forest use and management in the era of globalisation (Inoue, 2011; Lockwood et al., 2010). Collaboration among various stakeholders who have a range of interests in local forest use and management can be achieved through collaborative governance (Inoue, 2004). A mutually supportive and cooperative relationship among various stakeholders is needed in order to ensure good collaborative governance (FAO, 2011).

Collaborative governance requires 'open-minded localism' where local people consent to open their resources and environment to outsiders (e.g., project implementing authority) for sustainable management (Inoue, 2011). This open-minded localism may lead to 'graduated membership' in the executive committee. 'Graduated membership' refers to local people acting as core members (first-class members) of the executive committee; core members have the strongest authority and cooperate with other graduated members (second-, third-class members and so on) who have relatively weaker authority (Inoue, 2011). This guideline implies that all villagers cannot be core members, only selective persons who are responsive, honest and acceptable by the majority should be included in the executive committee. As we saw in the above cases, the leader in village B seemed irresponsive, and thus, most of the villagers had shown their dissatisfaction with his activities. However, the majority of the villagers of village A reported that their leader is responsive and helped villagers when needed. Therefore, under this guideline it is imperative to have provisions to change the leadership and even the core members by promoting second-class members. Previously, local people did not allow outsiders to implement any activities in VCFs, but recently they have been convinced to invite NGOs to execute forest conservation and livelihood-related functions in VCFs. The condition of these invitations is that the outsiders cannot take any benefit from their operations. The NGOs helped to establish VCF management committees, when necessary, or helped to strengthen the existing committee. This means that local people had become open to entertaining NGO activities for their own welfare.

Core members, who are the most responsible members in the society and take care of their resources, should have greater power over the decision-making processes than the second- and third-class members of a society. This is the 'commitment principle', which states that the authority to make decisions should correlate with the degree of commitment to relevant activities of the stakeholders (Inoue, 2011). If we consider this guideline with our case studies, we can see that members of the VCF have high autonomy in making and implementing decisions for VCF management. We observed that indigenous people have a strong commitment towards VCF management and conservation, not only for their current generation, but also for future generations. As a result, they have been sustainably managing VCFs for a long time. In the case of USP, the local executive committee has no autonomy to make any project management decisions, and, thus, they did not participate effectively in project activities, like plantation management. Therefore, it is important that project authority should have commitment to engage community members in decision-making processes. It can be hoped that if community members are genuinely involved in the decision-making process, a sense of community ownership would be instilled in them, and they would participate in project activities effectively.

*Building social capital* (e.g., trust) with outsiders is a pre-condition for 'graduated membership' and the 'commitment principle' (Inoue, 2011). Currently, donor agencies and development practitioners encourage building social capital among stakeholders for the sustainability of development projects. The formation of social capital in the form of local organisations is an effective approach to mobilising local resources for the conservation of natural resources and the improvement of rural livelihoods. The project village committees need to be involved truly in project activities. Where necessary, committees need to be reformed so that honest and responsive leaders lead the committees. The project authority requires ensuring proper collaboration with the participants and motivating the participants to create a sense of ownership among them. If this occurs, it is likely that participants will work together for the sustainability of the project activities. However, it should be kept in mind that social capital and governance are not independent entities but are interrelated. While social capital generates a foundation for collective action, the practice of good governance ensures its continuation, which facilitates the sustainability of programme outcomes.

# References

ADB (2007). Bangladesh: Forestry sector project (Completion Report. Project Number: 26623 Loan Number: 1486), Asian Development Bank. Retrieved from http://www.adb.org/Documents/PCRs/BAN/26623-BAN-PCR.pdf.

AF (2010). Conserving forests for the future (Annual Report 2009). Dhaka, Bangladesh: Arannayk Foundation,. Retrieved from http://www.arannayk.org/docs/af_annual-report_2009.pdf.

Ahmed, F.U. & Akhtaruzzaman, A.F.M. (2010). *A Vision for Agricultural Research in Bangladesh: Sub-sector—Forestry 2030.* Dhaka, Bangladesh: Bangladesh Agricultural Research Council (BARC). Retrieved from http://www.barc.gov.bd/documents/Final-Mr.%20Farid.pdf.

Arnold, J.E.M. (1998). Managing forests as common property. FAO Forestry Paper 136. Rome: Food and Agriculture Organization of the United Nations.

Baten, M.A., Khan, N.A., Ahammad, R., & Misbahuzzaman, K. (2010). *Village Common Forests in Chittagong Hill Tracts, Bangladesh: Balance between Conservation and Exploitation.* Dhaka, Bangladesh: Unnayan Onneshan-The Innovators.

BFD (2011). The official website of the Bangladesh Forest Department. Retrieved December 2011, from www.bforest.gov.bd.

Biswas, S.R. & Choudhury, J.K. (2007). Forests and forest management practices in Bangladesh: The question of sustainability. *International Forestry Review,* 9(2), 627–640.

Broderick, K. (2005). Communities in catchments: Implications for natural resource management. *Geographical Research,* 43, 286–296.

CHTDB (Chittagong Hill Tracts Development Board). (2001). Ministry of Chittagong Hill Tracts Affairs project proforma (revised) of Upland Settlement Project (2nd phase). Rangamati, Bangladesh: Chittagong Hill Tracts Development Board.

Dolom, B.L. & Serrano, R.C. (2005). The Ikalahan: Traditions bearing fruit. In P.B. Durst, C. Brown, H.D. Tacio, & M. Ishikawa (eds), *Search of Excellence: Exemplary Forest Management in Asia and the Pacific* (pp. 83–92). Bangkok: RAP Publication.

Dendi, A., Shivakoti, G.P., Dale, R., & Ranamukhaarachchi, S.L. (2005). Evaluation of the Minangkabau's shifting cultivation in the west Sumatra highland of Indonesia and its strategic implications for dynamic farming systems. *Land Degradation and Development,* 16, 13–26.

Ernesto, A. & Aguirre, J.A.N. (2005). Forest from the mud: The Kalibo experience. In P.B. Durst, C. Brown, H.D. Tacio, & M. Ishikawa (eds), *In Search of Excellence: Exemplary Forest Management in Asia and the Pacific* (pp. 39–48). Bangkok: RAP Publication.

FAO (2011). *State of the World's Forests.* Rome: Food and Agriculture Organization of the United Nations.

FAO (2000). Forest resources of Bangladesh: Country report (Working Paper 15). Rome: The Forest Resources Assessment Programme, Forestry Department, Food and Agriculture Organization of the United Nations.

——— (2010). Global forest resources assessment 2010: Main report. FAO Forestry Paper 163. Rome: Food and Agriculture Organization of the United Nations.

FMP (Forestry Master Plan). (1992). *Forestry Master Plan: Forest Production.* Asian Development Bank (TA No. 1355-BAN). Dhaka, Bangladesh: Ministry of Environment and Forests, Government of Bangladesh.

Gray, I., Williams, R., & Phillips, E. (2005). Rural community and leadership in the management of natural resources: Tensions between theory and policy. *Journal of Environmental Policy and Planning,* 7, 125–139.

Halim, S. & Roy, R.D. (2006). Lessons learned from the application of human rights-based approaches in the indigenous forestry sector in the Chittagong Hill Tracts, Bangladesh: A case study of the village common forest project implemented by *Taungya.* Retrieved from http://hrbaportal.org/wp-content/files/1233223431_8_1_1_resfile.pdf.

Halim, S., Roy, R.D., Chakma, S., & Tanchangya, S.B. (2007). Bangladesh: The interface of customary and state laws in the Chittagong Hill Tracts. In H. Leake (ed.), *Bridging the Gap: Policies and Practices on Indigenous Peoples' Natural Resource Management in Asia*. Chiang Mai, Thailand: UNDP-RIPP, AIPP Foundation.

Hettige, H. (2006). *When do Rural Roads Benefit the Poor and How? An In-depth Analysis Based on Case Studies*. Manila, Philippines: Asian Development Bank.

Hong, H.N. (2005). Can Gio: Turning mangroves into riches. In P.B. Durst, C. Brown, H.D. Tacio, & M. Ishikawa (eds), *Search of Excellence: Exemplary Forest Management in Asia and the Pacific* (pp. 49–60). Bangkok: RAP Publication.

Iftekhar, M.S. (2006). Forestry in Bangladesh: An overview. *Journal of Forestry*, 104(3), 148–153.

Inoue, M. (2004). *In Search of the Principle of Commons*. Tokyo, Japan: Iwanami-shoten (in Japanese).

———— (2011). Prototype design guidelines for 'collaborative governance' of natural resources. A paper presented at the Thirteenth Biennial Conference of the International Association for the Study of the Commons held in Hyderabad, India from 10 to 14 January 2011.

Islam, M.A., Marinova, D., Khan, M.H., Chowdhury, G.W., Chakma, S., Uddin, M., Jahan, I., Akter, R., Mohsanin, S., & Tennant, E. (2009). *Community Conserved Areas (CCAs) in Bangladesh*. Dhaka, Bangladesh: Wildlife Trust of Bangladesh.

Jashimuddin, M. (2011). *Drivers of Land Use Change and Policy Analysis in Bangladesh: Theory and Policy Recommendations*. Germany: Lap LAMBERT Academic Publishing, GmbH & Co. KG.

Jashimuddin, M. & Inoue, M. (2012a). Community forestry for sustainable forest management: Experiences from Bangladesh and policy recommendations. *FORMATH*, 11, 133–166.

———— (2012b). Management of village common forests in the Chittagong Hill Tracts of Bangladesh: Historical background and current issues in terms of sustainability. *Open Journal of Forestry*, 2(3), 121–137.

Kessler, C.A. (2007). Motivating farmers for soil and water conservation: A promising strategy from the Bolivian mountain valleys. *Land Use Policy*, 24, 118–128.

Khisa, S.K. & Hossain, A.T.M.E. (2002). Upland settlement programme with small holding rubber based agroforestry farming systems in the CHT. In N.A. Khan, M.K. Alam, S.K. Khisa, & M. Millat-Emustafa (eds), *Farming Practices and Sustainable Development in the Chittagong Hill Tracts* (pp. 123–134). Bangladesh: CHTDB and VEFP-IC.

Lockwood, M., Davidson, J., Curtis, A., Stratford, E., & Griffith, R. (2009). Multi-level environmental governance: Lessons from Australian natural resource management. *Aust. Geogr.*, 40(2), 169–186.

———— (2010). Governance principles for natural resource management. *Soc. Natur. Resour.*, 23(10), 986–1001.

Millat-e-Mustafa, M. (2002). A review of forest policy trends in Bangladesh. Bangladesh forest policy trends. *Policy Trend Report 2002*, 114–121. Retrieved from http://pub.iges.or.jp/modules/envirolib/upload/371/attach/08_Bangladesh.pdf.

Muhammed, N., Koike, M., & Haque, F. (2008). Forest policy and sustainable forest management in Bangladesh: An analysis from national and international perspectives. *New Forests*, 36, 201–216.

Muhammed, N., Koike, M., Sajjaduzzaman, M., & Sophanarith, K. (2005). Reckoning social forestry in Bangladesh: Policy and plan versus implementation. *Forestry*, 78(4), 373–383.

Murali, K.S., Murthy, I.K., & Ravindranath, N.H. (2006). Sustainable community forest management systems: A study on community management and joint forest management institutions from India. *International Review for Environmental Strategies*, 6, 23–40.

Nayak, P.K. (2002, June). Community-based forest management in India: The issue of tenurial significance. Paper presented at the 9th Biennial Conference of the IASCP (International Association for the Study of Common Property), Victoria Falls, Zimbabwe.

Pini, B. & McKenzie, F.H. (2006). Challenging local government notions of community engagement as unnecessary, unwanted and unproductive: Case studies from rural Australia. *Journal of Environmental Policy and Planning*, 8, 27–44.

Prokopy, L.S. (2005). The relationship between participation and project outcomes: Evidence from rural water supply projects in India. *World Development*, 33, 1801–1819.

——— (2008). Ethical concerns in researching collaborative natural resource management. *Society and Natural Resources*, 21, 258–265.

Rahman, M.A. (2005). Chittagong Hill Tracts peace accord in Bangladesh: Reconciling the issues of human rights, indigenous rights and environmental governance. *Journal of Bangladesh Studies*, 7(1), 46–58.

Roy, R.C.K. (2000). *Land Rights of the Indigenous Peoples of the Chittagong Hill Tracts, Bangladesh* (231 pp.). Copenhagen, Denmark: International Work Group for Indigenous Affairs (IWGIA).

Roy, R.D. & Halim, S. (2002). Valuing village commons in forestry. *Indigenous Perspectives*, 5(2), 9–38.

Saha, P.S. (2010). Parbattya Chattagramer Mouza Ban: Prachin Praggyar Arek Rup. In P. Gain (ed.), *Dharitri*, 11th issue, an occasional SEHD magazine (Bangla). Dhaka, Bangladesh: Society for Environment and Human Development (SEHD). Retrieved from http://www.sehd.org/publications/magazines/dharitri.

Tiwari, S. (2003). Chittagong Hill Tracts: A preliminary study on gender and natural resource management, IDRC, Ottawa, ON, CA. Retrieved from http://hdl.handle.net/10625/30490.

# 3

# India: Determinants and Challenges of Sustainable Forest Governance

*Madhusudan Bandi and P.K. Viswanathan*

## Introduction

Community-based natural resources management (NRM) policies presume that communities are willing to manage natural resources collectively because of the latter's utilitarian and/or intrinsic benefits or because the communities are promised a reward for such management. An extensive body of theoretical and empirical literature explains the conditions under which collective action occurs to manage natural resources (Agarwal, 2001; Baland & Platteau, 1996; Dasgupta, 2008; Ostrom, 1990). Many of these conditions influence the success (or failure) of NRM policies. Even if governments are willing to decentralise, success depends on the resource under consideration, the community dependence on the resource and the type of institutions created to govern the management of resource (Shyamsundar & Ghate, 2011).

Forest governance captures the problems or challenges associated with collective action in the broader context of developing societies, including India. Forest governance in India has changed substantially in the past century. Large areas of forest land in India remained under a communal regime until the end of the 19th century (Singh, 1986). Colonial (mainly British) rulers confiscated communal rights over forest land using regulations/legislation,[1] only to expropriate India's forest wealth in the pretext of conserving it to facilitate the tremendous expansion of railways (Agarwal, 1999; Guha, 1983). As late as the 1880s, the Indian Forest Department (FD) was entertaining repeated requests from the British Navy for the supply of Madras and Burma teak ships built in the dockyards of Surat and along the coast of Malabar to meet such demands (Saikia, 2011).

Empirical evidence demonstrates that even after independence, forest governance in India has not changed much in terms of obliterating the colonial legacy of state control and regulations. The forest management system that India inherited from the British perpetuates the notion that forests constitute a distinct territory that must be governed repressively to extract a profit. Furthermore, the Wildlife (Protection) Act of 1972, India's main wildlife law, remains rooted in and modelled on the Indian Forest Act of 1927. More importantly, the domain of conservation and management of forests in India has been overwhelmingly dominated by a top-down bureaucratic approach cherished by the Indian Forest Service's civil servants. As some scholars argue, this trend had perniciously affected both the understanding and practice of forest conservation in the country (Gopalakrishnan, 2010).

The unification of forest laws and the extension of scientific management became the most important considerations of forest administration immediately after India's independence, when the states also enacted legislations to consolidate the privately owned forest areas under the control of the state FDs (Kashyap, 1990 as cited in Bandi, 2011). Three important policy pronouncements were considered instrumental to streamlining the fundamentals of forest governance in India, specifically, the 1952 Forest Policy, the National Commission on Agriculture (NCA) 1976 and the 1988 Forest Policy (Saxena, 2000). The 1952 policy affirmed forestry as an important land use category and insisted on keeping one-third of the country's land area as forests. But, commercial exploitation of forests was given top priority, depriving the needs of local communities. This policy continued till 1976, with heavily subsidised forest land given to industries and businessmen in the pretext of 'national interest' and

---

[1] The Forest Act of 1865 and the Forest Policy of 1894 de-recognised communal property and restrictions were placed on forest dwellers' collection of forest products (Guha, 1983). Conservation programmes resulted in progressive encroachment of the rights enjoyed by tribals for centuries over fuel wood, timber, non-timber forest products and hunting (Shyamsundar & Ghate, 2011).

industrial development, which had significant negative impact on forests and forest dwellers (Bandi, 2011, p. 79).

The NCA introduced 'social forestry' in 1975 to develop forestry in the unproductive non-forest government and community lands after realising the importance of locals' support for protection of forests. Simultaneously, farmers were encouraged to plant trees in their private lands, resulting in the overlapping of the terms 'social' and 'on-farm' forestry. In 1988, the National Forest Policy effected a paradigm shift, as it sought to involve 'tribal people closely in the protection, regeneration and development of forests as well as provide employment to people living in and around forests' (MoEF, 1988).

However, the conservation of forests and their ecosystems were always skewed against the forest-dependent people. Even in the earlier conservation policies the control of forests and its resources also remained with the state, thus granting limited rights to local communities to use and manage resources. The seriousness of the conflicts between the state and communities is significant, notably in a situation in which local communities have limited options of income and employment. Spatial and socio-cultural factors spurring forest dependence also intensified the conflicts, as the communities' livelihoods have been historically rooted in the forests and its resources.

All these factors, including an excessive population, livestock dependence and the requirements of forest products to support livelihoods, had generated pressure on forest resources such as fuel-wood, fodder, timber, lumber and paper, causing massive deforestation in India. Notwithstanding these pressures, the early decades of planning and development had also intensified the massive destruction/degeneration of forests because of a spurt in large projects, from large dams and thermal power projects to huge mines and massive industrial complexes (GOI, 2009).

Thus, the unbridled deforestation and degradation of forests causing a decline in forest cover have long been sources of concern for policy makers in India. These concerns and the near failure of earlier policies and legislations figured as backlashes, inducing the nation-state to evolve a new governance paradigm and institutional arrangements for overcoming the impasse. It was in this historical context that India had devised a governance model for the forestry sector referred to as joint forest management (JFM) to develop strategies and action plans for sustainable community-based forest management.

A discourse on the underlying principles and the outcomes of implementing forest governance in the form of JFM in India does not have sanctity on its own because of the rich and diverse empirical perspectives on this topic. Nevertheless, a critical assessment of JFM and its current status of implementation across India merits attention in the emerging context of climate change risks, as forests and forest-based communities represent the most vulnerable segments of the society. Hence, it is important to consider how future JFM policies and action plans would integrate conservation and management goals with climate change mitigation and adaptation strategies (including 'Reduce Emissions from Deforestation and Forest Degradation' [REDD+]), while serving the interests of forest-dependent communities.

Against this backdrop, this chapter critically assesses the forest governance system, that is, JFM that India launched in 1990 alongside a broader vision of sustaining India's forests and forest-based ecosystems, while appreciating the rightful claims of forest-dependent communities. Following a critical overview of the status and implementation of JFM across Indian states, it discusses the major determinants and challenges confronting the sustainability of forest governance, drawing on empirical evidence from the south Indian state of Andhra Pradesh (AP).

## JFM in India: An Overview of Its Implementation and Current Status

In India, JFM has emerged as a landmark intervention in the management of forest resources. While West Bengal first introduced JFM as early as in the 1970s, the programme has been launched nationally since 1990, following the 1988 National Forest Policy. JFM specifies a concept of managing and improving forest resources by forging partnerships between forest user groups (local communities) and the FD. Yet, the JFM policy is not based on a constitutional legislation and is only being implemented through Government Orders (GOs) of the respective states within the framework of the 1990 National Guidelines.

JFM recognises the livelihood and sustenance needs of the people through the principle of 'conservation by participation'. The concept of JFM has been interpreted in various ways, but its basic element is to establish grass-roots community-based institutions to protect and manage forests. JFM aims at empowering locals in their active participation as partners in the management of forest resources and sharing the benefits derived from forests. The JFM optimises returns, minimises conflicts and links forestry development with the overall development of land-based resources. It also aims to build technical and managerial capability at the grass-roots level (GOI, 2009).

Following the announcement of national JFM guidelines in 1990, all of India's provincial (state) governments had adopted and started implementing it in their respective states with appropriate resolutions (Bahuguna, Mitra, Capistrano, & Saigal, 2004). The states of West Bengal, Haryana and Odisha (formerly Orissa) have already completed two decades of the JFM programme (as they had initiated it on their own much before), while others, such as Assam, Sikkim and Mizoram, issued enabling orders in 1988. As of March 2008, there were 113,036 JFM committees in 28 Indian states. The area co-managed by these committees is measured at 22.02 million hectares (Mha). Approximately 8.3 million families are involved in JFM efforts, while the number of families indirectly benefited by it is much higher (Bahuguna et al., 2004 as cited in Bandi, 2011, p. 85). Table 3.1 presents an overview of the trends and the status of forest cover, as well as the implementation of JFM across major states in India.

## JFM and the Formulation of Design Guidelines in India

A closer examination of the important policy directives for JFM in India helps reveal the extent to which the design guidelines have been incorporated into the framework for implementation at the grassroots level. In this respect, Table 3.2 presents the chronology of policy directives and legislative incorporation vis-à-vis a description of the important features of JFM policy directives adopted by the Government of India over time. The policy directives seem to have been concerned by integrating the design guidelines with respect to the (a) involvement of local communities and NGOs in the regeneration of degraded forests; (b) principles of benefit distribution; (c) multi-level implementation and monitoring of systems and (d) conflict resolution.

## Outcomes of JFM in India: Some Evidence

An analysis of the impact of policy directives is important for determining whether the design guidelines on JFM have been properly incorporated into the policy directives to generate beneficial outcomes for

**Table 3.1:**
*Trends and status of implementation of JFM in major Indian states*

| States | Per cent share in TFA | Forest area as % of TGA, 2011 | No. of JFMCs | | Area under JFM ('000 ha) | | JFM area as % of total forest area |
|---|---|---|---|---|---|---|---|
| | | | 1990 | 2008 | 1990 | 2006 | |
| 1. Andhra Pradesh | 8.3 | 16.9 | 7,606 | 8,498 | 1,679 | 2,566 | 40.2 |
| 2. Bihar | 0.8 | 7.3 | 296 | 615 | 505 | 385 | 59.5 |
| 3. Chhattisgarh | 7.8 | 41.2 | 6,412 | 7,887 | 3,391 | 3,276 | 54.8 |
| 4. Gujarat | 2.5 | 7.5 | 1,237 | 2,578 | 138 | 273 | 14.4 |
| 5. Haryana | 0.2 | 3.6 | 471 | 1,831 | 66 | 60 | 38.5 |
| 6. Himachal Pradesh | 4.8 | 26.4 | 914 | 1,749 | 111 | 425 | 11.5 |
| 7. Jharkhand | 3.1 | 28.8 | 1,379 | 10,903 | 430 | 2,190 | 92.8 |
| 8. Karnataka | 5.0 | 18.9 | 2,620 | 4,849 | 185 | 303 | 7.9 |
| 9. Kerala | 1.5 | 44.5 | 32 | 571 | 5 | 173 | 15.4 |
| 10. Madhya Pradesh | 12.3 | 25.2 | 9,203 | 14,428 | 4,126 | 5,947 | 62.8 |
| 11. Maharashtra | 8.0 | 16.5 | 2,153 | 12,473 | 687 | 2,685 | 43.3 |
| 12. Orissa | 7.6 | 31.4 | 12,317 | 10,647 | 783 | 880 | 15.1 |
| 13. Rajasthan | 4.2 | 4.7 | 3,042 | 4,882 | 309 | 770 | 23.6 |
| 14. Uttaranchal | 4.5 | 45.8 | 7,435 | 13,523 | 607 | 545 | 15.7 |
| 15. West Bengal | 1.5 | 14.6 | 3,545 | 4,192 | 488 | 625 | 52.6 |
| Sub-total | 72.1 | 22.2 | 58,662 | 99,626 | 13,511 | 21,103 | 38.0 |
| All India ('000 ha/%) | 76,954 | 21.1 | 62,890 | 113,036 | 14,255 | 22,018 | 28.6 |

TFA, total forest area; TGA, total geographical area; JFMC, joint forest management committee.
**Source:** GOI, MoEF.

**Table 3.2:**
*Important features of policy directives and design guidelines on JFM in India*

| Policy directive | Main features of design guidelines |
|---|---|
| 1. The Circular (first) Concerning JFM, 1990 | • Involvement of village communities and NGOs in regeneration of degraded forests<br>• Benefits of participation should go to village communities |
| 2. JFM Cell Creation Notification, 1998 | • For monitoring impact of JFM carried out by state governments |
| 3. Standing Committee Notification, 1988 | • Advise on all operational aspects of JFM for its expansion to non-forest areas |
| 4. Terms of Reference Notification, 1999 | • Sharing of experiences of JFM implementation as each state passed its own resolution<br>• Monitoring JFM programmes |
| 5. Notification for JFM Network, 2000 | • To act as regular mechanism for consultation between various agencies involved in JFM at the national level |
| 6. Guidelines for Strengthening JFM, 2000, 2002 | • Present latest JFM policy directives and broad framework for its implementation<br>• Measures such as legal support for JFM committees, promotion of women's participation and conflict resolution<br>• Memorandum of understanding between forest department and JFM committees outlining short and long-term roles and responsibilities, and pattern of sharing of usufructs<br>• Capacity building for management of non-timber forest products for providing remunerative prices for users |
| 7. Operational Guidelines for Tenth Five Year Plan, 2002–2007 | • Formulation of National Afforestation Programme to facilitate sustainable forest development<br>• Implementation of afforestation schemes via a two-tier system consisting of Forest Development Agencies and JFM Committees<br>• Transfer of funds to JFM Committees through Forest Development Agencies |

*Source:* Balooni and Inoue (2009, p. 7).

forest-dependent communities. In this regard, a comprehensive review of empirical literature on various aspects of JFM and its performance across Indian states (Adhikari, 2005; Agarwal & Chhatre, 2006; Bawa, Joseph, & Setty, 2007; Dutta, Roy, Saha, & Maity, 2004; D'Silva & Nagnath, 1999; Ghate & Ghate, 2010; Ghate, Mehra, & Nagendra, 2009; Murali, Murthy, & Ravindranath, 2006; Patel, Subhash, Tripathi, Kaushal, & Mudrakartha, 2006; Pandolfelli, Meinzen-Dick, & Dohrn,

2007; Sahu, 2008) highlights several of its positive outcomes, viz., (a) increase in employment and household incomes; (b) increase in biomass production of fuel wood and fodder; (c) reduction in distance travelled and time expended by households (particularly women) for collecting fuel wood and fodder; (d) rise in income from non-timber forest products; (e) creation of long-term incentives through transfer of net income from sale of forest produce; (f) provision of livelihood security through forest regeneration; (g) women's autonomy and greater representation in JFM and (h) increased representation of marginalised sections in JFM executive committees. In the case of 'g' and 'h', women's increased representation occurred because of the reserved quotas for females and other under-represented sections in the JFM of their respective states.

At the same time, the JFM as implemented in India, has been facing innumerable problems as revealed by several studies. The regenerated forests themselves have become valuable assets that became sources of conflict among communities, threatening sustainability of JFM (Joshi, 1999). A major contentious issue highlighted by many studies pertains to sharing of benefits among communities. For instance, in Karnataka, only one-third of the community-assigned forest produce was distributed by the FD to members in 18 JFMs in the Uttar Kannada district (Damodaran, 2000). Moreover, significant variations exist across states in the benefit-sharing arrangements between village communities and FDs, though the states have passed their own resolutions to resolve these variations.

Studies also differ with respect to JFM's beneficial outcomes in India in terms of (a) poverty reduction; (b) welfare and (c) social security. A study from Jharkhand reveals that the wealthier sections benefited from JFM at the expense of the poorer sections (Kumar, 2002). Similarly, the principle of equity in benefit sharing has been a major casualty (Ghate, 2004), and marginalised groups such as shifting cultivators and head loaders were denied access to the forests in the JFM programme based on claims of forest protection (Carter & Gronow, 2005). Matta's (2006) study in Tamil Nadu revealed inequity in the participation in JFM activities and benefit sharing and a lack of adequate provisions for extending individual assistance to the poor and erstwhile forest users. Furthermore, a review by Matta and Kerr (2004) of 278 forest communities suggests that in most states, the forest committees created under JFM have not lasted long. In many cases, the forest protection committees (FPCs) became dysfunctional after either the initial enthusiasm evaporated or the incentive money was exhausted (Ghate & Nagendra, 2005; Kumar, 2002; Matta & Kerr, 2004).

JFM was launched in India under a variety of names, such as community forest management (CFM) or community based forest management (CBFM), participatory forest management (PFM), and joint forest programme (JFP), and a plethora of empirical evidence underscores that even after more than two decades of continued promotion and scaling up, such community-based forest management initiatives have not fulfilled their goals (Bandi, 2013). A prominent reason cited for the sub-optimal performance of JFM has been the colonial legacy (Saikia, 2008) and the 'command and control' principle that dominated the forest management regimes in most states over the past six decades under the planning process.

While there are apprehensions concerning the overall beneficial outcomes of JFM in India, an analysis of the changes in forest cover over the past two decades reveals noteworthy trends in major states. For instance, while the national forest cover declined at a rate of 0.6 Mha between 1991 and 1997, the period henceforth has shown significant growth in forest cover. National forest cover increased from 67.55 Mha in 2001 to 69.20 Mha in 2011, resulting in a net increase of 0.16 Mha per annum during the period. The largest increase in forest cover occurred in dense canopy forests, while medium density forests have declined. To a large extent, this growth in forest cover has been attributed to afforestation programmes and the forest protection policies pursued by the national and state governments under the JFM (FSI, 2009).

The trends in forest cover of India's major states are presented in Table 3.3. Between 2001 and 2011, the forest cover improved in all states except Gujarat, Haryana, Karnataka, Rajasthan and Chhattisgarh.

**Table 3.3:**
*Trends in forest cover in major Indian states, 2001-2011*

| States | Forest cover 2001 (km²) | % TGA | Forest cover 2011 (km²) | % TGA | % change between 2001 and 2011 |
|---|---|---|---|---|---|
| Andhra Pradesh | 44,637 | 16.2 | 46,389 | 16.9 | 3.9 |
| Bihar | 5,720 | 6.1 | 6,845 | 7.3 | 19.7 |
| Gujarat | 15,152 | 7.7 | 14,619 | 7.5 | −3.5 |
| Haryana | 1,754 | 4.0 | 1,608 | 3.6 | −8.3 |
| Himachal Pradesh | 14,360 | 25.8 | 14,679 | 26.4 | 2.2 |
| Karnataka | 36,991 | 19.3 | 36,194 | 18.9 | −2.1 |
| Kerala | 15,560 | 40.0 | 17,300 | 44.5 | 11.2 |

*(Table 3.3 Continued)*

*(Table 3.3 Continued)*

| States | Forest cover 2001 (km²) | % TGA | Forest cover 2011 (km²) | % TGA | % change between 2001 and 2011 |
|---|---|---|---|---|---|
| Madhya Pradesh | 77,265 | 25.1 | 77,700 | 25.2 | 0.6 |
| Maharashtra | 47,482 | 15.4 | 50,646 | 16.5 | 6.7 |
| Orissa | 48,838 | 31.4 | 48,903 | 31.4 | 0.1 |
| Rajasthan | 16,367 | 4.8 | 16,087 | 4.7 | −1.7 |
| Tamil Nadu | 21,482 | 16.5 | 23,625 | 18.2 | 10.0 |
| Uttar Pradesh | 13,746 | 5.7 | 14,338 | 6.0 | 4.3 |
| West Bengal | 10,693 | 12.0 | 12,995 | 14.6 | 21.5 |
| Jharkhand | 22,637 | 28.4 | 22,977 | 28.8 | 1.5 |
| Chhattisgarh | 56,448 | 41.8 | 55,674 | 41.2 | −1.4 |
| Uttarakhand | 23,938 | 44.8 | 24,496 | 45.8 | 2.3 |
| Sub-total | 473,070 | 20.3 | 485,075 | 21.0 | 2.5 |
| All India | 675,538 | 20.6 | 692,027 | 21.1 | 2.4 |

**Source:** GOI, MoEF, Forest Survey of India, Reports, 2001 and 2011.

While the average decadal increase in forest cover was closer to 2.5% nationally, a few states, viz., West Bengal, Bihar, Kerala, Tamil Nadu and Maharashtra, reported notable increases in forest cover because of conservation efforts.

## Community Forest Management in Andhra Pradesh: Determinants and Challenges

Following the discussion of the scenario on the status of implementation and outcomes of JFM, this section examines JFM's major determinants and challenges based on an empirical case study conducted in the south Indian state of AP.

AP is the fifth largest of the 29 Indian states in terms of area and population. According to the Forest Survey of India (2011), 63,814 km² of AP's total geographic area of 275,069 km² is forest area (i.e., 23.2% of the state's total area). AP ranks second (after Madhya Pradesh) in terms of forest area in India, with a relative share of 8.3% of the country's total forest area.

The FD in AP began implementing JFM in 1993 with the World Bank's (WB) support. *Vana Samrakshana Samithi*s (VSSs) or the FPCs are institutions established to protect the respective village forests. In 2003, JFM was modified into CFM by GO 13[2] issued in 2002, to make it 'more democratic' in the interest of the 'community' by relegating the FD to the role of mere 'facilitator'.

As of 2006, 8,412 VSSs were functioning in AP. Of these, 5,000 committees are sponsored by the WB, with the rest under centrally sponsored schemes such as the Forest Development Agency, Employment Assurance Schemes, the Rural Infrastructure Development Fund and the National Bank for Agriculture and Rural Development (APFD, 2006). These projects require funding to undertake various forest regeneration activities such as plantation and labour payment and 'entry point programmes' such as building 'village community halls', 'buying irrigational equipment' or other activities that encourage village development, thus keeping the community interested in the programme until its forests can yield sustained produce for the community.

## Empirical Site

For empirical analysis, the study selected one district from each of the three geographical regions of AP, viz., the Adilabad district of Telangana, the Chittoor district of Rayalaseema and Visakhapatnam district of Coastal Andhra. These three districts were selected because the number of VSSs functioning in these districts was the highest among other districts within their respective regions. The final sample comprised 11 VSS villages each from the Adilabad and the Visakhapatnam districts and 8 villages from the Chittoor district, thus totalling 30 villages. Among these VSS villages, 20 were sponsored by APCFM and the remaining 10 were non-APCFM supported.

Primary data were gathered using 'Focus Group Discussions' (FGD) and Household (HH) surveys using structured questionnaires. Data were also obtained using direct observation and informal conversations with the officials and the people concerned, individually and in groups. One FGD was moderated in each of the 30 VSS villages. The respondents comprised VSS General Body (GB) and Managing Committee (MC)

---

[2] The GO 13 is a comprehensive document detailing the constitution of the VSS. It also defines the duties and responsibilities of communities and the FD.

members, including the chairperson and vice-chairperson of the respective VSS. In addition, 360 VSS members (12 HH in each village) were individually interviewed at the HH level. Accordingly, the study conducted 360 HH interviews, comprising 132 interviews each in Adilabad and Visakhapatnam and 96 interviews in Chittoor. The field survey was conducted between 1 March and 31 August 2006.

Providing for an institutional mechanism to manage 'forests' does not indicate in itself that the objectives of an institution such as CFM will be achieved as desired. Many different factors beyond the purview of governance principles also determine the course of community-based institutions. The remaining section assesses the causes of such decisive issues that can influence the performance of the CFM programme in the AP districts.

Among the sample of 30 VSS villages, there were reports of conflicts of one form or the other, though the frequency of conflicts varied from one VSS to another. Most members in their respective VSS identified the major sources of conflicts as theft, smuggling, grazing, the conduct of VSS members and hostilities from non-VSS HHs in the same village (Table 3.4).

As evident, theft tops the list of sources of conflict, indicated by approximately 36% of VSS and the offenders reportedly were neighbouring villagers. Theft occurred as the neighbouring village HHs would suddenly find themselves barred from entering a forest from where they were collecting firewood for domestic use. In some cases, these HHs secured their livelihoods by selling firewood, while artisans such as

**Table 3.4:**
*Sources of conflict in VSS (%)*

| Sources of conflict | Adilabad | Chittoor | Visakhapatnam | Overall |
|---|---|---|---|---|
| 1. Theft | 38.97 | 16.7 | 50.0 | 35.7 |
| 2. Smuggling | 22.2 | 5.6 | 0 | 8.9 |
| 3. Grazing | 16.7 | 27.8 | 10.0 | 17.9 |
| 4. VSS member's conduct | 16.7 | 38.9 | 30.0 | 28.6 |
| 5. Hostilities between VSS and non-VSS HHs | 5.6 | 11.1 | 10.0 | 8.9 |
| Total (*n*) | 100 (18) | 100 (18) | 100 (20) | 100 (56) |

*n* = this is higher than the sample number of VSS in their respective districts due to multiple responses from respective VSS.
**Source:** Field survey (FGD).

ironsmiths and washer men collected firewood to sustain their traditional occupations. In the opinion of the VSS members, most offending neighbours were finally convinced to not enter their forests, though some time was required to convince them. Now, they are reportedly using alternatives to meet their firewood needs. Some adamant villages were also awarded with a VSS to pacify them. However, people who exploit the forest for smuggling continue to clash with VSS members (two villages in Adilabad) quite often. In one of the Visakhapatnam villages, some VSS members narrated how they were beaten up by their neighbouring villagers while passing through the latter's village as a revenge of their opposition to permit them to steal wood from their forest.

Conflicts arising from smuggling were reported by 9% of the VSS, with Adilabad reporting the highest incidence (22%). Smuggling is a more serious offence compared with theft with respect to social outcomes. All the VSS reporting problems of smuggling are located in districts known for timber trade, which fetches good remuneration in open and black markets. Smuggling was not reported by VSS villages in Visakhapatnam because of their remoteness, poor road connectivity and absence of priced timber. Although one mixed caste village in Chittoor does not grow high-valued timber in its forest, its VSS members claim to suffer from smugglers' offences because the offenders use a route that bypasses their village to smuggle red sanders from the forest situated on the other side of the hill adjacent to their forest. They are threatened by dire consequences if their activity is reported to the FD or the police.

Grazing is yet another problem that creates a conflicting situation in 18% of the VSS villages (Table 3.4). It happens when cattle from neighbouring villages stray unintentionally or sometimes are herded deliberately into the VSS forests. The magnitude of the problem in one mixed village (in Chittoor) was severe, as thousands of goats from neighbouring villages were herded into its forest every day for grazing, thus destroying its vegetation in an unprecedented manner. Countless efforts were made by the VSS members in containing this menace unleashed by their neighbours. Consequently, the VSS members have not been able to salvage even 10% of the plantation. They were disillusioned after fighting unending and hopeless battles with the goat herders and not receiving support from the FD, despite lodging complaints repeatedly.

The most discouraging aspect of all conflicts concerns the internal fights amongst the VSS members reported in approximately 29% of the

VSS villages (Table 3.4). These fights occur approximately uniformly across tribal and mixed villages in all three districts. The cause of the internal fights among members is the non-payment of wages on time and even non-payment of expected wage-rate by their respective chairpersons. The internal fights related to non-payment of wages happen because ground-level FD officials cheat them and deprive them of their rightful and anticipated share. Invariably, the quarrels flare up when members are inebriated. Overall, the conflicts appear to be occurring more frequently among the mixed villagers compared with tribal villagers, to some extent because of the dynamics in the composition of the members.

Though minuscule in presence, a smaller segment of the sample VSS (8.9%) reported clashes between members and non-members for reasons listed in Table 3.4. The non-VSS HHs alleged that their chairperson and MC members deliberately barred them from VSS membership. Personal rivalry is cited as the reason in some VSS villages, while party politics laced with caste discrimination is cited in others. In some villages, VSS members accused non-VSS members of being greedy, wanting to share the forest benefits enjoyed by them. This problem arises occasionally in one of the study villages, while the issue has reached the court in another village. The hostility-induced battle continued for over a year, thus hampering regular VSS activities because the FD had kept the alleged VSS under suspension since then. In another mixed caste village (Chittoor), though there was no direct conflict between VSS and non-VSS members, there existed uneasy relations between members, probably due to caste/community differences.

## Podu/*Encroachment and the CFM Interface*

The term *podu* is elucidated differently by different people. Generally, *'podu'* refers to 'slash and burn' cultivation, a lengthy procedure implying the clearance of forests on hill slopes, burning trees and growing crops in the ashes, and after a certain period, shifting the cultivation to a new hill slope to allow soil fertility and vegetation on old plots to regenerate. However, because of the increasing tribal population and reduced forest cover over the past century, *podu* rotations have been reduced to 2 or 3 years from 10 years of fallow period. It was observed that the communities indulging in shifting cultivation are the poorest sections. Additionally, in the absence of inputs such as bullocks, agricultural implements and manure for *podu* cultivation, the tribals grow different types of cereals and vegetables in a single plot and often, this results in insecure and uncertain output and yield along with attendant legal consequences.

Given APFD's stance on *podu*, that is, treating it as an encroachment (APFD, 1999), and the accusations against it for seizing tribals' traditional lands justified by J/CFM, the VSS respondents were asked to share their experience in this regard. Members in 11 of these VSS acknowledged cultivation as having implications for J/CFM (Figure 3.1).

Only 11 of 30 VSSs reported a *podu*/encroachment problem in their villages. Of the 132 VSS members in these 11 villages, 67.4% were affected by the initiation of CFM for merging their land with VSS forest. The mode of acquisition differed in each case. While some VSS members claimed to have surrendered willingly, others faced forceful evacuation. The most affected district was Visakhapatnam, notably the tribal villages.

When the affected respondents were further probed regarding how the loss of land affected them, the reasons cited included (a) reduced income, (b) food insecurity and (c) dependence on another's field for work as a labourer. The worst affected were those suffering from food insecurity, because they had little alternatives for their livelihoods, as they either were not assured wage-works from the VSS or could not find work.

**Figure 3.1:**
*Member responses about the influence of merging lands with CFM*

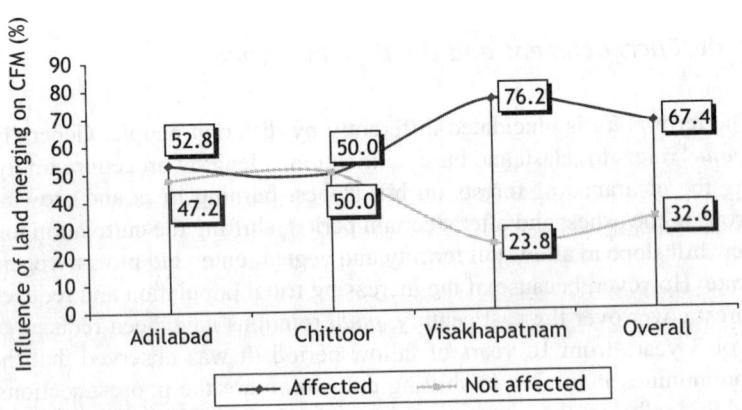

*Source:* Field survey (HH).

## Methods Used by the FD to Acquire *Podu/Encroached Lands*

Concerning the methods used by the FD to recapture *podu* lands from the tribals, VSS members revealed that the FD employed convincing, compensatory and coercive tactics selectively with the communities, based on their background. In the case of more aggressive communities, the FD used convincing methods or provided faster compensation. Naïve communities such as the primitive tribal groups were rarely treated respectfully, and their lands were simply seized and merged into the VSS forest. Conversely, the more compassionate the officer was, the better the alternatives provided to the community.

A section of VSS members were furious for not receiving the promised compensation. When the issue was raised to the FD in the jurisdiction of the aggrieved VSS members, the officials denied any such promises of compensations, instead maintaining that the land belonged to the FD, and hence, compensation to reclaim it was unnecessary. Some officers even demanded a penalty from the offenders to send strong signals to prevent others from such encroachments.

Thus, it emerges that the *podu*/encroachment continues to have negative impact on J/CFM performance, because of the rigid and indifferent approach of the FD towards the tribals practising *podu*, that is, ignoring their traditional rights and economic implications. Hence, many tribals resorted to the cultivation of their previous land (officially now controlled by the VSS), while some tribals are considering returning to their former lands or wish to occupy new lands.

## Coordination between CFM and Other Departments, Including NGOs as Determinants

### Interdepartmental Coordination

The Revenue Department, Tribal Welfare Department, Livestock Department, Department of Rural Development, Integrated Tribal Development Authority (ITDA), Defence Research and Development Organisation, Drought-Prone Areas Programme (DPAP), Girijan Cooperative Corporation (GCC), *Velugu*,[3] NGOs and self-help groups are some of the

---

[3] The Indian Government has issued GO MS No.78, EFS and T (For III) Department: 17-10-2003 on the convergence of the Velugu project with CFM to ensure greater

government departments, cooperatives and organisations that influence the outcomes of the CFM programme. All these organisations work in the same jurisdiction as the CFM. Hence, interdepartmental coordination becomes vital. The role of many of these line departments is limited to an advisory level, while certain bodies, organisations and cooperatives such as the panchayat, ITDA, GCC and NGOs have a significantly greater role to play in the day-to-day operations of the CFM.

In this section, we examine the role of some important organisations that influence the outcomes of the CFM in AP. The analysis is based on the views expressed by VSS members during the FGD. Table 3.5 presents the inter-departmental coordination and role of few organisations in influencing the CFM process in the study villages. It also describes the role of various line departments/agencies in influencing the CFM activities. For instance, the Revenue Department assumes a supposedly key role in addressing land rights and the sensitive issue of encroachment. However, it is also being criticised by the FD for issuing *patta*s to people without verifying their land credentials.

Furthermore, when the VSS members were asked to reflect on the role of NGOs, members of only one of the 30 VSS villages acknowledged that NGOs played an active role in creating awareness among members about the CFM programme. By contrast, members of nine VSSs reported that NGOs limit themselves to the preparation of a micro plan in the initial stages of J/CFM. Moreover, only a few of the members are said to have conducted a participatory rural appraisal to prepare the micro plan document. Members in most VSS (67%) referred to NGO's presence as negligible. Of these, 12 VSSs (40%) are based in remote tribal areas, where the presence of NGOs is most necessary. Sharing their bitter experience with an NGO, some tribal VSS members in Visakhapatnam recounted how they were tired of paying ₹750 to their NGO every time they received their VSS wages.

---

convergence between these two projects and the Tribal Welfare Department. Convergence with DPIP and APRPRP projects, popularly called Indira Kranthi Pathakam (formerly Velugu), is being implemented in Andhra Pradesh through financial assistance from the World Bank. Because the development objectives and most of the target groups of these projects and the Andhra Pradesh Community Forest Management (APCFM) project overlap, a mechanism for convergence has been devised to ensure an effective implementation strategy and the non-overlap of investments, and orders in this regard have been issued in GO MS No.78 EFS and T (For III) Department: 17-10-2003.

**Table 3.5:**
*Inter-departmental coordination and CFM outcomes*

| S. no. | Department | Role and influence on CFM outcomes |
|---|---|---|
| 1 | Revenue Department (RD) | • Plays a key role while managing land rights, particularly the sensitive issue of encroachment |
| | | • As alleged by some FD officials, the Revenue Department issues '*pattas*' to people without verifying their credentials on lands that are deep in the reserved forest |
| | | • According to VSS, this lack of verification not only creates bad relationships between people and the FD but also generates mistrust between them |
| 2 | Integrated Tribal Development Authority (ITDA) | • VSS members maintain positive relations with the ITDA, as they benefit from the ITDA schemes |
| | | • In the Visakhapatnam tribal belt, many VSS members are grateful to the ITDA not only for helping them with coffee, silver oak and pepper plantation and throughout the stages of these plants' growth but also for their enhanced economic condition |
| | | • As per the respondents, the ITDA has an active presence only in the Visakhapatnam district |
| 3 | Girijan Cooperative Corporation (GCC) | • The VSS members have no faith in the GCC |
| | | The members now consider the GCC an exploiter identical to the *Sahukars* |
| 4 | Gram panchayat (GP) | • The role of panchayats, according to GO 13 (2002), commences from marking of boundaries of the VSS forests when the Forest Range Officer (FRO) consults the sarpanch of the particular village in completing the task |
| | | • The sarpanch is a member of the advisory council constituted by the FRO and chairs its meetings |
| | | • Although each VSS has one or two members from the panchayat, sometimes even from ranks such as the sarpanch, they have never observed interactions between the VSS and the panchayat |
| 5 | Non-government organisations (NGOs) | • CFM envisages a positive role for the NGOs in its set-up through the CFM GO 13 (2002). They are expected to prepare micro plans, train VSS members on marketing and value addition of the forest produce, micro credits, accounting procedures, awareness creation/capacity building and conflict resolution |

*Source:* Authors' compilation.

Similarly, many other NGOs were not too enthusiastic and claimed that they were no longer interested in VSS because of non-cooperation from the FD and because they felt the programme had lost its purpose and objective. When asked why NGOs do not like to work in remote VSS villages where their presence is important, most of the NGO staff questioned why government teachers dreaded to go to such remote villages/hamlets.

### Training and Capacity Building

We also consider training and capacity building as an important determinant of CFM performance in the villages studied. Under the CFM project, the FD claimed to have undertaken capacity building and awareness building programmes by training the VSS members, particularly the chairpersons, vice-chairpersons and MC members. Emphasis was placed on training in financial management and book-keeping activities for the selected VSS members to improve their skills in record maintenance, minutes' preparation, forest management, orientation and demonstration. Training was also provided on specific forest-based micro-enterprises to provide VSS members with synergy and sustainable dependence on forests. The use of a global positioning system (GPS), map interpretation, silviculture nursery raising and grafting of high-yielding varieties were some of the other areas in which the FD set agendas to train VSS members.

The members in the sample VSS were asked whether they had attended a training programme and whether it benefited them and their VSS. The response was far from satisfactory, as only 19% of the 360 sample VSS members opined that they had benefited from it. Most such members were chairpersons, vice-chairpersons or MC members and on the utility of the training, half of them acknowledged it as beneficial, while the other half did not agree. Those reporting 'not benefited' reflected that the training was more of a tourist exposure, as they were taken to places of interest in Hyderabad city. However, some members expressed that they personally benefited from the training programme, as they fully understood the VSS concept after attending a couple of training camps. Women members in a few VSSs were happy to have been trained in making incense-sticks and leaf-plates.

None of the members in the sample VSSs reported receiving higher training such as for the usage of a GPS, map interpretation or micro-enterprises, as envisaged by CFM training. The trainings, in most cases, were limited to the basics of plantation techniques, book-keeping and account maintenance. Many members confessed that they did not relay to their GB what they learnt in the training, because their GB was uninterested; the chairperson cared little to perform the necessary functions, or in some cases, the incompetence of the chairperson in imparting the training was very much evident.

### External Funding for CFM as a Determinant

In the context of the emerging presence of wage-works as the backbone to sustain the CFM programme (Reddy, Reddy, Saravanan, Bandi, & Springate, 2004), we enquired 'whether the VSS members agree with such a notion wherein the CFM lasts as long as it provides wage-works and not after it?' Approximately 33% of respondents apprehended that VSS (and thus, CFM) will lose its meaning in the absence of monetary incentives (Table 3.6). Already, some VSS members have moved away from the CFM. In such VSS, the forest has neither reached a sustainable level, nor is likely to reach a beneficial stage because of the poor management by the FD and the VSS, because of financial and other irregularities in implementation.

About 33% of the VSS members were hopeful of continuing with CFM even in the absence of external funds. Some of them had evolved a unique arrangement of planting cashew in individual plots, to sustain their members' interest in the VSS. Hence, they are individually motivated to reap definite benefits from these plots, even if the benefits from the larger portion of the VSS forest require more time to yield sustainable benefits.

**Table 3.6:**
*Response of VSS members as regards the sustenance of CFM with external funding for wage employment (%)*

| Response | Adilabad (n = 132) | Chittoor (n = 96) | Visakhapatnam (n = 132) | Overall (n = 360) |
|---|---|---|---|---|
| 1. CFM sustain only through external support for wage employment | 18.5 | 25.0 | 54.5 | 33.3 |
| 2. Not necessarily | 36.4 | 37.5 | 27.3 | 33.3 |
| 3. Neither | 45.1 | 37.5 | 18.2 | 33.3 |
| Total | 100 | 100 | 100 | 100 |

**Source:** Field survey (household).

Another 33% members responded neither way, thus indicating that for all practical purposes, their VSSs are defunct. In Adilabad, 45% of the members concur, followed by Chittoor (37%) and Visakhapatnam (18%).

## Climate Change Adaptation and Mitigation Strategies for India's Forests: Can JFM Model the Way?

The ensuing climate change risks are most likely to cause dramatic changes in the forest landscape of India. An assessment of the impact of climate change on India's forest ecosystems based on climate projections of the Regional Climate Model of the Hadley Centre (HadRM3) shows that nationally, approximately 45% of the forested grids might change. A vulnerability assessment showed that while such vulnerable forested grids are spread across India, their concentration is highest in the upper Himalayan stretches, parts of central India, the northern Western Ghats and the Eastern Ghats. Low tree density, low biodiversity status, higher levels of fragmentation and climate change increase the vulnerability of these forests.

Preliminary assessments at the national level indicate major changes, such as (a) shifts in forest boundaries; (b) changes in species assemblages or forest types; (c) changes in net primary productivity; (d) possible forest extinction in the transient phase and (e) potential losses of or changes in biodiversity. Even in a relatively short span of approximately 50 years, most of India's forest biomes seem highly vulnerable to projected climate changes. Approximately 70% of the vegetation in India is likely to be sub-optimally adapted to its existing location, thus making it more vulnerable to the adverse climatic conditions as well as to the increased biotic stresses. These impacts on forests will bear adverse socio-economic implications for forest-dependent communities, as approximately 173,000 villages are classified as forest villages (Gopalakrishnan, Jayaraman, Bala, & Ravindranath, 2011).

These eventualities necessitate effective conservation plans for sustaining forests, given the dependence of livelihoods on them and protecting them against the adversities of climate change. Because forests have immense potential for mitigating climate change through their carbon sequestration, the UN Framework Convention on Climate Change launched a major initiative, REDD+, in 2008. In the forest sector, mitigation strategies entail reducing emissions from deforestation and forest degradation; increasing the role of forests as carbon sinks and substituting products, such as using wood rather than fossil fuels for energy, and forest products rather than materials whose manufacture involves high green house gas emissions.

Within the REDD+ framework, initiatives (including financial transfers) are underway at the global level for conservation and sustainable management of forests and the enhancement of forest carbon stocks. India launched the National Action Plan on Climate Change in 2008, called the National Mission for a Green India (GIM). The mission aims to address climate change by (a) enhancing carbon sinks in sustainably managed forests and other ecosystems through afforestation of 6 Mha and the revival of degraded forestlands; (b) enhancing the resilience and ability of vulnerable species/ecosystems to adapt to changing climates and (c) enabling the forest-dependent communities to adapt various measures to overcome the climatic risks. The implementation of REDD+ in India is being proposed through the JFM, the FD and the forest-based communities.

However, the critical issue is how the forest-dependent communities and the JFM beneficiaries can be effectively integrated with the REDD+ initiatives in India such that their stake is enhanced with mechanisms and arrangements for the proper distribution of welfare gains. REDD+ initiatives must also incorporate the VSS (under the J/CFM) to act as custodians of the forest conservation funds to sustainably manage forest resources and forest ecosystems. Thus, developing and implementing adaptation strategies to minimise the possible adverse impacts of climate change risks are necessary. In this regard, the framework described in Figure 3.2 shows how the JFM can be effectively integrated with climate change adaptation and mitigation strategies encompassing four major components, viz., (a) carbon sequestration; (b) conservation of forest carbon stocks; (c) strengthening of adaptive capacities at the system level; and (d) strengthening adaptive capacities at the community level.

## Sustainable Community Forest Management in India: A Road Map for the Future

The issues discussed in this chapter are of a basic nature and ground-level and pertain to people living around the forests. The determinants were not entirely unforeseen, but the magnitude of their dimensions and nature

**Figure 3.2:**
*Framework for integration of JFM with climate change adaptation and mitigation strategies in India*

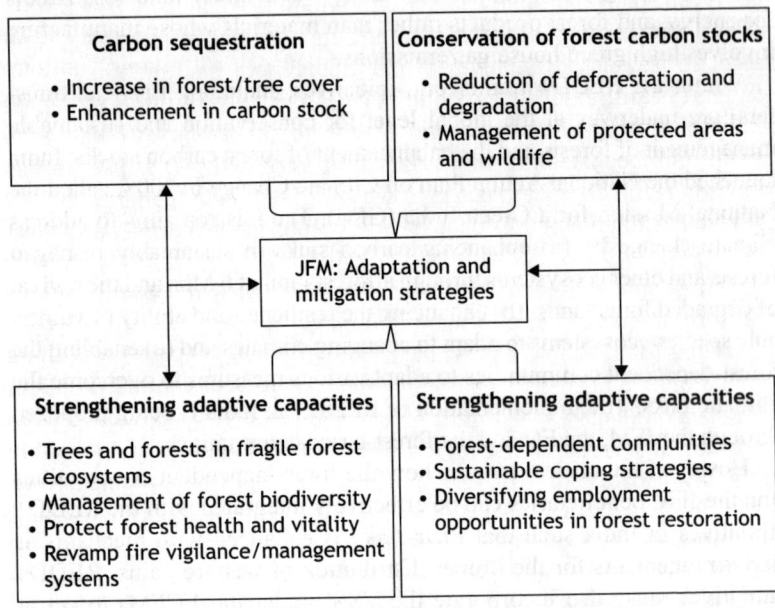

| Carbon sequestration | Conservation of forest carbon stocks |
|---|---|
| • Increase in forest/tree cover<br>• Enhancement in carbon stock | • Reduction of deforestation and degradation<br>• Management of protected areas and wildlife |

JFM: Adaptation and mitigation strategies

| Strengthening adaptive capacities | Strengthening adaptive capacities |
|---|---|
| • Trees and forests in fragile forest ecosystems<br>• Management of forest biodiversity<br>• Protect forest health and vitality<br>• Revamp fire vigilance/management systems | • Forest-dependent communities<br>• Sustainable coping strategies<br>• Diversifying employment opportunities in forest restoration |

*Source:* Modified by authors based on FAO (2012) framework for forests and climate change.

were unanticipated. However, addressing these challenges based on the presented case study could immensely benefit not only the forest-dependent people but also the forests in the larger context of the environment.

Forest governance in India according to JFM reveals that the implementation and success of the programme have changed substantially and manifested variably across states. While the programme had resulted in significant positive outcomes in several regions, the sustenance of J/CFM is thwarted by several challenges and issues at the grass-roots level. These challenges and issues emerge predominantly from the specific governance regime based on command and control principles followed by India over the past under the colonial and the post-colonial periods. Despite the tremendous achievements in forest management from changes in governance practices, the challenges of mobilising the local communities for collective action and protection of forests and forest ecosystems still persist.

Given the long tradition of community-based natural resource conservation practices that India has evolved over generations, sustainable forest management could have been possible, if the policies and governance institutions were efficient, proactive and more responsive in the post-independence period. Instead, the management regime continued to be greatly influenced by colonial policies and governance structures, without making inroads into the livelihoods of forest-dependent communities and forest resources. The contradictions in the current management system reinforces the conflicting interests between economic and ecological benefits or between the community and the state, particularly in situations in which the local communities are left with limited alternatives of income and employment, thus implying a situation of limited substitutability.

With rapidly changing global and local natural environments alongside resource-depleting development processes and climate change induced threats, the conservation of forests and their ecosystems should constitute the critical aspects of India's forest management policies and J/CFM regimes. As the forest biodiversity and ecosystems in India have been facing serious problems of degradation and exhaustion, it is important that alternative livelihood opportunities be explored for forest-based communities to reduce the current pressures on forests for agriculture and other livelihood demands. The efforts towards reducing the dependence on forests presupposes devising a long-term conservation strategy and agenda in which communities are integrated with the various types of conservation action programmes. Invariably, providing alternative sources of fuel wood should also form an integral part of the strategy. Alongside, the communities' contributions must be adequately compensated in terms of employment opportunities, incentives for conservation practices, and community involvement in the sustainable management and conservation of PAs and the ecosystem services they provide.

Ideally, shifting the pressure from the local communities should coincide with heightened vigilance of illegal extraction from the forests by strong social and political interest groups. The onus of promoting sustainable forest regeneration by simultaneously opening up new avenues for alternative sources of energy for local communities, therefore, may depend on various developmental schemes beyond forest management *per se*. The solution in terms of an alternative paradigm for addressing the growing need for fuel wood in general and for the local communities in particular thus requires immediate attention.

Finally, communities depending primarily on forests for their livelihoods must maintain this precious resource in a state of ecological balance. To do so, the communities must have an effective institutional mechanism to govern them. Such an institutional governance mechanism could also uphold conducive and proactive policies of the respective governments towards 'sustenance of ecosystem'. Even then, the type of institutional arrangements established that would constitute a foolproof system of governance of conservation or a similar system is a serious concern. Most likely, this chapter's proposed framework for the integration of JFM and the forest-dependent communities might facilitate the sustainable management of forests in India in the emerging context of climate change threats and communities' persistent dependence on forests for their livelihoods. The management framework suggested for the integration of JFM as the key institutional mechanism for implementing climate change adaptation and mitigation strategies would provide a sustainable road map for the conservation of forest-based biodiversity and ecosystems and the sustenance of the livelihoods of India's forest-dependent communities.

## References

Adhikari, B. (2005). Poverty, property rights and collective action: Understanding the distributive aspects of common property resource management. *Environment and Development Economics*, 10, 7–31.

Agarwal, B. (2001). Participatory exclusions, community forestry, and gender: An analysis for South Asia and a conceptual framework. *World Development*, 29(10), 1623–1648.

Agarwal, R. (1999). Van panchayats in Uttarakhand: A case study. *Economic and Political Weekly*, 37(38), 3881–3883.

Agarwl, A. & Chhatre, A. (2006). Explaining success on the commons: Community forest governance in the Indian Himalayas. *World Development*, 34(1), 149–166.

APFD (2006). Andhra Pradesh Community Forest Management Project Status Report for the Period up to 31-03-2006, for Mid Term Review. Retrieved from http://forest.ap.nic.in/JFM%20CFM/CFM/Status%20report_MTR.htm.

Bahuguna, V.K., Mitra, K., Capistrano, C., & Saigal, S. (eds) (2004). *Root to Canopy: Regenerating Forests through Community-State Partnerships*. New Delhi: Commonwealth Forestry Association.

Baland, J.M. & Platteu, J.-P. (1996). *Halting Degradation of Natural Resources: Is There a Role for Rural Communities?* Oxford, UK: FAO and Oxford University Press.

Balooni, K. & Inoue, M. (2009). Joint forest management in India: The management change process. *IIMB Management Review*, 21(1), 1–17.

Bandi, M. (2011). Forest governance in India with particular reference to Andhra Pradesh: A review of policy shift from state control to community participation. *Man and Development*, 33(4), 75–90.

—— (2013). *Tribals and Community Forest Management*. Jaipur: Rawat Publications.

Bawa, K.S., Joseph, G., & Setty, S. (2007). Poverty, biodiversity and institutions in forest-agriculture ecotones in the Western Ghats and Eastern Himalaya ranges of India. *Agriculture, Ecosystems and Environment*, 121, 287–295.

Carter, J. & Gronow, J. (2005). *Recent Experience in Collaborative Forest Management* (CIFOR Occasional Paper No. 43). Bogor Barat: CIFOR.

Damodaran, A.N. (2000). *Report on Economic Instruments for Biodiversity Conservation in India: An Assessment Exercise for the National Biodiversity Action Plan* (Technical Report No. 3). Bangalore: Indian Institute of Management Bangalore.

Dasgupta, P. (2008). Property resources: Economic analytics. In R. Ghate, N. Jodha, & P. Mukhopadhyay (eds), *Promise, Trust and Evolution Managing the Commons of South Asia*. Oxford, U.K.: Oxford University Press.

Dutta, M., Roy, S., Saha, S., & Maity, D.S. (2004). Forest protection policies and local benefits from NTFP. Lessons from West Bengal. *Economic and Political Weekly*, 39(6), 587–591.

D'Silva, E. & Nagnath, B. (1999). *Local People Managing Local Forests: Behroonguda Shows the Way in Andhra Pradesh, India* (Working Paper Series No. 3). California, USA: Asia Forest Network.

FAO (2012). *Global Forest Resources Assessment 2010 Main Report* (pp. xiii+340) (FAO Forestry Paper 163). Rome: FAO.

FSI (2009). *India State of Forest Report 2009*. New Delhi: Forest Survey of India (FSI), Ministry of Environment, and Forests (MoEF).

Ghate, R. & Ghate, S. (2010). *Joint Forest Management, Role of Communication, and Harvesting Behavior: Evidence from Field Experiments in India* (23 pp.) (SANDEE Working Paper No. 53–10) (October). South Asian Network for Development and Environmental Economics.

Ghate, R. Mehra, D., & Nagendra, H. (2009). Local institutions as mediators of the impact of markets on non-timber forest product extraction in central India. *Environmental Conservation*, 36(1), 51–61.

GOI (2009). *State of Environment Report, 2009*. Government of India, Ministry of Environment and Forests.

—— (2011). *State of Forests in India Report*. Forest Survey of India, Government of India, Ministry of Environment and Forests, New Delhi.

Gopalakrishnan, S. (2010). Blame the forest management system. *Economic and Political Weekly*, 45(49), 32–34.

Gopalakrishnan, R., Jayaraman, M., Bala, G., & Ravindranath, N.H. (2011). Climate change and Indian forests. *Current Science*, 101(3, August), 348–355.

Guha, R. (1983). Forestry in British and post-British India: A historical analysis. *Economic and Political Weekly*, 17(44), 1882–1896.

Joshi, A. (1999). *Progressive Bureaucracy: An Oxymoron? The Case of Joint Forest Management in India* (Network Paper 24a, Winter-98/99), ODI, London.

Kumar, S. (2002). Does "Participation" in common pool resource management help the poor? A social cost-benefit analysis of joint forest management in Jharkhand, India. *World Development*, 30(5), 763–782.

Matta, J.R. (2006). *Transition to Participatory Forest Management in an Era of Globalization Challenges and Opportunities.* Paper presented at the Eleventh Conference of the IASCP, Bali, Indonesia, June.

Matta, J.R. & John, K. (2004). *Reframing Joint Forest Management in Tamil Nadu through Payments for Environmental Services.* Presented at the International Association for the Study of Common Property Biennial Meeting, Oaxaca, Mexico, August 9–13.

MOEF (1988). National Forest Policy. No. 3-1/86-FP. Ministry of Environment and Forests (Department of Environment, Forests & Wildlife), New Delhi. Retrieved 15 August 2013, from http://moef.nic.in/downloads/about-the-ministry/introduction-nfp.pdf.

Murali, K.S., Murthy, I.K., & Ravindranath, N.H. (2006). Sustainable community forest management systems: A study on community forest management and joint forest management institutions from India. *International Review for Environmental Strategies*, 6 (1), 23–40.

Ostrom, E. (1990). *Governing the Commons: The Evolution of Institutions for Collective Action.* Cambridge, UK: Cambridge University Press.

Pandolfelli, L., Meinzen-Dick, R., & Dohrn, L. (2007). *Gender and Collective Action: A Conceptual Framework for Analysis* (CAPRi Working Paper No. 64). Retrieved from http://www.capri.cgiar.org/pdf/capriwp64.pdf.

Patel, R., Subhash, M., Tripathi, J.P., Kaushal, V., & Mudrakartha, S. (2006). Regeneration of teak forests under joint forest management in Gujarat. *International Journal of Environment and Sustainable Development*, 5(1), 85–95.

Reddy, R., Reddy, M.G., Saravanan, V., Bandi, M., & Springate, O. (2004). *Understanding Livelihoods Impact of Participatory Forest Management Implementation Strategies in India: The Policy Issue and Impacts of Joint Forest Management in India with Particular Reference to Andhra Pradesh: A Review* (Working Paper No. 62). Hyderabad, CESS.

Sahu, N.C. (2008). Socio-economic heterogeneity and distributional implications of joint forest management (JFM): An empirical investigation from Orissa. *Indian Journal of Agriculture Economics*, 63(4), 614–628.

Saikia, A. (2008). Forest land and peasant struggles in Assam, 2002–2007. *Journal of Peasant Studies*, 35(1), 39–59.

——— (2011). *Forest and Ecological History of Assam.* New Delhi: Oxford University Press.

Saxena, R. (2000). *Joint forest Management in Gujarat: Policy and Managerial Issues* (Working Paper 149). Anand: Institute of Rural Management.

Shyamsundar, P. & Ghate, R. (2011). *Rights, Responsibilities and Resources: Examining Community Forestry in South Asia* (SANDEE Working Paper No. 59–11) (May) 24pp.

Singh, C. (1986). *Common Property, Common Poverty: India's Forests, Forest Dwellers and Law.* New Delhi: Oxford University Press.

# 4

# Nepal: Evaluating Different Forest Management Regimes

*Ganesh P. Shivakoti, Birendra K. Karna,*
*Ambika P. Gautam and Makoto Inoue*

## Introduction

There is a growing trend towards decentralisation of forest management in the developing nations of South and Southeast Asia. Nepal is one of the leading countries in the region (and world) in terms of initiating innovative forest policies and management approaches. However, not all the decentralised forest management approaches in the country have been equally successful in meeting the objectives of reversing deforestation trends through sustainable forest management (see, e.g., Webb & Gautam, 2001, Schweik, Nagendra, & Sinha, 2003; Nagendra & Ostrom, 2011). These differences have led to a considerable debate in the country on the appropriateness of different forest management approaches. The debate has gained further momentum in recent years as the country has geared its efforts towards establishing a federal national governance structure through the Constituent Assembly.

This chapter analyses the forest governance arrangements in the centrally controlled national forestry, community forestry and community-based pro-poor leasehold forestry regimes and compares the effectiveness of the three forest management regimes in improving forest conditions. The objective is to provide inputs to the policy reform process that is underway in Nepal and many other Asian countries. The study is particularly important considering that Nepal is one of the leading countries in the world in terms of community-based forest management, which is currently undergoing political transition from a centralised system to a federal national governance system.

## Evolution of Forest Policies and Governance Systems in Nepal

In Nepal, serious public concern about managing forest resources began in the beginning of the 20th century. Initially, the focus of the government was the exploitation of quality forests for exports to India to earn revenue and the conversion of forests to farmlands to widen the tax base, which was followed by increased national control of forests. The nationalisation of forests in 1957, the formulation of stringent laws, the establishment of a separate Ministry of Forestry in 1959 and the expansion of the forest bureaucracy are evidence of increased national control of the resource. This approach to governance failed, and widespread deforestation and forest degradation occurred across the country from the 1960s to the 1980s (Gautam, Shivakoti, & Webb, 2004).

As a means to abate widespread deforestation and increase the supply of basic forest products for subsistence needs, the government initiated a participatory approach to forest management in the late 1970s. The community forestry programme that started in 1978 has continuously evolved under the aegis of supportive forest policies and legislations. The programme involves handing over of forest patches to local forest user groups (FUGs) for management and use according to a forest management plan mutually agreed upon by the user group and the concerned district forest office.

Since 1992, the government has also implemented a pro-poor leasehold forestry programme, which involves 40-year rent-free leasing of small areas (up to 10 ha.) of degraded national forest land to small groups (6–10 households) of local farmers living below the poverty line and with little or no private land resources.

More recently, a new policy of collaborative forest management has been introduced in eight *terai* (a narrow strip of lowlands extending east-west in the southern part of the country) districts with the objective of enhancing participation of distant forest users and local government bodies in forest management. According to this policy, contiguous large blocks of forests in the *terai* and inner-*terai* are to be managed as national forests under a collaborative management arrangement.

Despite the introduction and implementation of different types of innovative decentralised policies, the centralised forest management system introduced after the nationalisation of forests in 1957 still remains the dominant forest management system in the country, at least in terms of its spatial coverage.

Alongside the changes in policy, the process to expand forest bureaucracy and change the organisational structure of the Department of Forests continued. Significant among those processes were the changes of 1976, 1983, 1988 and 1993 (Gautam et al., 2004). The department, which was established in 1942 with 3 regional and 12 divisional forest offices under it, now has 74 district forest offices, 92 *ilaka* (sub-district) forest offices and 698 range posts. Currently, over 6,000 employees are working in these different-level offices, which represents a three-fold increase in staffing over the last five decades.

## Changes in Forest Conditions

Nepal's forest area has continuously decreased since 1965. An analysis of the results of two of the most comprehensive surveys (i.e., the Land Resources Mapping Project (LRMP), 1986 and National Forest Inventory 1987–1998) shows that the forest area in the country decreased by 24% during the period of 1979–1994, and the area under shrubs increased by 126% during the same period (DoFRS/FRISP, 1999; LRMP, 1986). The National Forest Inventory results of 1994, which were published in 1999, show a 4.3 million hectares (Mha) (29%) area under forest cover and an additional 1.6 Mha (10.6%) under shrubs. According to another analysis of land use changes carried out by the Central Bureau of Statistics (CBS), a total of 1,560,000 ha of forest area were lost during the period of 1991–2001, with an average annual rate of 2.7% (CBS, 2008). A more comprehensive survey of forest resources for the country is currently being conducted with financial and technical supports from the Government of Finland.

There are regional variations in terms of changes in forest conditions. For example, between 1965 and 1986, the highest rate of deforestation took place in the *terai* followed by the *siwalik*s. This trend of high rates of forest loss in the *terai* and *siwalik*s has continued since then. The middle mountains may have gained some forest cover in recent years due to the community forestry programme (e.g., see Niraula, Gilani, Pokharel, & Qamer, 2013).

## Methods

### Study Sites

This study is based on data collected from six forest sites in Nepal. Two of the forests (Karne and Nehit) were under the direct control of the forest department, two (Baramchi and Bhagawatisthan) were under the leasehold forestry programme and the remaining two were under community forestry regimes. Additional details about the sites are presented in Table 4.1.

### Data Collection

The data for all the sites were collected at two different times by teams composed of a forester, botanist, social scientist and local helpers using International Forest Resources and Institutions (IFRI) protocols. IFRI is a long-term, multi-country research programme that collects data on forests and users using standardised interdisciplinary data collection protocols. The main objectives of the IFRI programme are to examine linkages and interrelationships between and among (a) institutional arrangements, (b) forests or other resource base conditions, (c) harvesting and other management activities, (d) rules in use and (e) resource impacts over time. The fundamental question the programme attempts to answer is how diverse institutions affect forest conditions and forest sustainability (Wollenberg, Merino, Agrawal, & Ostrom, 2007).

Randomly selected circular plots composed of three concentric circles of 1, 3 and 10 m in radius were used to collect the botanical data necessary to measure forest conditions. In the innermost circle (1 m radius),

**Table 4.1:**
*Description of the study sites*

| District | Tanahu | Parbat | Sindhupalchowk | Kabrepalanchowk | Makawanpur | Jhapa |
|---|---|---|---|---|---|---|
| Forest name | Ahal Danda | Bhadkhore | Baramchi | Riyale | Karne | Nehit |
| Forest governance type | Community forest | Community forest | Leasehold forest | Leasehold forest | National forest | National forest |
| Forest size (ha) | 75 | 58 | 15 | 42 | 200 | 125 |
| Number of households in the FUG | 183 | 140 | 17 | 53 | N/A | N/A |
| Year of handover to FUG | 1994 | 1993 | 1994 | 1994 | N/A | N/A |
| Ethnic composition | Heterogeneous | Heterogeneous | Homogeneous | Heterogeneous | Heterogeneous | Heterogeneous |
| Date of first visit | 1997 | 2001 | 1994 | 1994 | 1994 | 1997 |
| Date of second visit | 2006 | 2007 | 1998 | 1998 | 2000 | 2003 |

N/A, not applicable.
*Source:* IFRI Nepal database.

woody seedlings and herbaceous ground cover was sampled. In the next circle (3 m radius), shrubs, saplings and climbers were identified and counted. In addition, the diameter at breast height (DBH) and heights of woody stems between 2.5 and 10 cm in diameter were recorded. In the largest circle (10 m radius), stems equal to or greater than 10 cm in DBH were identified and counted, and the DBH and height were measured. Each forest plot was also described in terms of its topography and soil condition. In addition to the quantitative data, the perceptions of local forest users of changes in forest conditions over time were also recorded for each forest and time period.

Qualitative data on the forest governance systems, including the characteristics of institutional arrangements, operational rules, property rights situations, the size and attributes of the FUGs and external agencies involved in the governance of forests were collected through rapid appraisal and traditional interview methods. Secondary data/ information available from the forest management plans, the constitution of the FUGs, office records, reports, existing forest acts and bylaws, research papers and other published sources were used to supplement the primary data.

## Data Analysis

Several approaches and methods can be used to assess the condition of a forest depending upon the research objective. In this study, we used measured values of dependent variables, including the DBH of trees ($\geq$10 cm DBH), the density of trees, the DBH of saplings (tree species with 2.5 to <10 cm DBH), the density of saplings, and species richness to quantitatively determine and compare the conditions of the forests under study. The significance of differences in the mean plot values of the dependent variables between the two study periods was analysed using a non-parametric Mann–Whitney $U$ test. The forest conditions were further compared qualitatively using the perceptions of local forest users of changes in forest conditions over time.

Institutions governing the forests, particularly those relating to the operational rules (who develops rules, including harvest rules and power to modify rules), property rights (the rights to withdraw, manage, exclude and alienate) and management rules (equity, monitoring, income and investments) were analysed and compared qualitatively.

# Results

## Changes in Forest Conditions

The changes in the biological conditions of the forests during the study periods differed substantially across the three management regimes. There was considerable improvement in the condition of community forests, moderate improvement in leasehold forests and deterioration in government-managed forests.

A non-parametric Mann–Whitney $U$ test revealed that there were significant increases in the tree and sapling density per plot and mean DBH in both of the community forests in between the two study periods. In one of the leasehold forests (i.e., Baramchi), the mean sapling density and saplings diameter increased significantly over the study period. In the other site, there was a significant increase in the mean saplings density and an increase in the mean saplings diameter. There were significant decreases in most of the parameter values between the two periods in both the national forests (Table 4.2). Assessments of the forest conditions by the members of the FUGs also indicated that the density of trees and saplings of community forests and leasehold forests increased, whereas in national forests, the tree and sapling values decreased. These perceptions of local forest users provide another avenue of support for the improved biological conditions of the community forests (Table 4.3).

## Institutions Governing the Forests

A detailed analysis of the existing property rights situation, forest governance and use rules for the six forests is presented in Table 4.4. The findings indicate that there are substantial variations among the three management regimes in terms of the property rights within which they function, the monitoring and harvesting rules, and the level of internal and external support. In general, the leasehold forest groups had limited permission for harvesting products and changing operational rules; community forest groups enjoyed the highest level of property rights and a better level of monitoring systems; and the government-managed forests had the weakest monitoring and management arrangements.

**Table 4.2:**
*Change in forest condition over time*

| Parameter | Community forests | | Leasehold forests | | National forests | |
|---|---|---|---|---|---|---|
| | Ahal Danda | Bhadkhore | Baramchi | Bhagawatisthan | Karne | Nehit |
| Number of forest plots sampled | 60 | 30 | 58 | 32 | 22 | 34 |
| Mean tree density (per plot) | 1997<2006* | 2001<2007* | 1994<1998 | 1995<1999 | 1994>2000* | 1997>2003 |
| Mean tree diameter per plot (cm) | 1997<2006* | 2001<2007* | 1994>1998 | 1995>1999 | 1994>2000* | 1997>2003 |
| Mean sapling density per plot (per ha) | 1997<2006* | 2001<2007* | 1994<1998* | 1995<1999* | 1994>2000* | 1997>2003* |
| Mean sapling diameter per plot (cm) | 1997<2006* | 2001<2007* | 1994<1998* | 1995<1999* | 1994>2000 | 1997>2003* |

*Difference between first and second visits in forest condition parameters. The year represents the year of visit.

**Source:** IFRI Nepal database.

**Table 4.3:**

*Changes in forest between first and second visits as evaluated by the user group members\**

| Forest | Density of trees | Density of saplings | Density of ground cover |
|---|---|---|---|
| **Community** | | | |
| Ahal Danda | Increased | Increased | Remained the same |
| Bhadkhore | Increased | Increased | Remained the same |
| **Leasehold** | | | |
| Baramchi | Increased | Increased | Increased |
| Bhagawatisthan | Increased | Increased | Increased |
| **National** | | | |
| Karne | Decreased | Deceased | Remained the same |
| Nehit | Decreased | Decreased | Decreased |

\*Evaluated in terms of three ordinal values—'increased', 'decreased' or 'remained the same'.
*Source:* IFRI Nepal database.

**Table 4.4:**

*Comparison of property rights, and governance and management situations in three forest management regimes*

| | Community forest | Leasehold forest | National forest |
|---|---|---|---|
| **Property right situations** | | | |
| Rights to withdrawal | Users have the right to withdraw any forest product according to the operational plan | Users are not allowed to withdraw timber products | No one is allowed to withdraw any product from the forest; however, just opposite in practice |
| Rights to manage | Users have high degree of control over modification to management system | Very little or no control over modification to management system | Rigid modification of management system |
| Rights to exclude | Users have the right to decide membership. No authority to transfer rights to exclude | The District Forest Office decides membership in consultation with community. No authority to transfer rights to exclude | No user group |

*(Table 4.4 Continued)*

*(Table 4.4 Continued)*

|  | Community forest | Leasehold forest | National forest |
|---|---|---|---|
| Rights to alienate | No right to sale and transfer of the forest land, which is owned by the state | No right to sale and transfer of the forest land, which is owned by the state | State has all right to sell or transfer the land |
| **Governance and management** | | | |
| Rules formulation | Users' coordination with District Forest Officer (DFO) to formulate the operational rules | DFO develops operational rules for users | State has rigid rule formation system |
| Power to modify rules | Users have capacity to modify some management rules but need to get approval from DFO | The initial set of rules can be modified by DFO after request from users | State has capacity to modify the rules |
| Forest management | Regular plantation, silvicultural operations done every year | Extensive plantation of non-wood forest species conducted with the help of DFO-supported programme | Rarely done |
| Harvesting rules | Users are allowed to harvest any forest products according to the operational plan but in case of timber, prior permission from DFO is required | Limited to minor forest products; standing trees are property of the state | Strict restriction to harvest any forest product in rules but very poor implementation |
| Monitoring and evaluation | Generally through hired forest guards when users have enough fund and occasionally by local volunteers | Mostly by users | Rarely |
| Income source | From sale of forest products, membership fee and donations | Limited fund collection | – |
| External support | Limited support from DFO and donor agencies directly | High degree of external support in terms of technical and financial through DFO | Rare external support to manage national forests |

*Sources:* Adopted from Nagendra, Karmacharya, and Karna (2005) and Nagendra and Gokhale (2007).

# Discussion

The findings that there was consistent positive growth in the community forests over time as compared to the forests under the other two management regimes indicate that the decentralised forest management system, with clearly defined property right situations, operational rules and governance structure motivated the users to conserve and utilise the community forests in a more sustainable way, which ultimately led to improved forest conditions. Relatively favourable property rights arrangements might also have motivated the community forests users for some longer-term investments (e.g., plantation of tree species). These findings are, in general, consistent with some other past studies that found that community forestry programmes have contributed to improvements in the condition of forests in Nepal (e.g., Gautam, 2006; Jackson, Tamrakar, Hunt, & Shepherd, 1998; Schweik et al., 2003; Virgo & Subba, 1994).

There are two possible factors that contributed to the improvements in the condition of community forests. First is the establishment of plantations in barren or severely degraded forest patches within the community forests. Plantation in community forests, especially of pine species, with or without external assistance is common across the middle hills of Nepal, which has contributed to incremental forest cover. Second is the self-regeneration of native broadleaf species in degraded forestland after the areas came under FUG's protection. Past studies (e.g., Gautam, Webb, & Eiumnoh, 2002; Webb & Gautam, 2001) and our own experience indicate that most community FUGs implement strict protectionist strategies over their resources, thereby improving the chances of plantation success or natural regeneration, which ultimately leads to improved forest conditions.

In the case of leasehold forests, the users mainly focused on activities that can provide them with immediate benefits. The shorter horizon of leasehold forest users can be related to the two fundamental aspects of leasehold forestry. First, the economic status of the group members (poorest of the poor) might have compelled them to focus on activities that help fulfil their day-to-day subsistence needs rather than planning for longer-term benefits. Second, the uncertainty about the tenure over the forests after the expiry of the lease agreement might have caused reluctance among the group members to invest in activities that require waiting long term before receiving returns from the investment.

The condition of the national forests, which were under direct control of the district forest office and without any local collective action, generally deteriorated during the period mainly due to the lack of effective monitoring and enforcement by the forestry staff. In fact, the forests under direct control of the government (particularly in the hills and mountains) are virtually open access because the limited capacities of district forestry offices are mostly utilised by the community forestry.

In general, the community FUGs had more decentralised endowments, stronger property rights and greater ability to craft or modify the operational rules of governance and management than semi-decentralised leasehold FUGs and the centralised national forest management system. In leaseholds, the FUGs had limited power to create and modify the operational rules of governance and management of forest resources and were permitted to harvest only minor forest products. In national forests, although the state possesses all the governance and management rights and local people are barred from extracting forest products and participating in management activities without the permission of the forest department, the rules generally are not followed. The widespread 'illegal' harvesting of forest products led to deterioration of the forest conditions.

Thus, we can observe through the lens of design guidelines that 'graduated membership' is in its maximum form with community forestry, followed by leasehold forestry, in which some of the local people act as 'core members' with the strongest authority cooperating with other graduated members having relatively weaker authority, which is more or less absent in the case of government-managed forest. If we evaluate these forests from the perspective of the 'commitment principle', which recognises an authority to make decisions in a capacity that corresponds to a degree of commitment to forest use and management, both community forestry and leasehold forestry have higher degrees of commitment in management and use of forests than the national forest. This observation suggests that the 'commitment principle' might be effective in community and leasehold forestry, even though the internal decision-making process within each group was not examined in the study. In terms of the 'fair benefit distribution', in which the benefit distribution is not necessarily equal but is fair in accordance with cost-bearing, leasehold forestry, with 40-year lease ownership to manage the forest and reap benefits from it, is clearly the best. By contrast, community forestry presents some doubts about the fair benefit distribution due to the fears of the elite capturing the resources and the core members of the FUG paying attention to the timber and other prime forest benefits rather than the poor households' dependence issues on NTFP and other day-to-day livelihood needs from the forest. Because forest products cannot be harvested in the centrally government-managed forests, there is least possibility of a fair benefit distribution. We can conclude that these design guidelines are the most critical factors in the success of collaborative forest governance, as evidenced by the conditions of the forest and users and the perceived evaluation of the fair benefit distribution and the different property rights regimes in governing and managing the forest resources (Inoue, 2011). Based on the comparative perspectives and analysis of these design guidelines, we can suggest an effective forest governance mechanism to bridge multi-level outcomes.

## Conclusions

The results of this study show general improvements in forest conditions under community forestry, followed by leasehold forestry. With limited downward accountability and state-forced institutions that are relatively inflexible and unable to adapt to change, the governance system leads neither to the betterment of forests nor to the strengthening of local communities. In many, if not most instances, decentralisation reforms need to be louder in counties like Nepal. Thus, flexibility for policies and the willingness of governments to allow local actors to experiment, test and explore new potential solutions are absolutely essential for long-term success. Rigid governmental structures that do not fully decentralise the authority and, inherently, the opportunity to craft local solutions are destined to be less successful than those allowing—and supporting—local creativity and local institutions to thrive.

## References

CBS (2008). *Environment statistics of Nepal, 2008*. Kathmandu: Central Bureau of Statistics (CBS), Government of Nepal.

Department of Forest Research and Survey/Forest Resource Information System Project (DoFRS/FRISP) (1999). *Forest Resources of Nepal (1987–1998)* (Publication No. 74). DoFRS, His Majesty's Government of Nepal and FRISP, The Government of Finland.

Gautam, A.P. (2006). Combining geomatics and conventional methods for monitoring forest conditions under different governance arrangements. *Journal of Mountain Science*, 3(4), 325–333.

Gautam, A.P., Shivakoti, G.P., & Webb, E.L. (2004). A review of forest policies, institutions, and changes in resource condition in Nepal. *International Forestry Review*, 6(2), 136–148.

Gautam, A.P., Webb, E.L., & Eiumnoh, A. (2002). GIS assessment of land use/land cover changes associated with community forestry implementation in the Middle Hills of Nepal. *Mountain Research and Development*, 22(1), 63–69.

Jackson, W.J., Tamrakar, R.M., Hunt, S., & Shepherd, K.R. (1998). Land-use changes in two Middle Hill districts of Nepal. *Mountain Research and Development*, 18(3), 193–212.

Land Resources Mapping Project (LRMP) (1986). *Land Utilization Report*. Kenting Earth Sciences Limited, His Majesty's Government of Nepal and Government of Canada.

Nagendra, H. & Gokhale, Y. (2007). Management regime, property rights, and forest biodiversity in Nepal and India. *Environmental Management*, 41(5), 719–733.

Nagendra, H., Karmacharya, M., & Karna, B. (2005). Evaluating forest management in Nepal: Views across space and time. *Ecology and Society*, 10(1), 24. Retrieved from http://www.ecologyandsociety.org/vol10/iss1/art24/.

Nagendra, H. & Ostrom, E. (2011). The challenge of forest diagnostics. *Ecology and Society*, 16(2), 20. Retrieved from http://www.ecologyandsociety.org/vol16/iss2/art20/.

Niraula, R.R., Gilani, H., Pokharel, B.K., & Qamer, F.M. (2013). Measuring impacts of community forestry program through repeat photography and satellite remote sensing in the Dolakha district of Nepal. *Journal of Environmental Management*, 126, 20–29.

Schweik, C.M., Nagendra, H., & Sinha, D.R. (2003). Using satellite imagery to locate innovative forest management practices in Nepal. *Ambio*, 32(4), 312–319.

Virgo, K.J. & Subba, K.J. (1994). Land-use change between 1978 and 1990 in Dhankuta District, Koshi Hills, Eastern Nepal. *Mountain Research and Development*, 14, 159–170.

Webb, E.L. & Gautam, A.P. (2001). Effects of community forest management on the structure and diversity of a successional broadleaf forest in Nepal. *International Forestry Review*, 3, 146–157.

Wollenberg, E., Merino, L., Agrawal, A., & Ostrom, E. (2007). Fourteen years of monitoring community-managed forests: Learning from IFRI's experience. *International Forestry Review*, 9, 670–684.

# 5

# Sri Lanka: Forest Governance of Community-based Forest Management

*Mangala De Zoysa, W.D. Lakmi Saubhagya and
Makoto Inoue*

## Introduction

Community forestry has become neither a government programme nor a foreign aid-driven activity; rather, it has become a complex governance regime for a forest-dependent social–ecological system. Community forestry may be defined as 'the governance and management of forest resources in designated areas, by communities for commercial and non-commercial purposes to further their livelihoods and development' (Brown, 2009). The complexity of community forestry can be explained in terms of the range of actors involved, the scale of resources mobilised, the diversity of the processes involved in conflicts and collaboration, and the policy and practical issues encountered (Ojha, Persha, & Chhatre, 2009). Community-based forest governance opens new spaces for communities to exercise political control of their territories and resources through horizontal decision-making mechanisms with community transparency and accountability (Baltodano, 2011). Forest governance practices may alter the following central activities: learning; planning; decision making; mobilising marginalised groups to create pressures on the elites; developing clearer visions and indicators, and altering the purposes of the forest user groups and related community forestry organisations; monitoring; promoting transparency; electing executive committees; creating ownership through organisational change processes; improving communication and promoting public hearings and auditing. Deliberative forest governance relies on the quality of deliberative interactions among local communities, political activists, development organisations and forest officials. The popularity of participation in forest governance demonstrates that it has generated procedural gains that include democratic deliberation, the building of procedures and institutions and substantive gains that include the creation of livelihood opportunities, ecological conservation, and social justice and equity. The governance systems of community forestry advance with the evolution of strong local institutional mechanisms and the clarity of the definition of property rights (Ojha et al., 2009). Community-based forest governance allows communities to do the following: live in a manner that is integrated with the local ecosystem; satisfy their needs while simultaneously conserving and enriching resources; maintain equal and just relationships within the community itself and with other communities; promote the horizontal integration of decision making and take advantage of traditional knowledge to help a large number of communities fulfil their essential needs.

The derivation of supportive policy reforms, the decentralisation of governance systems and the devolution of authority to local administrations are ongoing trends in the South Asian region that allow communities greater access to public forest lands. The National Forestry Policy clearly lays out the major reorientations of the forestry sector that are required to successfully address the challenges facing the country. To address the issues of deforestation, forest degradation, and the direct and indirect effects of forest health on the livelihoods of the local communities and the natural environment, the government of Sri Lanka has launched various community-based forest development programmes. The centralised management system and the lack of coherence of the capacities of provincial and local authorities to manage the forest resources have resulted in satisfying the interests of a few people while marginalising the majority of local communities and have also resulted in resource depletion (De Zoysa & Inoue, 2008). Sri Lanka has one of the highest rates of forest loss in the world and is losing forests at rates that exceed 1.4% per year (Fabie, 2009). The loss of forest cover continues in Sri Lanka due to agricultural encroachment, illegal exploitation of forest resources, damage from wild fires and the conversion of forestland for other uses (Dangal & De Silva, 2009). Hence, the study of community-based forest management and forest governance in Sri Lanka has become practically significant. This chapter reviews the available literature using a search methodology that was designed based on a conceptual framework and discusses community-based forest management and forest governance in Sri Lanka in light of related concepts such as formal recognition, devolution of authority and the setting of prototype design guidelines.

A special analytical emphasis is placed on the two recently implemented community-based forestry projects. The Small Grants Program for Operations to Promote Tropical Forests (SGPPTF) was an innovative, community-based forest-related project that was implemented by the UNDP in Sri Lanka from 2004 through 2007 (Escobin, Gonsalves, & Eduardo Queblatin, 2011). The SGPPTF supported the following activities with the aim of protecting tropical forests and contributing to poverty reduction: demonstrating community-based forest management and resource use; facilitating dissemination of innovative community practices; and the building of grass-roots capacities for localised management through partnerships and networks (Velasco & Lim, 2012).

The Farmer's Woodlot (FWL) was a model developed by the Community Forestry Project (CFP) and introduced in 1982, and the Participatory Forest Project (PFP) was launched in 1992; these programmes sought to motivate farmers to plant timber trees together with their agricultural crops in degraded shifting cultivation lands (De Zoysa & Inoue, 2009). The main objective of the FWL was to encourage poor and marginal farmers to grow trees on degraded government land using an agro-forestry approach to increase wood supply and improve the livelihoods of the farming households (ADB, 2003). The FWL models used degraded lands that were being used for shifting cultivation to set up 3-year crop rotation strategies that were based on an average 0.4-ha plot size and a predominant tree species (teak or eucalyptus) or mixed (teak/ margosa) tree species (ADB, 2003). The physical target of the FWLs component of the PFP of 4,000 ha was substantially exceeded; 9,808 ha was utilised by the end of the project (ADB, 2003).

## Formal Recognition

### *Implementation of Supportive Policy Reforms*

The effective community participation in decentralised forest governance practices in developing countries is still limited due to lack of proper policy guidance. Community forestry requires effective policy and programmatic approaches for building natural and community capital and considerable community capacity to overcome the resistance to bottom-up approaches (Baker & Jonathan, 2003). The policy recognises that the role of government will have to change to substantially improve forest governance and to meet the demand for greater public participation in the forest sector. Good governance and policies are needed to complement, support and encourage community participation in working effectively towards sustainable forest management (Keller, 2009).

The ancient historical chronicles reveal that village communities in Sri Lanka were well organised and lived in harmony with the forest environment as long ago as 543 BC (Maddugoda, 1991). The rights and responsibilities of communities to manage the forests were shifted to bureaucratic agencies run by the British colonial authority after the 1820s (Poffenberger, 2000). However, even after achieving independence in the 1950s, forest resources were administered by officers who were not trained to recognise indigenous rights, appreciate local knowledge or understand the economic dependency of the communities on forest resources. The National Forest Policy in 1980 considered community forestry to be a promising strategy for local control over, and maximisation of the monetary and non-monetary benefits offered by, local forest resources, which led to sustainable rural development (De Zoysa & Inoue, 2008). Market-oriented forest policies prevailed, were amended and revised in the National Forest Policy of the late 1980s and created provisions for community involvement in forest management (EU–UNDP, 2004). The National Forest Policy of 1995 aimed to conserve forests for posterity and specifically emphasised biodiversity, soils, water and historical, cultural and religious values (Keller, 2009).

The government owned and managed more than 98% of the forest resources in Sri Lanka due to the command and control approach (Bandaratillake, 2001). The ancient resource use strategies and traditions of community involvement became necessary and required strengthening of the country's management of the forest resources (Poffenberger, 2000).

The forest policy in 1995 emphasised the recognition of the important role of traditional knowledge and the experiences of rural people in sustainably meeting the needs of forest-dependent communities and ensuring environmental services and the supply of forest products (Escobin et al., 2011). Communities should be able to play a major role in the management of forests, and this should be a part of the broader process of local government reform. Community forestry approaches must be focused on the communities' collective ownership and custodial care of their traditional lands and forest resources, and the communities must manage their lands according to high standards of governance (Brown, 2009). The National Forestry Policy of 1995 accepted 'joint forest management' and 'leasehold forestry' as promising strategies (Forestry Planning Unit, 1995).

Good governance and policies are needed to complement, support and encourage community participation in working effectively towards the implementation of community forestry development programmes in Sri Lanka (Keller, 2009). The 'Caring for the Environment 2003–2007 path to sustainable Development (CFE)' is the most crucial policy document in Sri Lanka and stresses the involvement of all relevant stakeholders at different levels of authority in the management of forest resources (MENR, 2003). Good governance that promotes the rule of law, transparency, accountability and the meaningful participation of people in local decision-making processes is necessary for the development and implementation of community-friendly national forest policies, programmes and regulatory frameworks (RECOFTC, undated). However, there is a huge gap between policy and practice due to the insufficient institutional capacity for forest management in Sri Lanka (Kiyulu, 2011).

## Acceptance of Forest Governance in Government Legislation

The recognition and normalisation of community forest governance facilitates and legalises the promotion of sustainable forest resource use. Community forest governance of forest resources has to be considered as a decentralising action within the politically defined administrative structure (Baltodano, 2011). Nearly all forestlands in Sri Lanka are government-owned, and traditional resource management practices are not officially recognised (*Tropenbos International*, 2009). A cross-country review of Sri Lankan policy stated that forest management could mainly be improved with legal frameworks that secure private, community-based land tenure. Farmers began raising FWLs on long-term lease bases in the 1980s under a partnership with the Forest Department. The Forest Ordinance of 1995 was amended with legal provisions for leasehold forestry, which is an effective form of partnership (Nanayakkara, 2001).

The current local institution is sufficient to balance responsibilities if legal land tenures are disseminated to local communities (Lynch & Talbott, 2010). The Sri Lanka Australia Natural Resource Management

Project (SLANRMP) mobilises target communities to form groups and institute formal registered community-based organisations (CBOs) while providing the necessary support for institutional capacity building. Forest Department and community representatives signed 25-year legal contracts for the management of community forests (Dangal & De Silva, 2009). The improvement of community forest governance and the recognition of community rights are politically feasible and cost-effective strategies for rural poverty alleviation in Sri Lanka (De Zoysa & Inoue, 2007). Weak governance in community forest management at the community level causes people to feel that community forests are no man's lands and that they are not responsible for managing their forest resources in a sustainable manner. The implicit intention of including communities in forest governance and management is the redistribution of power to mobilise all poor and marginalised groups within the community (Lema, 2011). The legality of community forests and their connections to forest management and the rights and benefits of the local communities are crucial when conflicts arise and villagers are required to protect their rights regarding the forests (Tan, Thanh, & Hoang Huy Tuan, 2009).

## Devolution of Authority

### Use of Community Knowledge of Forest Resources Management

The traditional or acquired knowledge of the community regarding the climate, geography, biology and the use of biodiversity elements is vital for the planning and governance of forest resources. This knowledge of the forest and its elements is useful for planning and regulating its use and guarantees respect for the different components of the forest (Baltodano, 2011). The National Forest Policy of Sri Lanka of 1995 recognised the crucial importance of involving local communities in forest management because that arises from the valuable traditional knowledge and experience those communities possess. The participation of communities optimises the benefits to those communities while ensuring the sustainability of the forest's resources (Keller, 2009). SGPPTF has recognised the possibility of combining local and scientific knowledge to support resource management and livelihoods for environmental benefits and the commercial development of specific resources (Velasco & Lim, 2012). The SGPPTF included the planting of kiriala as an immediate cash crop; kiriala is currently being grown traditionally in organic conditions. The programme signed a sales agreement with exporters to produce considerable cash income (Escobin et al., 2011). Local communities are not sufficiently included in the planning of forest resource management in Sri Lanka. In the FWL programmes in the Badulla and Kandy districts, the initial decisions, such as the selection of the project's location, land distribution and planting design, are primarily made by project officials. Further, project officials make operational decisions related to such issues as weeding and maintenance regimes and the application of fertilisers. Only decisions related to the selection of participants and indigenous knowledge and inputs for intercropping were made by the local community based upon indigenous knowledge. The involvement of local communities in the administration and coordination of projects at the village level is limited (Dissanayake, 1998). Some parts of the FWLs initiated under the CFP have been abandoned by farmers because the FWL schemes were incompatible with the traditional farming systems that are managed with based on the farmers' indigenous knowledge and experiences (Wickramasinghe, 1997). Sri Lanka needs more assistance in the collection of community knowledge data and needs to conduct fairly extensive forest inventories to reduce deforestation and forest degradation (*Tropenbos International*, 2009).

## Community-based Designing of Resource-use Regulations

Forest-dwelling communities are formally constrained by national laws and regulations regarding which resources they can legally access and use. It has been realised that community participation is needed to regulate forest resource use. Community members develop their community forest protection regulations and manage the forest in their community's interest (Tan et al., 2009). The success of FWLs could also be attributed to the flexibility of farmers in accepting land- and tree-tenure agreements in which the farmers have enjoyed usufruct rights for their traditional shifting cultivation lands for generations. The establishment of FWLs alone is not enough to offset illegal encroachment and logging until the farmers generate a continuous flow of direct values in the form of timber and other forest products from FWLs (De Zoysa & Inoue, 2009). The community requires collecting information and making agreements with respect to the norms through effective forms of education and communication (Baltodano, 2011). The 'Caring for the Environment 2003–2007 path to sustainable Development (CFE)' policy document stresses the involvement of the general public and communities in designing plans for resource use and sustainable management of forest resources in Sri Lanka (MENR, 2003). The SGPPTF has made efforts to strengthen the roles of local people, CBOs and non-governmental organisations (NGOs) in the formulation of community regulations for forest management. The SGPPTF has promoted conservation and improved sustainable forest management through active partnerships with the state, rural communities and the private sector (EU–UNDP, 2004). In 2007, the SGPPTF expected sustainable forest management with community participation to ensure sustained delivery of better forest services, better delivery of forest goods, multi-sector partnerships and the promotion of the development of an alternative forest resource base (Keller, 2009).

## Establishment of a Vigilance and Flexible Monitoring System

The economy, which is central to the success of community forestry efforts, recognises the wide array of natural and social capital across which investment is needed. The environment, economy and equity are equally important components of community forestry. The SGPPTF has assisted resource base assessment through resource inventories to establish baseline information concerning forests in the area to ensure effective planning processes. Existing village land use plans were considered in the preparation of project activity plans that mainly concentrated on individual home gardens. Further, degraded and unutilised areas have been identified for future interventions (Velasco & Lim, 2012). Sets of 0.4 ha degraded shifting cultivation land plots for each farm household were leased for a period of 25 years within a 20–30 ha block of degraded forestland under the FWLs. An annual renewal permit was issued for the first five years of the lease period under the condition that the land must be maintained with forest cover (Philips, 2003). The ADB conducted one inception, seven reviews and one mid-term and three special project administration missions to evaluate the performances and activities of the FWL projects (ADB, 2003). Inadequate monitoring and inadequate assistance from the Forest Department has led to weak maintenance of woodlots by farmers in some parts of the FWLs.

The farmers have successfully completed the planting of seedlings and the maintenance of woodlots over the first 3–4 years in response to the incentives provided. The farmers have ignored timber production, which is the long-term or final benefit (Wijewarnasuriya, 2009). Sustainable forest health, ecosystem function and biodiversity reflect the commitment of community forestry practitioners. Equity refers to the distributions of power, knowledge and economic benefits and addresses the claims of related stakeholders to forests and forest resources (Baker & Kusel, 2003). Flexible and effective monitoring systems are required to comply with forest-use regulations that ensure accountability in the implementation of community governance processes (Baltodano, 2011). Members of the community need to organise themselves and receive the support from local authorities to make a collective decision to take action to stop encroachment into their community forest (Tan et al., 2009). Sri Lanka commonly lacks monitoring; data regarding the socio-economic factors involved in local communities' abilities to properly monitor forest governance has not been aggregated (*Tropenbos International*, 2009). The successful collective outcomes of community forest management vary with the strength of the local compliance monitoring arrangements. The community, under the SGPPTF, has successfully uprooted invasive species and protected regenerating forests and cattle from fire through effective monitoring systems (EU–UNDP, 2004). Motivated members of the community forestry system contribute their time and effort to the monitoring and enforcement of compliance rules related to the access to, and use of, commonly held forest resources (EU–UNDP, 2004).

## Promotion of the Capacity for Conflict Resolution

Weak governance of forest resources has increased tension among local communities over access to, and the use of, forest resources. Conflicts in community forest management commonly arise over disagreements about tenure and use rights, harvest regulations, competition with other users for limited resources, unsustainable use and unfair distributions of benefits (Lema, 2011). One of the main policy statements highlighted in the National Forest Policy of 1995 was the building of partnerships with local communities for the protection of natural forests. A national programme of forest boundary demarcation has been undertaken by the Forest Department to address the degradation caused by agricultural encroachment into forestlands. Through skilled facilitation and participatory mapping processes, the SGPPTF grantees have helped to mediate conflicts over boundaries and settlement rights as part of their community forest management activities. Community members have planted boundary fences that are mutually agreed upon (Velasco & Lim, 2012). The policing efforts that have attempted to prevent unauthorised exploitation and promote the prosecution of offenders by the Forest Department have not been sufficiently effective due to limited human and other resources and the lack of community support. Reducing deforestation requires extensive institutional reform and capacity building. Community-based patrolling is the most widespread forest protection strategy of the SGPPTF for the monitoring of illegal forest activities. Vigilant committees in the villages that have been established by the recipients of SGPPTF micro-capital grants patrol the forest. The youth, particularly the unemployed village youths, have volunteered to become members and are strongly involved in the vigilant committees in their villages (Velasco & Lim, 2012).

Deforestation and the poor management of forests, together with the risk of forest fires in Sri Lanka, threaten the natural regeneration of the forest (Keller, 2009). The bond between people and forest is not strong as many communities still depend, directly or indirectly, on forests (Dangal & De Silva, 2009). Community forest governance is helpful for the development of mechanisms that enable the community to resolve internal conflicts in creative and transparent manners. These mechanisms should encourage dialogue about monitoring and accountability within the community to resolve internal conflicts (Baltodano, 2011). The SLANRMP, the local community and the forest department enter into agreements at the forest range level regarding the management of forest areas, especially those areas with elevated risks of encroachment, illegal harvesting activities and annual wildfires (Dangal & De Silva, 2009). Forest governance systems initiated by the local communities have emerged to confront the forces responsible for forest degradation and livelihood deprivation. Communities have their own methods of protecting their forests with norms and sanctions. In traditional shifting cultivation, approximately 10–12 farm families cultivate around 8–10 ha as a single site through mutual understanding and shared benefits (Perera, 2001). These groups solve the important issues of the village through their activated indigenous village institutions (Nurse & Hitinayake, 2000). However, the management patterns of the FWLs are poorly defined, and the newly created indigenous institutional structures are latent, weak or dysfunctional (De Zoysa & Inoue, 2009).

## Improvement of the Capabilities for Resource Governance and Administration

Forest governance empowers the community to successfully interact with the powerful commercial interests of the forest industry to address local, regional and national concerns (Brown, 2009). Under community forest governance, the community and organisations have to develop basic tools for administration, decision making, governance, innovation and creativity in the face of change. The forest lands employed for traditional shifting cultivation in Sri Lanka were previously commonly owned, while the crops were individually owned. The Forest Department released degraded forest lands to the farmers to grow trees as FWLs and manage the lands for their own benefit. The farmers planted timber trees along with their agricultural crops after the establishment of FWLs. The success of FWLs can be attributed to the flexibility of farmers in accepting land- and tree-tenure agreements. The farmers preferred that individual blocks of land be allocated to each family to enable each family to reap the undivided benefits. The legal provisions for leasehold forestry, which is an effective form of partnership, were included in the amendment to the Forest Ordinance in 1995 (Nanayakkara, 2001).

The PFP of 1993 not only consulted communities during the process of planning the FWLs but also sought to maximise the farmers' individual capacities to operate the FWLs by providing seedlings, advice and technical training (FAO, 1998). The FWL programmes empowered farmers by increasing land possession rights, which provided the farmers with future security, increased farmers' access to loans and improved farmers' attitudes towards soil conservation on their land and forest conservation and management (Philips, 2003).

In local forest governance, the quality of the decision-making process is more important than the forest management plan. The forest sector's limited human and financial resources require that community organisations achieve national reforestation objectives. Social mobilisation through the community forest management of the SGPPTF produces lasting positive effects that extend beyond environmental goals. The cost-sharing mechanisms of the SGPPTF that employ revolving fund schemes that are linked community forest protection responsibilities have minimised dependency and encouraged self-reliance (Velasco & Lim, 2012). The community takes ownership of the forest, and the entire community understands that current negotiations determine future well-being; thus, the community will manage resources decisively and carefully (Baltodano, 2011). The communities take more active roles in the management, processing and utilisation of forest and other resources due to the devolution of management and decision-making authority to the local level (Baker & Kusel, 2003). The government-led individually leased land tenure policy of the FWLs limits the customary rights enjoyed by farmers regarding their traditional shifting cultivation lands (De Zoysa & Inoue, 2009).

The Forest Department, with the help of the Environment Protection Foundation, has successfully leased a 100-ha pine plantation to the community of Matara district for 30 years with the aim of re-establishing natural forest patches through inter-planting with indigenous species. This lease allows the community to sustainably collect non-wood forest products (mainly medicinal plants) for sale to drug manufacturing companies (Velasco & Lim, 2012). The SGPPTF promotes the conversion of exotic monoculture plantations into endemic, natural and multiple-use forests. The state-owned exotic pine (*Pinus carebaea*) plantations in the Matara district have been rehabilitated with endemic medicinal trees (*Garcenia quessita*), endemic trees (e.g., Pihimbia) and fast-growing timber species such as mahogany (*Sweteinia macrophilla*) with the participation of community members (Velasco & Lim, 2012). The members of community forestry systems rely on their own, self-organised sets of rules-in-use that establish pragmatic and mutually beneficial arrangements regarding forest use. Communities develop and practice a set of informal rules regarding the conduct of community members that seek the common interest, mutual respect and the development of institutions necessary for self-governed forest resource management (Tan et al., 2009). Villagers in the Kegalle district of Sri Lanka formed small self-help groups and introduced alternative livelihoods. These villagers formed a social fence to actively protect the forest from intruders and no longer encroach upon, or cut valuable timber from, the forest. The implementation of the SGPPTF promoted participatory forest biodiversity conservation, the development of alternative livelihood initiatives, and the enhancement of the skills and capacities of local communities (Keller, 2009). The construction of fire belts, planting of fire resistance species and employment of vigilant committees to patrol the forest are examples of the actions taken by the communities of Nilgala, Sri Lanka. The SGPPTF initiated viable and competitive rural forest-based enterprises by exposing communities to private and state enterprises and disseminating environment-friendly technical knowledge (EU–UNDP, 2004). The SLANRMP commenced its operations in February 2003 and implemented its programmes with the overall goal of contributing to poverty reduction by improving natural resource management (Dangal & De Silva, 2009). This jointly funded project assisted in improving the capacity of the Forest Department and other service providers to mobilise communities for participatory forestry management programmes that are in accordance with the National Environmental Action plan for 2008–2012.

## Setting Prototype Design Guidelines

### Recognition of Graduated Membership

Although the forest governance approaches are decentralised at the local level, local forest-dependent communities have not yet successfully participated in forest governance policies and practices (Ojha, Subedi, Dhungana, & Paudel, 2008). Although the right to manage forests has been transferred to local communities under the community forestry programmes of Sri Lanka, community participation in governance continues to face techno-bureaucratic challenges from forest officials. Foresters still need training to manage social and political matters and manage the technical and economic aspects of forestry to implement community-based forest governance. In the community forestry sector, all community members do not have equal access to power over forest resources. Marginalised groups within the community, particularly the poor and women, have few leadership roles. Although the marginalised groups participate in many activities and in the decision-making process, their voices are not heard to the same degree as those of the elite in the community.

Inoue (2011) argues that the 'graduated membership' design guidelines of the design principles of the Common Pool Resources recognises that some of the local people who act as 'core members' have the greatest authority and cooperate with other graduated members who have relatively less authority. The SGPPTF strengthened community organisations and formed local committees and interest groups to build human and social capital. Involvement of the youth is a strategy that is common to 11 of the 18 SGPPTF projects because this strategy stabilises current CBOs. Their skills in project management, financial accounting, documentation, strategic negotiations, working with local authorities and paralegal techniques have been developed (Velasco & Lim, 2012). The delivery of many of the SGPPTF small grants directly to CBOs was successful because these organisations were crucial actors in community forest management and community development (Velasco & Lim, 2012). The active participation of local community members may not be a prerequisite for the decentralisation of forest governance. To effectively benefit the constituents, local community groups must actively pursue opportunities that become available through the creation of decentralisation reforms (Agrawal & Ostrom, 2001). However, the participation of elite members of the community improves governance relative to the government management of forests, and ensuring that marginalised members of the community can participate equally in the process is an ongoing challenge (Ojha et al., 2008). Discussions between community members could be carried out by representatives of key groups. The conflict management processes of forest governance should establish reliable communication between community representatives and their constituencies (Lema, 2011). One approach that promotes mutual learning among community forest management team members is to require that information be shared among all members throughout the entire CFP. Involvement in the projects was based on individual experience and commitment to improving forest governance structure for the benefit of the forest community (Tan et al., 2009). The target beneficiaries of FWLs were to be rural households whose members were farmers, agricultural labourers and share-croppers. The participants in the FWL programmes were selected based on the set criteria. Participants were selected based on surveys of household profiles that were conducted by the motivators. The participants should not have family incomes exceeding 1,500 SLRs per month (in 1992 prices) and should be committed to active involvement in the project (ADB, 2003).

## Promotion of Commitment Principle

There are often major gaps between decentralisation policies and their implementation that may not be overcome by political will and commitment to community forestry in Sri Lanka. Political commitment to the strengthening of forest law enforcement and governance has become vitally important. Approaches for decentralising forest governance need to be adapted to specific situations in different provinces, districts and locations depending on the capacities and commitments of the local governments, local institutions and other actors. The authority to make decisions corresponding to the degree of commitment to forest use and management is recognised under the 'commitment principle' (Inoue, 2011). The success of international forest governance depends on international commitments to achieving globally sustainable forest management. The development of international forest-related policy and obligations has been rapid since the United Nations Conference on Environment and Development, which was held in 1992 in response to strengthened commitments to international actions that facilitate sustainable worldwide forest management. In addition to effective planning, the implementation of the international commitments to forests will depend on political commitment, the availability of adequate institutional, technological and human capacities, and the availabilities of financial resources in different countries (Braatz, 2003). Sustained decentralisation and the guarantee of its long-term viability require sufficient political support and commitment from governments in the form of committing funds sufficient to make decentralisation work. The strong commitment of the central government to the forestry department and stakeholder consultations will facilitate the development of an integrated plan for community forest management at the district level. Commitment building among stakeholders is vitally importance for the decentralisation of forest governance to the district level. The central government should provide incentives and financial assistance to support infrastructure development. A national commitment that extends down to the local governance system and partnerships with local stakeholders is indispensable for sustainable community forest management. In a decentralised forest governance system, the extension of a national commitment to sustainable community forest management at the level of local governance requires intensive stakeholder consultations and the consultative planning of the management of forest resources (Efransjah et al., 2008). One of the main objectives of the SGPPTF was to empower the traditional forest people and related stakeholders to develop and practice traditional and novel forms of sustainable forest use (European Commission to Thailand, 2006). The SGPPTF moved, with the vision of conservation and improved sustainable forest management, through active partnership with the state, rural communities and the private sector (EC–UNDP SGPPTF, 2004).

In some cases, multi-stakeholder approaches do not afford sufficiently clear authority to the community to promote local commitment to conservation (Grundy et al., 2003). The government's commitment to the involvement of local communities in forest governance is theoretical when the forestry sector contributes little to the national economy (Kiyulu, 2011). Hence, village governments are obliged to appoint members to the natural resource committees that represent citizens' interests in any particular forest or environmental issue. The reduction in the annual water yield in the Rajjuruwagala Forest Reserve and the effects on its watershed functions were mainly due to the encroachment of small-scale tea plantations. The SGPPTF formed small community groups of six to eight people and adopted a participatory forest conservation model that involved the community in all aspects of project planning and implementation. The community used the traditional practice of shared labour or '*aththam*' to protect the watershed (Escobin et al., 2011). The SGPPTF supported matching commitments from forest communities to develop micro-hydro schemes for electrification in rural communities in the Kegalle district. The households serviced by this scheme had contributed financially (Velasco & Lim, 2012). Participation in the FWL programmes was limited to selected families of the villages who had shown interest. Some who had not participated in the programme earlier were later motivated by the benefits gained from the project (Philips, 2003).

## Persuasion of Fair Benefit Distribution

Community forest users face the challenge of the uneven distribution of benefits among different groups whose livelihoods depend on the forests of Sri Lanka. The fair and equitable sharing of benefits should be assured for the success of community-based forest management. Increased and diversified sustainable income-generation opportunities and the guarantee that benefits will be shared equitably are needed to help reduce poverty and motivate the active participation of local people in forest governance and management. Farming households participating in FWLs were provided with 25-year lease agreements that included ownership of the trees in their woodlots (ADB, 2003). The financial benefits for farmers participating in FWLs (in the form of food coupons and intercropping) were significant. These farmers grew crops for three consecutive years during the establishment of the woodlots and received revenue from pre-commercial thinning of the trees. The FWLs have been well maintained and have shown satisfactory yields, which the farmers regarded as long-term savings and insurance (ADB, 2003). Inequities in the distribution of benefits are common in community forest management systems due to differences in the power, assets and capacities of the participants. Inequities in benefit-sharing quite often contribute to conflict (Acharya & Yasmi, 2008). Unless the governance of local institutions is strengthened, mechanisms that ensure the accountability of decision making are established, and monitoring systems that review outcomes are implemented, inequities in benefit-sharing in community forest management systems will continue in the future (Mahanty, Guernier, & Yasmi, 2009). Communities need real benefits in return for the time and energy they expend in forest management to make long-term commitments to sustainable forest management. The SGPPTF exposed the rural communities to private and state enterprises and environment-friendly technical knowledge that enabled the rural communities to run viable and competitive rural forest-based enterprises. Marginalised groups should be involved in enterprise planning, their interests and skills should be addressed, and equity in the sharing of enterprise benefits should be ensured. Cooperative funding that benefits the poor and underprivileged forest user groups, such as the indigenous peoples and other forest dwellers, should be developed (EC–UNDP SGPPTF, 2004).

'Fair benefit distribution' is the most critical factor for the success of collaborative forest governance in which the distribution of the benefit is not necessarily equal but is fair in terms of the bearing of costs (Inoue, 2011). Decision-making processes in well-governed community forestry include the determination of a benefit distribution system that is favourable to all members of the community. Members of the community are supposed to organise themselves and distribute the benefits of the community forest in an equitable way for the good of the entire community. Although NGOs offered higher prices, collectors in the SGPPTF sold some of their products to community-based traditional buyers to raise the purchase price of their products and to manage risk by maintaining these networks. Support organisations provide valuable market information that allows the communities to benefit more fully from these transactions (Velasco & Lim, 2012). With good governance, all community members, particularly the most vulnerable, have opportunities to improve or maintain their well-being through the equitable sharing of forest benefits (Lema, 2011). Sustainable improvement of community-based forest governance requires tenured security to secure and use forest resources, investments to stimulate enterprise development, long-term political commitment to capacity building, and institutional reform to make decision making less centralistic and more democratic. Community forest management generates a workable scale of local benefits that include livelihood contributions and incentive for communities to make long-term investments in sustainable forest management. The successful achievements of the FWLs and the increasing realisation of the market value of the trees have provided strong incentives for many farming households, which have planted trees on other's land at their own cost (ADB, 2003). Under good forest governance, the revenue collected from the harvest of the forest products is used to improve the quality of community life through the provision of better services such as education and health services (Lema, 2011).

## Conclusion and Policy Implication

The government of Sri Lanka has formally recognised community-based forest management and forest governance through the implementation of supportive policy reforms and the acceptance of forest governance in government legislation. Community-based forest management and forest governance have, to some extent, transferred authority to the community and allowed the community to use its knowledge for forest resource management, to design forest resource-use regulations, to establish vigilant and flexible monitoring system, to promote conflict resolution capacities, and to improve their resource governance and administration capabilities. The creation of prototype design guidelines that recognise graduated membership, and promote commitment principles and fair benefit distribution is required for sustainable community forest management and governance.

## References

Acharya, G.R. & Yasmi, Y. (2008). Conflict management strategy adopted in community forestry of Nepal: A study of four community forests in Midwestern region. *Banko Jankari*, 18(2), 44–52.

ADB (2003). Project performance audit report on the participatory forestry project in Sri Lanka, Manila.

Agrawal, A. & Ostrom, E. (2001). Collective action, property rights and decentralization in resource use in India and Nepal. *Politics and Society*, 29(4), 485–514.

Baker, M. & Kusel, J. (2003). *Community Forestry in the United States: Learning from the Past, Crafting the Future* (pp. 238–247). Washington: Island Press.

Baltodano, J. (2011). *Friends of the Earth Costa Rica. Friends of the Earth International Secretariat.* The Netherlands: Amsterdam. Retrieved August 12, 2011, from www. foei.org.

Bandaratillake, H.M. (2001). ADB Forestry sector strategic framework: Forest Department perspective. Forest Department, Battaramulla, Sri Lanka.

Braatz, S. (2003). *International Forest Governance: International Forest Policy, Legal and Institutional Framework.* Proceedings of the World Forestry Congress, Quebec City, Canada; 21–28 September 2003.

Brown, D. (2009). *Building national capacity for forest governance reform: The role of institutions.* Keynote paper, 'Governance and Institutions', *World Forestry Congress,* Buenos Aires, October 2009.

Dangal S.P. & De Silva, P.M.A. (2009). *Community Forest Management in Sri Lanka Lesson Learnt and Future Direction.* Proceedings the Community Forestry International Workshop, Pokhara, Nepal, September 2009.

De Zoysa, M. & Inoue, M. (2007). Community forest management in Sri Lanka: Concepts and practices. Proceedings of International Symposium on Forest Stewardship and Community Empowerment: Local Commons in Global Context. Kyoto International Community House, Japan, 11–12 October 2007.

———— (2008). Forest governance and community based forest management in Sri Lanka: Past, present and future perspectives. *International Journal of Social Forestry (IJSF),* 1(1), 27–49.

———— (2009). *Farmer Woodlots Development in Sri Lanka: Gains, Losses and Remedies.* Proceedings Small Scale Forestry Symposium; Morgantown, West Virginia, USA, 7–11 June 2009.

Dissanayake, M.W.M.W.T.B. (1998). *Evaluation of the Farmers' Woodlot Component of the Participatory Forestry Project in Sri Lanka.* Proceedings of Forestry Symposium 1998, University of Sri Jayewardenepura, Sri Lanka.

Escobin, R., Gonsalves, J., & Eduardo, Q.E. (2011). Forest management through local level action small grants programme for operations to promote tropical forests (SGPPTF), Sri Lanka. Retrieved 28 December 2011 from http://www.searca.org/ ptf/temp/docs/SriLanka_CH.pdf.

EC–UNDP SGPPTF (2004). EC UNDP SGP PTF Mid-Term Review. The EC-UNDP small grants programme for operations to promote tropical forests in south and south-east Asia, Manila, Philippines.

Efransjah, G.P., Hassan, C.H., & Santosa, K.D. (2008). Consultative planning for effective forest governance: Case studies from Malaysia and Indonesia. *In* C.J.P. Colfer, G.R. Dahal, & D. Capistrano (eds), *Lessons from Forest Decentralization: Money, Justice and the Quest for Good Governance in Asia-Pacific* (pp. 133–147). London, UK: Earthscan Publications.

European Commission to Thailand (2006). European Union–Thailand co-operation activities report 2005, Bangkok 10330, Thailand.

EU–UNDP (The European Commission, United Nations Development Program) (2004). Small grants program for operations to promote tropical forests (SGP PTF): Country guideline paper for Sri Lanka (2004–2007), Colombo, Sri Lanka.

Fabie, P. (2009). Political economy in the natural resources sector. Regional Seminar on Political Economy of Corruption 9–10 September 2009, ADB Headquarters, Manila, Philippines.

FAO (1998). *Woodfuel in Sri Lanka—Production and Marketing.* Bangkok: FAO of the ▪ited Nations.

Forestry Planning Unit (1995). *Sri Lanka Forestry Sectors Master Plan: National Forestry Policy and Executive Summary*. Sri Lanka: Ministry of Agriculture, Lands and Forestry.

Grundy, I., Campbell, B., White, R., Prabhu, R., Jensen, S., & Ngamile, T. (2003). Towards participatory forest management in conservation areas: The case of Cwebe, South Africa. *Forests, Trees and Livelihoods*, 14, 149–165.

Inoue, M. (2011). *Prototype Design Guidelines for 'Collaborative Governance' of Natural Resource*. Presented at 13th Biennial Conference of the International Association for the Study of the Commons, Hyderabad, India, 12 January. Janvanbodegom, A., Savenije, H., & Wit, M. (2009). *Forests and Climate Change: Adaptation and Mitigation*. Wageningen, The Netherlands: Tropenbos International.

Keller, D. (2009). Community participation in sustainable forest management, Sri Lanka. In Arend Jan van Bodegom, Herman Savenije, & Marieke Wit (eds), *Forests and Climate Change: Adaptation and Mitigation*. Wageningen, The Netherlands: Tropenbos International.

Kiyulu, J. (2011). *Forest Governance in the Democratic Republic of Congo* (SAIIA Policy Briefing 33). Norway: Norwegian Ministry of Foreign Affairs.

Lema, L. (2011). *Training Manual on Good Forest Governance at Community Level, Envirocare*. Retrieved from www.envirocaretz.org

Lynch, O.J. & Talbott, K. (2010). *Balancing Acts: Community Based Forest Management and National Law in Asia and the Pacific*. Washington, DC: Natural Resources Management & Development Portal, U.S. Agency for International Development.

Maddugoda, P. (1991). *Experience of Community Forestry in Sri Lanka*. Proceedings Second regional workshop on multi-purpose trees. Kandy, Sri Lanka, 5–7 April 1991.

Mahanty, S., Guernier, J., & Yasmi, Y. (2009). A fair share? Sharing the benefits and costs of collaborative forest management. *International Forestry Review*, 11(2), 268–279.

MENR (2003). *Caring for the Environment 2003–2007: Path to Sustainable Development*. Battaramulla: Ministry of Environment & Natural Resources.

Nanayakkara, V.R. (2001). *Regional Study on Forest Policy and Institutional Reforms: Final Report of the Sri Lanka Case Study*. Colombo, Sri Lanka: Asian Development Bank.

Nurse, M. & Hitinayake, G. (2000). Change on the horizon: Participatory forest management in Sri Lanka—recent experiences from the dry zone. Retrieved on 5 February 2009 from http://www.livelihoods.org/post/Docs/nurse1.rtf.

Ojha, H., Persha, L., & Chhatre, A. (2009). *Community Forestry in Nepal: A Policy Innovation for Local Livelihoods* (IFPRI Discussion Paper 00913). Washington, DC: The International Food Policy Research Institute.

Ojha, H.R., Subedi, B.P., Dhungana, H., & Paudel, D. (2008). *Citizen Participation in Forest Governance: Insights from Community Forestry in Nepal*. Proceedings of Conference on Environmental Governance and Democracy, 10–11 May 2008, Yale University, New Haven.

Perera, G.A.D. (2001). Secondary forest situation in Sri Lanka: A review. *Journal of Tropical Forest Science*, 13(4), 768–785.

Philips, B.R. (2003). *Integrated Approach in Watershed Management and Poverty Reduction*. Manila: Forests & Water/PRSNT/03, Agriculture, Natural Resources & Social Sectors Division, Regional and Sustainable Development Department Asian Development Bank.

Poffenberger, M. (2000). Communities and forest management in south Asia a regional profile of WG-CIFM: The working group on community involvement in forest management. Asia Forest Network, USA.

Tan, N.Q., Thanh, T.N., & Hoang Huy Tuan, H.H. (2009). *Community Forestry and Poverty Alleviation: A Synthesis of Project Findings from Field Activities.* Vietnam, Hanoi: Forest Governance Learning Group (FGLG).

Velasco, M.T. & Lim, A. (2012). *Forest Lives: Lessons on Sustaining Communities and Forests from the Small Grants Program for Operations to Promote Tropical Forests (SGPPTF).* Retrieved December 2012 from http://www.recoftc.org/site/uploads/content/pdf/Forest_Lives_81.pdf.

Wijewarnasuriya, A. (2009). *Establishment of Farmers' Woodlots in Sri Lanka; Research Studies on Forest Management in Sri Lanka.* Retrieved from http://www.sjp.ac.lk/~upul/forestmanagementlanka.

Wickramasinghe, A. (1997). Women and social forestry in Sri Lanka. *ENERGIA News*, April, Issue 2.

# 6

# Bhutan: Forest Resources Management and Conservation in and outside Protected Areas

*Om N. Katel and Dietrich Schmidt-Vogt*

## Introduction

Bhutan is a small country located in the eastern Himalayas (Figure 6.1) with a population of about 700,000 (RGOB, 2011). Bhutan was also isolated from the rest of the world for a long period of time, and became officially open to outsiders only after 1960 (Dorji, 2003). Documented history dating back to the 1600s provides ample evidence for a close link between the state and its abundant natural resources. Bhutan is especially famous for its extensive forests, which cover more than 80% of the total land area (RGOB, 2012). Forests are an important component of the Bhutanese farming system as farmers obtain a variety of products and services from forests, such as leaf litter for animal bedding and for the production of organic manure. Forests also provide timber and fuel wood as well as non-timber forest products (NTFPs) such as mushrooms and edible ferns to supplement the diets of local individuals and provide a source of cash income.

Bhutan is a predominantly agricultural country practising traditional mixed farming, mainly for subsistence. The agricultural sector consisting of crop farming, horticulture and livestock employs more than 69% of the total population. Crop farming, livestock rearing and the use of forest

**Figure 6.1:**
*Location of Bhutan (shaded)*

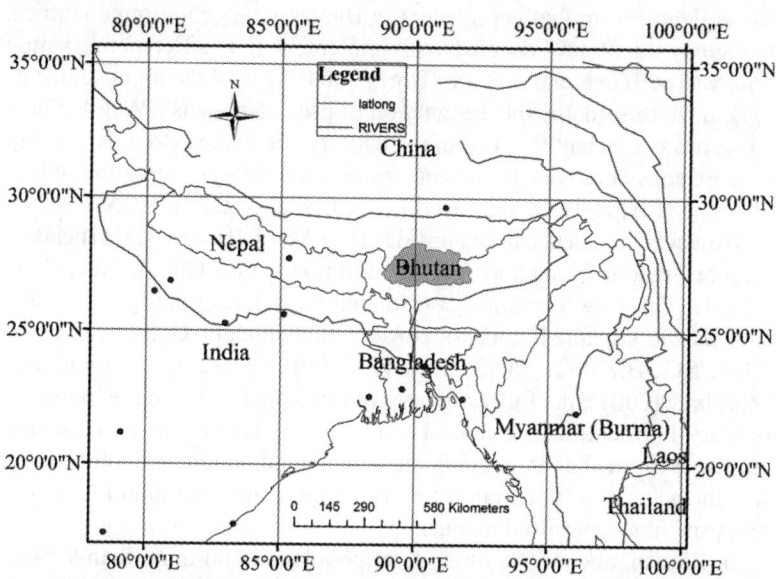

*Source:* http://www.diva-gis.org/Data.
(Disclaimer: This image has been redrawn by the author (Om Katel) and is not to scale. It does not represent any authentic national or international boundaries and is used for illustrative purposes only.)

products are integral to the farming system that has evolved over a long period of time within a variety of ecological conditions. Although agriculture is one of the most important sectors of the Bhutanese economy, its contribution to GDP has declined from 26% in 2001 to 25% in 2003 and 19% in 2008 (Gurung, 2012; NSB, 2009). Recently, the Royal Government of Bhutan has been encouraging farmers to practice organic farming and to use leaf litter collected from the forest as manure instead of chemical fertiliser.

The importance of forests as primary sources of organic manure adds another dimension to the environmental goal of protecting and maintaining an extensive forest cover in Bhutan. Environmental conservation represents one of the four pillars of gross national happiness and is one of the top priorities for sustainable development (RGOB, 2003). The emphasis on conservation is further reinforced by the constitution of Bhutan (RGOB, 2008), which stipulates that a minimum of 60% of the total land area of the country must be maintained under forest cover at all times (RGOB, 2002b, 2004). To achieve this aim and to avoid the severe forest degradation that is prevalent in Bhutan's neighbouring countries, the government has enacted and implemented a strict conservation policy to curb overexploitation (Dorji, 2003). One of the main strategies to achieve this goal is the designation of protected areas (PAs) that now cover more than half (51.4%) of the country's land area (RGOB, 2010a). PAs are considered one of the best strategies for the conservation of biodiversity (Chape, Harrison, Spalding, & Lysenko, 2005; IUCN, 2003).

However, residents living inside the PAs of Bhutan are dependent on forest resources such as fuel wood, timber and NTFPs for their livelihoods. These residents are constrained by a conservation policy that follows the so-called 'fortress conservation model' (Galvin & Haller, 2008; RGOB, 2002a, 2003; Seeland, 2000). According to Hulme and Murphree (2001), the fortress conservation model refers to the 'police or military-like central state control of a PA', in which human use of natural resources from the PA is completely restricted. While decentralisation was introduced in Bhutan in the early 1990s, conservation in PAs is still practised in a centralised manner.

In Bhutan, about 69% of the total population and more than 90% of people living inside PAs are dependent on forests (RGOB, 2003). Therefore, the perceptions of park residents towards PAs and PA management are likely to be influenced by their ability to access and use forest resources. This chapter presents a case study of the forest dependence of park residents in the context of the current conservation policy and in comparison to the situation outside PAs. The use of forest resources by park residents is discussed, along with the perceptions and attitudes of park residents about conservation management. The PA governance system is then compared with community forestry (CF) programmes using design guidelines. To provide an overview of resource management inside and outside of PAs, two case studies were carried out, one of park residents in Jigme Singye Wangchuck National Park (JSWNP) and the other of CF in Mongar *Dzongkhag* and Lhuentse *Dzongkhag* in eastern Bhutan. JSWNP was selected for three reasons: (a) JSWNP is located in central Bhutan and is surrounded by settlements on all sides, (b) a large number (about 6,000) of local people live inside the park and along the park boundary, and (c) the park is located in the Inner Himalaya, an area of transition between the monsoon-exposed southern side and the dry, cold northern side of the Himalayas (Katel & Schmidt-Vogt, 2011; Ohsawa, 1987; Schweinfurth, 1982).

Information for the case study of the CF programme in the villages of *Markuling* in the *Dzongkhag* of Lhuentse and *Yakpugang* in the *Dzongkhag* of Mongar was obtained from primary data, including published and unpublished literature. *Dzongkhag* stands for district in the national language of Bhutan, Dzongkha. *Yakpugang* CF (YCF), founded in 2001, was chosen because it is the second oldest CF established in Bhutan, and *Markuling* CF (MCF) was chosen because it is also among the oldest CFs in Bhutan.

## Overview of Biodiversity Conservation in Bhutan

Conservation and conservation behaviour have a long history in Bhutan, which is closely linked with the history of religion, and with the establishment of the state and its institutions. There is also a long tradition of local decentralised forest management, which was interrupted by the nationalisation of forests, but which may provide a basis for the recent emphasis on decentralisation.

At around 600 AD, the Bon religion spread from Tibet into Bhutan. The Bon religion is characterised by deities manifested in nature in forms such as mountains, trees and lakes. In the eighth century, when Buddhism was introduced in Bhutan, some Bon elements such as the deification of nature were incorporated into the new religion (Dorji, 2003). These elements still provide the spiritual basis for conservation behaviour and conservation efforts.

The introduction of Tibetan Buddhism to Bhutan brought with it the establishment of a formal government headed by religio-political rulers. The first of these was Shabdrung Ngawang Namgyal, who came to Bhutan from Tibet in 1616. Shabdrung Ngawang Namgyal ruled Bhutan through a theocratic administrative system (called the Choesid system) that vested all political and religious authority in him until his death in 1652 (Mathew, 2012). After Shabdrung Ngawang Namgyal, spiritual and temporal powers were divided and vested in two authorities called the Je Khempo (spiritual head) and the Druk Desi (temporal head). With the establishment of religio-political rule, a taxation system was introduced to materially sustain the rulers. Local people paid their taxes mainly with resources derived from forests. At that time, forests were managed by local residents with *de facto* access rather than through regulations issued by authorities. Communities defined the boundaries of the forests and developed local arrangements to manage them for subsistence and for meeting taxation requirements. In this way, the establishment of the formal government was paralleled by the development of local decentralised forest management. The theocratic system remained in place until it was replaced by a hereditary monarchy based on the Wangchuck dynasty in 1907 (Dorji, 2003).

Very little information is available on the taxation system before the 1950s, when the feudal system was abolished by the third hereditary monarch (Dorji, 2003). In 1952, the Department of Forest was established as the first central institution for the management of forest resources in Bhutan. This was followed by a change in the land tenure system that differentiated between private and public property. In 1959, *Thrimsung Chenmo* or the Supreme law was passed. This was followed by the enactment of the Bhutan Forest Act (BFA) in 1969. The BFA of 1969 brought all forests under state control and focused on traditional forest protection with the introduction of user permits (Penjore & Rapten, 2004; RGOB, 2004). The nationalisation of forest management replaced all oral and unwritten customary laws with written formal laws, ultimately transferring citizens' rights to the central state. State control of forests persisted until the 1990s, when a decentralised system of forest management system was introduced.

Until the 1960s, the absence of roads and other means of communication forced Bhutan to remain isolated from the rest of the world. The opening of Bhutan in 1969 heralded the implementation of conservation-related programmes and policies, such as the National Forestry Policy of 1974 (NFP, 1974), which established the guiding principle of maintaining a minimum of 60% of the total land area under forest cover in perpetuity. The NFP also consolidated the nationalisation of logging operations in 1979 and designated the PA network in 1983 (Dorji, 2003).

In 1995, new national parks were established comprising 26% of the country's total land area. In 1999, the addition of biological corridors increased the protected land area to about 35% (RGOB, 2003, 2004, 2010a). The establishment of Wangchuck Centennial Park in 2008 contributed so much land that PAs (parks, sanctuaries, reserves and the additional 9.53% representing biological corridors) now make up more than half (51.44%) of the country's total land area (Katel & Schmidt-Vogt, 2011; RGOB, 2010a; see Figure 6.2).

Although the nationalisation of forests in the 1960s led to more effective forest protection, forest resource users lost customary ownership (MOAF, 2002 cited in Brooks & Tshering, 2011). In the recent past, the

**Figure 6.2:**
*PAs in Bhutan*

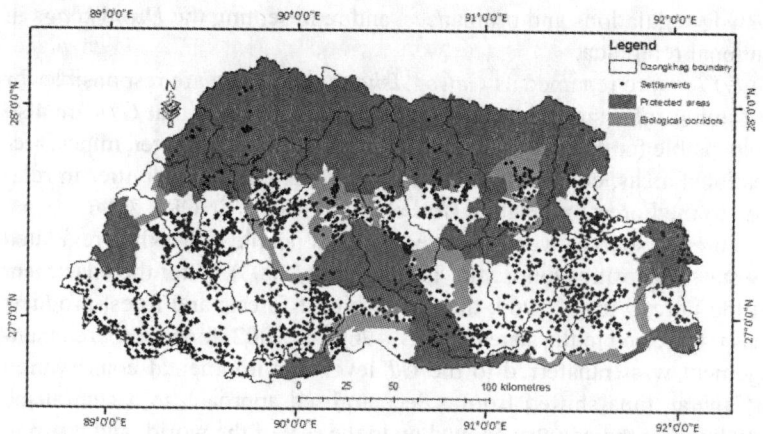

*Note:* Protected areas reflected here include national parks, wildlife sanctuaries and nature reserve.
*Source:* GIS Lab, College of Natural Resources, Royal University of Bhutan.
(Disclaimer: This image has been redrawn by the author (Om Katel) and is not to scale. It does not represent any authentic national or international boundaries and is used for illustrative purposes only.)

government realised the importance of local people's participation to manage natural resources more effectively, and placed more emphasis on decentralisation.

The origins of decentralisation can be traced back to the early 1980s. In 1981, the fourth king of Bhutan initiated a process of decentralisation by dividing the country into 20 districts (*Dzongkhag*s) and established *Dzongkhag* development committees called *Dzongkhag Yargye Tshogchung* (*DYT*) in each *Dzongkhag*. In 2002, county-level committees were also formed, known as *Gewog Yargye Tshogchung* (*GYT*). These committees (*DYT*s and *GYT*s) are represented by elected members at the local level. *DYT*s in each *Dzongkhag* comprise two elected members representing *GYT*s and one elected member representing the municipality from within that *Dzongkhag*. *GYT*s are formed at the county level as secondary administrative units and comprise at least seven members chosen by local people through direct election.

Since the enactment of the Local Government Act in 2009, the *DYT*s have been renamed *Dzongkhag Tshogdu* (District Councils), and were established as the highest decision-making body in the *Dzongkhag*.

These bodies are responsible for balancing socio-economic development, coordinating the activities of government agencies, reviewing
*Gewog* regulations and ordinances, and representing the *Dzongkhag*s in
national referenda.

*GYT*s were renamed as *Gewog Tshogde* (*GT*) and are responsible for
enforcing rules similar to DTs. The only difference is that *GT*s are also
responsible for the provision of drinking and irrigation water, mines, recreational areas, construction, land use and agricultural activities in relation to the Forest and Nature Conservation Rules (FNCR), 2006.

Forest resources were managed by local people rather than regulated
by state authorities before the 1950s (Dorji, 2003). After the enactment
of the Forest Act of 1969, the ownership of forests and forest products
was transferred to the state (RGOB, 1969). In 2002, forest resource management was transferred to the *GT* level. Environmental conservation
in Bhutan thus shifted from a decentralised approach to a centralised
system after the country opened up to the rest of the world, and again to
decentralisation in the recent past.

However, the recent move towards decentralisation does not apply to
PAs, although park residents living within the PAs are under the jurisdiction of the respective *Dzongkhag*. Over the years, the government of
Bhutan has enacted several acts, regulations and orders that provide a
legal framework for park management (Table 6.1).

Each park has its own management plan. Park management plans
include the zoning of park areas into the three management categories
of core zone, multiple use zone and buffer zone. The basic protection
regimes differ from one zone to another. The core zones provide strict
protection, preventing access by anyone except government officials on
duty and researchers conducting ecological monitoring. Multiple use
zones accommodate the needs of park residents for timber and other
forest resources, and buffer zones are demarcated as a protective layer
against development activities.

Bhutan's park policies allow residents to remain inside PAs (Wang,
Lassoie, & Curtis, 2006a) and do not displace them from the parks.
However, as some residents of PAs live in the core zone, further zoning
of land use within the core zones of the parks is required to address
the communal use of forests and the sustainable use of common pool
resources.

Protection of forests inside PAs is complemented by the conservation
of forests outside PAs through CF. The first CF was established in 1997.
However, there was at that time no strong legal framework supporting

**Table 6.1:**
*Summary of acts, regulations and orders*

| S. no. | Acts, regulations and orders | Subjects dealt with |
|---|---|---|
| 1 | Forest and Nature Conservation Act, 1995 | PAs, Conservation of wildlife, Soil and water conservation, enforcement and penalties |
| | Gazette Notification of PAs, 1995 | Gazette of four priority PAs |
| 2 | Land Act of Bhutan, 1998 | Encroachment of land, grazing land, government land procedure for allotment |
| 3 | Mines and Mineral Management Act, 1995 | Protection of environment from commercial mining operations |
| 4 | Pastureland Act of Bhutan, 1979 | Livestock grazing |
| 5 | Livestock Act of Bhutan, 2001 | Livestock breeding, health and production |
| | DYT Act, 2002 | *Dzongkhag*, park/sanctuaries, farm feeder road, forest management units, forest protection, grazing land monitoring and land encroachment |
| | GYT Act, 2002 | Protection and harvest of edible NTFPs, crop damage by wildlife, grazing land monitoring and enforcement, conservation and management of water resources |
| 6 | Forest and Nature Conservation Rules of Bhutan, 2006 | Timber transit within the park, fishing rules and regulations, allotment of government land for development of village basic facilities |

*Source:* JSWNP Conservation Management Plan (2003–2007).

the CF programme, and many restrictions on forest management were in place. CF was legally established only in 2000 (Temphel & Beukeboom, 2006; Wangdi & Tshering, 2006). The FNCA, 1995 provided the primary authority to establish CFs in Bhutan by including chapters on CF programmes that were not included in previous Acts (RGOB, 1995). The FNCA (RGOB, 1995) states that the ownership of forest produce can be transferred to appropriate groups of inhabitants and that the 'group to which CFs have been transferred shall manage them for sustainable use in accordance with the rules for community forests' (RGOB, 1995). This arrangement is further supported by the Forest and Nature Conservation Rules of 2006 (FNCR, 2006). The rules for the establishment of CF require that 'the forest area allocation shall not exceed more than 2.5 ha per household'. The FNCA (RGOB, 1995) vested powers in the Ministry of Agriculture and Forests (MOAF) regarding the establishment of CFs and the formation of rules. The main aim of the introduction of the CF programme was to conserve and promote sustainable utilisation of forest resources (RGOB, 2010b). The second CF was established in 2001. There were 31 CFs in 2006, 46 in 2007 and over 300 in 2010 (Chhetri, Schmidt, & Gilmour, 2009; Phuntsho, Schmidt, Kuyakanon, & Temphel, 2011).

## Case Studies

This study is based on data collected from JSWNP located in central Bhutan (Figure 6.3), and from MCF and YCF located in eastern Bhutan (Figure 6.5). Forests in JSWNP are under the direct control of the Department of Forests and Park Services (DOFPS), while YCF is under the CF regime.

**Figure 6.3:**
*(a) Map of Bhutan showing protected areas (shaded) and (b) JSWNP with Gewog boundaries*

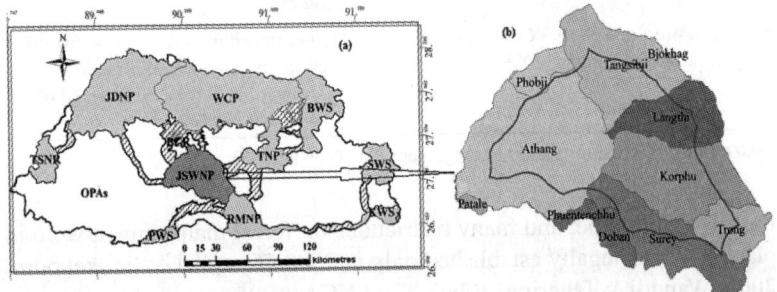

Grey line on the map indicates demarcation of periphery and inner areas. JDNP, Jigme Dorji National Park; WCP, Wangchuck Centennial Park; TNP, Thrumsingla NP; JSWNP, Jigme Singye Wangchuck NP; RMNP, Royal Manas NP. Wildlife sanctuaries: BWS, Bomdeling WS; SWS, Sakteng WS; KWS, Khaling WS; PWS, Phibsoo WS; OPAs, Other Protected Areas and the Toorsa Strict Nature Reserve (TSNR). Shading with line represents biological corridors connecting PAs.
(Disclaimer: This image has been redrawn by the author (Om Katel) and is not to scale. It does not represent any authentic national or international boundaries and is used for illustrative purposes only.)

The two cases were studied with different objectives. While the first case study was carried out to assess forest resource use by local communities living inside and at the periphery of JSWNP, the second case study was conducted to understand people's perceptions towards forest resource use and their participation in the governance and management of forest resources in CFs. As the research for these two case studies was not specifically carried out for the purpose of this book, there is an imbalance between the two case studies in terms of scope and intensity.

# A Case Study from JSWNP

## Study Sites and Methodology

JSWNP is the third largest national park in the kingdom of Bhutan. It covers over 1,723 km$^2$ in central Bhutan. JSWNP encompasses the administrative jurisdictions of five *Dzongkhags* (districts): *Tsirang, Sarpang, Zhemgang, Trongsa* and *Wangdue-phodrang* and of 11 *Gewogs* or sub-districts (Figure 6.3).

The data for communities inside the national park were collected during six months of fieldwork from May to December 2008 (Katel, 2012) in eight *Gewogs* of JSWNP (Figure 6.3) covering 39 villages (Table 6.2). The villages selected for this study are dispersed across the park. Stratified random sampling was used to collect forest resource use data.

## Data Collection and Analysis

Research for this case study was conducted in 8 of the 11 *Gewogs* in JSWNP. Three *Gewogs* were excluded because two *Gewogs* have no settlements and because one *Gewog* is located above the timber line. We classified the eight selected *Gewogs* as belonging either to the inner area (*Ada, Korphu* and *Doban*) or to the peripheral area of the park (*Tangsibji, Langthel [Langthi], Trong, Jigmechhoeling [Surey]* and *Patale*) based on the travel distance to the nearest roads. People living in the peripheral areas have direct access to roads or are less than 3 hours away from a road, whereas residents of the inner areas must travel for more than 3 hours on foot to reach the nearest road.

**Table 6.2:**
*Number of households per* Gewog *in JSWNP*

| Dzongkhag | Gewogs* | Number of villages in the park | Ecological zones, forest types and elevation (m) |
|-----------|---------|-------------------------------|--------------------------------------------------|
| Sarpang | Doban | 4 | Sub-tropical, warm temperate, broad leaved, 1,800 m |
| | Jigmechhoeling (survey) | 7 | Sub-tropical, warm temperate, broad leaved, 1,900 m |
| Tsirang | Patale | 2 | Sub-tropical, mixed forest (broad leaved + coniferous), 1,200 m |
| Trongsa | Korphu | 3 | Cool temperate, broad leaved, 2,200 m |
| | Langthel | 7 | Warm temperate, broad leaved + coniferous, 1,800 m |
| | Tangsibji | 4 | Cool temperate, broad leaved + coniferous, 2,200 m |
| Wangdue | Ada | 9 | Sub-tropical, warm temperate, broad leaved + coniferous, 1,800 m |
| Zhemgang | Trong | 3 | Sub-tropical, broad leaved + coniferous, 500 m |

*Village information from *Gewog*s: Phuentenchu and Drakteng are not included as no households are believed to be located inside park area and Phobji is located in higher elevations and patterns of forest resources use is different.
**Source:** Authors.

Research was carried out by means of pre-tested semi-structured questionnaires with structured and open-ended questions on the perceptions of park value and management, and on forest resource use and constraints. The questionnaires were supplemented with informal meetings and discussions, and with the direct observation of peoples' activities in their villages. The categorical demographical variables were transformed into dichotomous variables and entered into logistic regression and chi square models.

## Results

### Patterns of Forest Resource Use

Forest resource use is closely linked to subsistence farming in JSWNP. Forest resources are used for fuel wood, grazing, shifting cultivation, leaf litter, shingles to build roofs, kindling, medicine and dyes. Park residents

**Table 6.3:**
*Forest product use ranking by park residents*

|  | Responses (%) | |
| --- | --- | --- |
| *Forest product use ranking* | *Inner* | *Periphery* |
| Fuel wood | 100.0 | 96.1 |
| Timber | 73.3 | 94.4 |
| Leaf litter and fodder | 84.9 | 76.4 |
| Medicinal plants | 69.8 | 23.6 |
| Weaving and basketry | 37.2 | 16.3 |
| Tool handles | 33.7 | 1.70 |
| Poles for fencing and prayer flags | 3.5 | 14.6 |
| Thatch materials | 10.5 | 1.1 |

Total responses are more than 264 (*N* = 264) as multiple answers were recorded.
*Source:* Katel and Schmidt-Vogt (2011).

ranked fuel wood, timber and NTFPs as the three most important uses of the forest in JSWNP (Table 6.3). Fuel wood is the main source of energy for these communities, as only a few of the households in the peripheral area had access to electricity, and none of the households in the inner area had access as of the end of 2008. Fuel wood is used for heating and cooking purposes. In terms of household use, timber was ranked higher in the peripheral areas (94.4%) and fuel wood was ranked higher in the inner areas (100%) (Table 6.3). *Castanopsis* sp. is the most important fuel wood species in both the inner and peripheral areas (Table 6.4) and is common in all the zones.

Forest resources like fuel wood are harvested illegally in some areas mainly to avoid lengthy administrative procedures. Individuals, who require fuel wood must obtain the fuel wood requirement form, fill it in and submit it to the park head office, which then approves the requisition. The trees to be felled for fuel wood are marked with hammers by officials as prescribed by the Forest and Nature Conservation Rules of Bhutan (RGOB, 2006). Park residents must pay a royalty fee to obtain fuel wood or timber as per the guidelines stated in the rules (RGOB, 2006).

Tree species are used for fodder and NTFPs. NTFPs are used for food, clothes dyeing, medicines, weaving, basketry and religious purposes (Table 6.5). In addition, farmers also use leaf litter as more than 95% of the park residents practice subsistence agriculture (Wang & Macdonald, 2006) and leaf litter is the primary source of manure for farming.

**Table 6.4:**
*Species used for fuel wood*

| Location | Gewog | Name of tree species |
|---|---|---|
| Inner area | Ada | *Schima wallichii, Callicarpa arborea, Castanopsis hystrix, Castanopsis tribuloides, Pinus roxburghii* |
| | Doban | *Alnus nepalensis, Macaranga denticulata, Macaranga pustulata, Castanopsis hystrix* |
| | Korphu | *Quercus griffithii, Castanopsis hystrix, Pinus roxburghii, Schima wallichii* |
| Periphery area | Jigmechoeling | *Castanopsis tribuloides, Castanopsis hystrix, Schima wallichii* |
| | Langthel | *Quercus griffithii, Castanopsis hystrix, Pinus roxburghii, Schima wallichii* |
| | Patale | *Castanopsis tribuloides, Quercus griffithii, Lithocarpus elegans, Schima wallichii, Macaranga* spp |
| | Tangsibji | *Quercus griffithii, Pinus roxburghii* |
| | Trong | *Castanopsis tribuloides, Altingia excelsa, Schima wallichii* |

**Source:** Authors.

**Table 6.5:**
*Species used for NTFPs*

| Species used as NTFPs | Purpose |
|---|---|
| Tender shoots of *Calamus flagellum, Dryopteris* spp, *Elastostemma* spp | Vegetables |
| *Rhus* sp, *Strobilanthes* spp, *Rubia* spp, *Prunus* spp | Dyes |
| *Calamus* sp, *Bambusa tulda, Bambusa alamii, Bambusa clavata* | Weaving and basketry |
| *Pinus roxburghii, Castanopsis hystrix, Ficus cordata, Bambusa* spp, *Musa* spp, *Oroxylum indicum, Rhus* spp, *Cymbopogon* spp, *Artemisia vulgaris, Quercus* spp | Religious and cultural purposes |
| *Bombax ceiba, Phyllanthus emblica, Artemesia vulgaris, Terminalia chebula, Rubia cordifolia, Zanthoxylum* spp, *Adhatoda vasica, Erythrina arborescence* | Medicinal |
| *Toddalia* spp, *Gynocardia* spp, *Aesandra* spp | Edible oil |

'spp' indicates different species of the same genus.
**Source:** Katel and Schmidt-Vogt (2011).

## Local Residents' Knowledge and Awareness of Park Establishment

The results of the household questionnaire survey revealed that 79.9% of all respondents were aware that JSWNP was established more than 15 years prior to the survey. More residents (87.2%) were aware of the establishment of the park in the inner areas than in the peripheral areas (76.4%) ($\chi^2 = 4.2189$, df $= 1, p < 0.05$).

Logistic regression results (Table 6.6) show that males who received some form of incentives from park management, who had families with more than 7 members in the household, and who belonged to households owning more than 11 heads of livestock were more knowledgeable about the establishment of JSWNP. Incentives included solar lighting and corrugated galvanised (CGI) sheets for roofing provided to park residents at a subsidised price. These items are provided as incentives to reduce the use of forest resources. Variables such as gender and livestock possession were indicators of knowledge of the park establishment and of the

**Table 6.6:**
*Results of logistic regression showing the effects of independent variables on dependent variables*

| Independent | Dependent |
| --- | --- |
| Gender (male) | Knowledge park establishment (yes/no)** |
| Incentives (yes) | -do-** |
| Family size (>7) | -do-** |
| Livestock (>11) | -do-* |
| Park location (periphery) | Awareness on types of benefits (yes/no)** |
| Gender (male) | -do-** |
| Marital status (married) | -do-* |
| Age (<40) | -do-* |
| Livestock (>11) | -do-* |
| Park location (inner) | Attitudes of park residents (satisfied/not satisfied)* |
| Gender (male) | -do-* |
| Education (literate) | -do-* |
| Age (>40) | -do-* |
| Family size (<7) | -do-* |
| Livestock (<11) | -do-* |
| Gender (male) | Park perception (important/not important)** |
| Religion (Buddhist) | -do-* |
| Family size (<7) | -do-* |
| Livestock (<11) | -do-** |

** $p < .01$; * $p < .05$; variables in parenthesis indicate their relationship to dependent variables.

*Source:* Katel and Schmidt-Vogt (2011).

**Figure 6.4:**
*Park residents' reasons for their compliance with park rules*

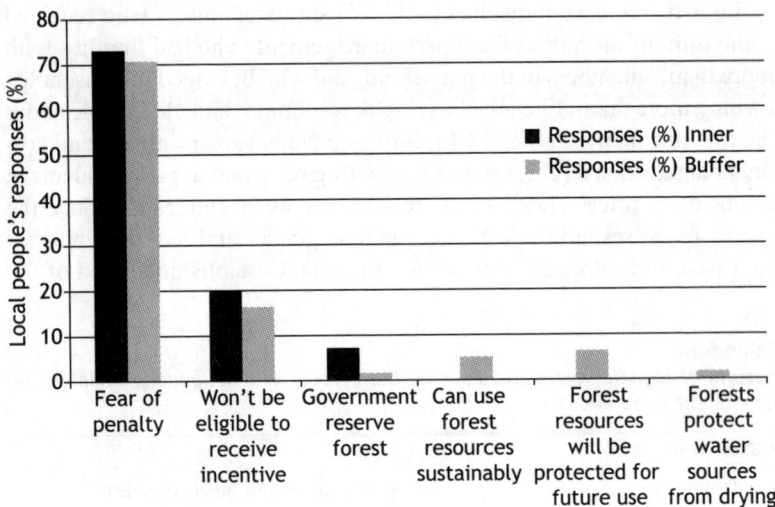

*Source:* Katel and Schmidt-Vogt (2011).

awareness regarding the benefits of the park for rural people. Other variables such as a resident's location (inner/periphery), marital status and age exhibited significant relationships with park residents' awareness of the park establishment.

Inner area respondents, male respondents, literate respondents, elderly respondents, respondents with households with fewer than seven members and respondents owning fewer than 11 heads of livestock all had positive attitudes towards the park.

Most respondents have heard about the park rules (93.0% in the inner area and 74.3% in the peripheral area), and more than 50% of the respondents from both areas said that they obey the rules, which are strictly enforced by the park management (Figure 6.4).

### Residents' Participation in Management

In this study, participation was assessed based on the involvement of park residents in silvicultural practices, planting seedlings and protecting plant species of conservation interest in the park. Results showed that park residents' involvement in managing forests was significantly lower ($\chi^2$ = 14.321, df = 1, $p$ < 0.001) in the inner area (28.2%) than in the peripheral area (53.2%).

# Case Studies of CF in Markuling and Yakpugang

## Brief History of CF Development in Bhutan

About 30% of the people living in the rural areas of Bhutan live under the poverty line. As 69% of the total population lives in rural areas, poverty in Bhutan is largely a rural phenomenon (NSB, 2007). Poor farmers are highly dependent on forest resources for their subsistence livelihoods. CF was introduced to Bhutan to provide poor farmers with better access to forests and a more sustainable supply of forest products through decentralised and sustainable forests (Phuntsho et al., 2011).

The concept of CF was introduced to Bhutan following a Royal Decree in 1979 (Chhetri et al., 2009). Although CF emerged in the international arena in the 1970s, it was established in Bhutan only in the 1990s when the importance of peoples' participation in the protection and management of forests was finally recognised (Wangchuk, 2010). The FNCA (RGOB, 1995) and FNCR (2006) provide the legal basis for the establishment of CFs in Bhutan. These rules prescribe steps for the initiation, planning, implementation and review of a CF and its Community Forest Management Groups (CFMGs). According to FNCR (RGOB, 2006, p. 27), a CF is 'any area of Government Reserved Forest, in and around villages and human settlements including government land situated in the interspaces between registered private lands, suitable for management by a Community Forest Management Group (CFMG)'. The forest area to be allocated as a CF may not exceed 2.5 ha per household, except for CFs involving the management of NTFPs, in which case more than 2.5 ha may be allocated per household.

In Bhutan, the management and governance of CFs is coordinated at the local (*Gewog*) level by *Gewog* (block) extension officers (GExO). The GExO submits recommendations to the *Dzongkhag* Forestry Office (DZFO) based on requests made by the local community through CFMGs. The DZFO then provides technical support to the local community for the establishment of the CF. DZFOs in turn receive administrative and other advice from the Social Forestry Division of the DOFPS.

When CFs became an important part of the national forest policy, the number of CFs increased. As of December 2012, 485 CFs had been established, accounting for 62,237 ha; 1.8% of the national forest cover and 24.2% of the rural households are registered CFMG groups (RGOB, 2012).

## Study Sites, Methodology and Data Analysis

The data for this section were collected from two CFs, MCF located in *Lhuentse Dzongkhag* and YCF located in *Monger Dzongkhag*, eastern Bhutan (Figure 6.5).

MCF covers an area of 70 ha and consists of 28 registered households. The CF was formally handed over to CFMGs in 2003. In the past, the CF area was heavily used for the extraction of timber for the renovation of the *Dzong*, a fortress that serves as an administrative, religious and social centre.

**Figure 6.5:**
*Map of Bhutan showing study sites*

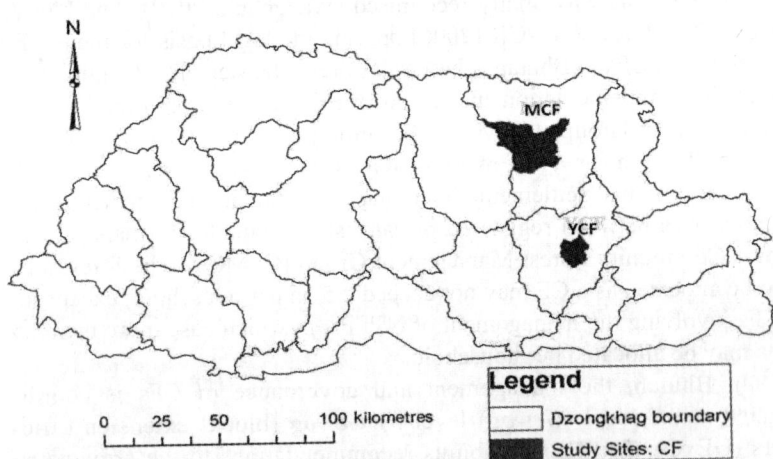

MCF, Markuling CF; YCF, Yakpugang CF; shaded in dark, eastern Bhutan
*Source:* College of Natural Resources, Royal University of Bhutan.
(Disclaimer: This image has been redrawn by the author (Om N. Katel) and is not to scale. It does not represent any authentic national or international boundaries and is used for illustrative purposes only.)

YCF covers an area of about 260 ha (Figure 6.5) and was formally established in May 2001. Currently, 115 households are registered under this CF.

The primary data for this study were collected through a household survey employing standard pre-tested questionnaires. In the two research sites (MCF and YCF), a total of 143 households were interviewed and questions were asked on the awareness of changes in forest conditions, households' involvement and participation in CF management activities.

The households were classified based on the criteria used by CFMG members to categorise rich, neither rich nor poor (*medium*) and *poor* (Table 6.7). These criteria were developed based on a focus group discussion with CFMG committee members, including the chairman, secretary, treasurer, village headmen (*Tsogpas*) and three CFMG members.

The forest products harvested by local communities were placed into six categories based on the CF management plan: *Drashing, Cham, Tsim,*

**Table 6.7:**
*Criteria and indicators to categorise households into different socio-economic groups*

| Criteria | Socio-economic grouping categories | | |
| --- | --- | --- | --- |
| | *Rich* | *Medium* | *Poor* |
| Land holdings | Large *Chhuzhing* and *Kamzhing* land | Sufficient land to feed family | Limited land mainly *Kamzhing* or landless |
| Livestock holdings | Large numbers of livestock (milking cows) | Few livestock (cattle heads) | Little or no cattle or small livestock |
| House conditions | Two-storied houses with tin roofs (CGI sheets) | Medium size houses, not in good conditions | Small houses and huts |
| Sources of income | Maximum income from sale of agricultural and livestock products | Limited income from sale of agricultural and livestock products | No income from sale of agricultural and livestock products |
| Food sufficiency | Sufficient food for 12 months with surplus for sale | Just or almost sufficient food for 9–12 months | 3–6 months or less food production |
| Farm machineries | Power tiller, chain saw, flour mills and the related farm machines | Limited to one or none | No farm machineries, lack of labour to work on farms |

**Source:** Authors.

*Dangchung,* flag poles and fencing posts. *Drashing* represents trees of at least 50 cm dbh (dbh stands for diameter of tree stem at breast height or at 1.3 m above the ground) that are used for sawed timber. *Cham* indicates trees of 30–50 cm dbh that are mainly used as beams, *Tsim* are trees of 20–30 cm dbh that are used as poles, and *Dangchung* are trees of 20–30 cm dbh that are used for small poles. These six forest products are all used for construction except for fencing posts and flag poles, which are used to fly religious flags.

## Results

### Socio-economic Groups

A total of 143 households were interviewed, and 69% of the total respondents were female (Table 6.8). In MCF, 61% were female respondents while in YCF 72% were female respondents. In Bhutan, women usually take leadership roles in social life such as attending meetings and conducting annual rituals. Although men make the important decisions, women take an equal part in almost all activities.

### Forest Resource Harvest

When the CF is established, user groups are identified and benefit entitlement is determined by the executive committee. Executive committee members and user group members are equally entitled to receive benefits from a CF. Results indicate that there was no statistical difference in forest resource harvest except that the percentage of trees harvested by executive committee members is higher in YCF (Table 6.9).

**Table 6.8:**
*Number of households under different categories of socio-economic groups*

| Name of CFMGs | Socio-economic categories | | | |
|---|---|---|---|---|
| | *Rich* | *Medium* | *Poor* | *Total* |
| *Markuling* | 08 | 17 | 03 | 28 |
| *Yakpugang* | 23 | 54 | 38 | 115 |
| Total | 31 | 71 | 41 | 143 |

**Source:** Field data collected by Mr Sonam Wangchuk, for his BSc project, College of Natural Resources.

**Table 6.9:**

*CF members who harvested trees by membership status and gender*

| CF | Size class | Membership status | | Sex | |
|----|-----------|--------------------------|-----------------|----------|-------------|
| | | Executive committee (%) | User groups (%) | Male (%) | Female (%) |
| MCF | *Drashing* | 8 | 25 | 27 | 18 |
| | *Cham* | – | – | – | – |
| | *Tsim* | – | – | – | – |
| | *Dangchung* | – | 4 | 9 | 6 |
| | *Flag poles* | – | 4 | 9 | 2 |
| | *Fencing posts* | – | – | – | – |
| YCF | *Drashing* | 22 | 27 | 27 | 27 |
| | *Cham* | 22 | 4 | 4 | 3 |
| | *Tsim* | 4 | 2 | 4 | 3 |
| | *Dangchung* | 4 | 2 | 4 | 1 |
| | *Flag poles* | 11 | 5 | 8 | 3 |
| | *Fencing posts* | – | – | – | – |

**Source:** Authors.

## Use of Forest Products and Distribution

Timber as a primary household forest resource is used periodically, while flag poles made from young trees are used annually. Based on the extension office records, communities harvested more timber trees from YCF than from MCF (Table 6.10).

## Fuel Wood and NTFP

Fuel wood is second in importance to timber. Fuel wood is used for cooking and heating purposes and includes two categories: standing trees and small branches and twigs. In both the sites, more trees are harvested from the CF for fuel wood than from the government reserve forest (GRF) and relatively more from MCF than from YCF (Table 6.11). About 71% of all respondents harvest NTFPs from the CF.

## Awareness of Changes in Forest Conditions

A total of 95% respondents in MCF and 93% respondents in YCF indicated that forest cover and tree density have increased over the 10-year period (Table 6.12).

**Table 6.10:**
*The percentage of CFMG members who received timber trees by communities with different socio-economic categories*

|  |  | Socio-economic categories | | |
| --- | --- | --- | --- | --- |
| *CF* | *Size class* | *Rich (%)* | *Medium (%)* | *Poor (%)* |
| MCF | *Drashing* | 17 | 2 | 11 |
|  | *Cham* | – | – | – |
|  | *Tsim* | – | – | – |
|  | *Dangchung* | 4 | – | – |
|  | *Flag poles* | – | – | – |
|  | *Fencing posts* | – | – | – |
| YCF | *Drashing* | 38 | 32 | 43 |
|  | *Cham* | 14 | 4 | 1 |
|  | *Tsim* | 2 | 1 | 2 |
|  | *Dangchung* | 3 | 3 | – |
|  | *Flag poles* | 3 | 8 | 2 |
|  | *Fencing posts* | – | – | – |

– indicates respondents did not harvest.
**Source:** Authors.

**Table 6.11:**
*Trees harvested for fuel wood (membership status and gender)*

|  |  | Membership status | | Sex | |
| --- | --- | --- | --- | --- | --- |
| *CF* | *Firewood source* | *Executive committee (%)* | *User groups (%)* | *Male (%)* | *Female (%)* |
| MCF | CF | 75 | 75 | 64 | 47 |
|  | GRF | 50 | 21 | 9 | 12 |
| YCF | CF | 22 | 46 | 46 | 43 |
|  | GRF | 22 | 19 | 19 | 19 |

**Source:** Field data collected by Mr Sonam Wangchuk, for his BSc project, College of Natural Resources.

Improvement in forest conditions is more pronounced in CFs than in GRFs (Table 6.12). Wangdi and Tshering (2006) report that the condition of the forest land now occupied by CFs was good until 1969 because communities had the right to manage, protect and utilise the

**Table 6.12:**
*CFMGs perception on the change in forest condition*

|  | Trend | MCF (%) (n = 28) | YCF (%) (n = 93) |
|---|---|---|---|
| Forest conditions of CF | Improved | 93 | 95 |
|  | Degraded | 0 | 5 |
|  | No change | 7 | 0 |
| Forest conditions of national forests | Improved | 32 | 51 |
|  | Degraded | 0 | 15 |
|  | No change | 68 | 34 |

**Source:** Fieldwork by Sonam Wangchuk, College of Natural Resources, Royal University of Bhutan.

resources. However, the conditions of these forests deteriorated after the nationalisation of the forests and the introduction of the permit system. In Bhutan, the government provides subsidised timber to rural people to enable them to construct a decent dwelling. Subsidised timber is provided to rural households after the timber requirement application submitted by the household has been scrutinised by the *Dzongkhag* and *Gewog* authorities (RGOB, 2011).

## Participation of Local Communities in CF Management

The involvement of CFMG members in CF administration was evaluated based on household members' meeting attendance and participation in other activities associated with CF management. About 70% and 50% of the total household members attended meetings and other CF activities.

## CF Governance

CF involves the governance and management of forest resources by communities for commercial and non-commercial purposes that hold social and even religious significance. In Bhutan, many divisions such as DZFOs and territorial forestry divisions under the DOFPs, including *Gewog* administration are involved in establishing CFs (Wangchuk, 2010). Interestingly, Wangchuk (2010) argues that in Bhutan DZFOs exert more influence over CF establishment and resource governance than the *Dzongkhag* and *Gewog* administrations (Figure 6.6).

**Figure 6.6:**
*Institutional influence and importance of CF development in Bhutan*

|  | Low Influence | High Influence |
|---|---|---|
| **Low Importance** | Outsiders (Non-CFMGs, Neighbouring villages) | *Dzongkhag* Adm* *Gewog* Adm* |
| **High Importance** | Donor agencies SFD, TFD | DZFOs |

*Adm, administration.
**Source:** Wangchuk (2010).

## Discussions

In JSWNP, forest resource use is closely linked to subsistence farming. The most important uses of the forest are grazing, shifting cultivation and collection of fuel wood and leaf litter. Fuel wood is the main source of energy, especially in the inner areas of JSWNP, as residents do not have access to electricity. The results show that respondents who have received incentives, and who have more than 11 heads of livestock and a family size larger than seven members are more likely to be knowledgeable about the park and its rules than others. The fact that families with more than seven members are more aware of the park than smaller families could be due to the fact that a larger household has more opportunities for interaction outside the home, including participation in awareness programmes. The ownership of more livestock could impose constraints on livestock rearing due to restrictions on the harvest of forest resources. Respondents aware of the establishment of the park were more likely to be male, which could be due to the fact that men have more time and more opportunities to interact socially and acquire information than women, who are busy with household work. This is consistent with the findings of Xu, Chen, Lu and Fu (2006) in Wolong Biosphere Reserve, China.

Respondents in the peripheral area are more likely to be aware of the benefits of the park because the park offices are located at the periphery of the park. The economic benefits provided to park residents affect their attitudes towards conservation favourably as reported for the Wolong Biosphere Reserve, China (Xu et al., 2006) and for the Annapurna Conservation Area (ACA) and Makalu-Barun Conservation Area (MCA) in Nepal (Mehta & Heinen, 2001). This is clearly shown by the fact that inner area residents, who receive compensation, hold a more positive attitude towards conservation than residents of other areas. Interestingly, more residents in the inner area follow the rules than in the peripheral area. There is evidence that the willingness of the majority of respondents to comply with park rules is due to strict enforcement by the park management (Katel & Schmidt-Vogt, 2011). This is consistent with the findings of Brooks (2010) and Brooks and Tshering (2011) and may reflect people's responses to the centralised approach to forest management practised since the 1960s, which remains in practice in PAs. Evidence from ACA and MCA in Nepal (Mehta & Heinen, 2001) and from Yunnan in China (Zhou & Grumbine, 2011) shows that a participatory approach can be more successful than the top-down approach in terms of improving local people's livelihoods and their attitudes toward conservation.

Families with fewer than seven members perceive the park as important because mostly smaller families receive incentives such as CGI sheets (Katel & Schmidt-Vogt, 2011). Residents with more livestock are particularly worried that park boundaries might entail additional restrictions on fodder collection and more risks towards the depredation of livestock by wildlife. This is consistent with the findings by Wang, Lassoie, and Curtis (2006b) that residents who have suffered losses due to livestock depredation by wildlife are usually opposed to conservation-related restrictions.

In the case of CF programmes, local communities hold a positive attitude towards CF because CF provides economic, social and environmental benefits that local communities perceive as having increased over the years (Chhetri et al., 2009; Wangdi & Tshering, 2006). Users benefit through ownership, empowerment, increased community participation and decreased conflicts among community members, while also deriving economic benefits from easy access to wood and more efficient procedures than in the national reserve forest. Several studies (Buffum, Tenzin, Dorji, & Gyeltshen, 2005; Chhetri et al., 2009; Temphel, Thinley, Wangchuk, & Moktan, 2005) show that the condition of the forest has improved as a result of controlled grazing and better forest management and fire prevention by community members. The assumption that CF is one of the best strategies to manage and sustainably use the forest resources in Bhutan (Chhetri et al., 2009; Wangdi & Tshering, 2006) is confirmed by the results from MCF and YCF.

The improvement in forest condition is due to the tree plantations established by local communities. Forest conditions may have also improved because gaining rights has imbued communities with a sense of stewardship, and motivated them to regulate the harvest of forest resources, limit grazing in CF areas and control forest fires (Temphel et al., 2005). Another reason for the improvement in forest conditions could be that CFMGs have been harvesting timber according to the CF management plan that limits the over-harvesting of forest products from CFs (Buffum et al., 2005; Temphel et al., 2005; Wangdi & Tshering, 2006).

CF provides benefits that increase social cohesion among people. Social cohesion in turn plays an important role in developing trust and reciprocity. While Mahanty, Gronow, Nurse, and Malla (2006) and Pandit and Thapa (2004) report that influential and rich people were able to obtain more benefits than poor people from CF in Nepal, no such discrepancy has been observed in Bhutan (Buffum, Lawrence, & Temphel, 2010; Chhetri et al., 2009), which could be one of the reasons why local communities hold positive attitudes towards CF.

## Implications for Design Guidelines

These results demonstrate that the residents of the inner area of JSWP are more dependent on forest resources than residents of the periphery, but that the degree of autonomy is minimal in both cases because the rules governing the use of forest resources are externally controlled. In the case of CF, on the other hand, local people hold authority, especially over the use of forest-derived products. From the perspective of forest conservation, CF provides withdrawal, management and exclusion rights to local communities, which are important factors allowing them to effectively manage the forest resources. These rights provide a sense of stewardship that can motivate communities to use and regulate forest resources accordingly. In the case of JSWNP, on the other hand, forest resources are directly managed by the Wildlife Conservation Division, reflecting the 'fortress model'.

Therefore, sustainable forest management and maintenance and the provision of property rights are important incentives for people to conserve forests (Dorji, 2003). CF can guarantee long-term tenure unlike the state-owned forests inside national parks. Findings indicate that local needs for timber and NTFPs can be met through community institutions, which are relatively effective in enforcing rules and regulations. Looking at the design guidelines, 'graduated membership' is maximal for CF and very weak for state forests such as national parks. Although the establishment, monitoring and even management of CFs can be influenced by the DZFO, local communities have more authority to influence decision making. In the case of JSWNP, user groups have less authority over resource use and management. Therefore, from the perspective of forest use and management, CFMGs have a higher degree of commitment than users do in the management and use of state forests. In terms of 'fair benefit distributions', benefits are equitably distributed in both MCF and YCF, while equitable distribution has been more difficult to attain in JSWNP. These results enable the conclusion that design guidelines are critical, especially in community-based natural resource management, where the two case studies clearly reflect the constraints in managing natural resources.

## Conclusion and Recommendations

In JSWNP, residents' perceptions and attitudes towards the park are closely associated with the lack of property rights over forest resource use because forest resource utilisation is closely linked to subsistence farming, which is their chief livelihood. Current park policies at JSWNP can be criticised for prioritising the maintenance of an extensive forest cover and not supporting traditional resource management, thus alienating park residents, who feel that their livelihood concerns are not sufficiently taken into account. Relaxation of rules, implementation through community-based approaches and improvements in compensation programmes are urgently needed to win support from park residents as a prerequisite for achieving conservation goals.

This is in marked difference to CF, where the implementation of people-based approaches has elicited positive responses from the community. In JSWNP, on the other hand, the centralised approach to park management has led to a situation where people comply with rules mainly because of enforcement. Providing access to forest resources through the support of traditional resource management could lead to a more positive attitude and more participation by park residents.

The results presented in this study demonstrate a high level of participation by local communities in CF. The institutions governing JSWNP are inflexible, with government institutions holding on to the centralised 'fortress model'. This model has not led to significant forest degradation, but the rigidity of this system of governance provides little scope for strengthening the participation of local communities.

Participation could be enhanced by decentralisation recognising the needs of the community for protection, conservation and sustainable utilisation of forest resources. Collaboration between park residents and the management authority through a local participation project can be conducive to a favourable environment for conservation in the national parks as is evident in the CF cases in Bhutan.

The application of good governance principles such as accountability, transparency and participation can develop into a viable approach contributing to sustainable forest management that can lead to rural development and improved local governance in Bhutan. The results of the PA case study and the comparison with CF show that a decentralised approach to forest management providing property and exclusion rights should also be extended to PAs in Bhutan.

People depending on forest resources must be incorporated into forest management and conservation strategies that can help local communities to make decisions and design strategies to meet their needs for forest resources through community institutions and through a mechanism of enforcing access and withdrawal rights.

Considering the limited human and financial resources available to protect forests, effective implementation of forest policy could be achieved despite the low capacity of the implementation agencies by relinquishing management rights to local communities. Thus, policy flexibility and the willingness of governments to allow for experimentation and explore new potential solutions to achieve a fair distribution of benefits are essential for successful forest management and governance.

# References

Brooks, J.S. (2010). The Buddha Mushroom: Conservation behaviour and the development of institution in Bhutan. *Ecological Economics*, 69(4), 777–795.

Brooks, J.S. & Tshering, D. (2011). Good Governance as an obstacle to community-based management of the matsutake mushroom in Bhutan. *Environmental Conservation*, 37, 336–346.

Buffum, B., Lawrence, A., & Temphel, K.J. (2010). Equity in community forests in Bhutan. *International Forestry Review*, 12(3), 187–199.

Buffum, B., Tenzin, Y., Dorji, S., & Gyeltshen, N. (2005). *Equity and Sustainability of Community Forestry in Bhutan: Analysis of Three Village Case Studies* (PFMP Report No. 21). Thimphu: Participatory Forest Management Project, Bhutan.

Chape, S., Harrison, J., Spalding, M., & Lysenko, I. (2005). Measuring the extent and effectiveness of protected areas as an indicator for meeting global biodiversity targets. *Philosophical Transactions of the Royal Society B*, 360, 443–455.

Chhetri, B.B., Schmidt, K., & Gilmour, D. (2009). *Community Forestry in Bhutan— Exploring Opportunities and Facing Challenges*. Paper for the Community Forestry International Workshop, Pokhara, Nepal, 15–18 September 2009.

Dorji, L. (2003). *Assessing the Evolution, Status and Future Implications of Forest Vegetation Resources Management in the Inner Himalayas of the Kingdom of Bhutan* (PhD dissertation). Natural Resources Management, School of Environment, Resources and Development, Asian Institute of Technology, Bangkok, Thailand.

Galvin, M. & Haller, T. (eds) (2008). *People, Protected Areas and Global Change: Participatory Conservation in Latin America, Africa, Asia and Europe*. Perspective of the Swiss National Centre of Competence in Research (NCCR) North South, University of Bern (3), Geographica Bernensia, 560 pp.

Gurung, T.R. (2012). Agricultural transformations in a remote community of Kengkhar, Mongar, Bhutan. *Bhu. J. RNR*, 8(1), 1–12.

Hulme, D. & Murphree, M. (eds) (2001). *African Wildlife and Livelihoods: The Promise and Performance of Community Conservation*. Oxford, UK: James Currey.

IUCN (2003). World conservation union (IUCN). World Parks Congress.

Katel, O.N. (2012). *Forest Resources Use by Park Residents and Conservation Management in Jigme Singye Wangchuck National Park, Bhutan* (PhD dissertation). Natural Resources Management, School of Environment, Resources and Development, Asian Institute of Technology, Bangkok, Thailand.

Katel, O.N. & Schmidt-Vogt, D. (2011). Forest resources use by local residents in Jigme Singye Wangchuck National Park, Bhutan: Practice and perceptions of constraint. *Mountain Research and Development*, 31(4), 325–333.

Mehta, J.N. & Heinen, J.T. (2001). Does community based conservation shape favorable attitudes among locals? An empirical study from Nepal. *Ecological Management*, 28(2), 165–177.

Mahanty, S., Gronow, J., Nurse, M., & Malla, Y. (2006). Reducing poverty through community based forest management in Asia. *Journal of Forest and Livelihood*, 5(1), 78–89.

Mathew, J.C. (2012). *China-South Asia Strategic Engagements—2 Bhutan-China Relations* (Institute of South Asian Studies Working paper. No. 157). National University of Singapore.

NSB (National Statistical Bureau) (2007). *Statistical Year Book of Bhutan*. Thimphu, Bhutan: National Statistics Bureau, Royal Government of Bhutan.

———— (2009). *Statistical Year Book of Bhutan*. Thimphu, Bhutan: National Statistics Bureau, Royal Government of Bhutan.

Ohsawa, M. (1987). Vegetation zones in the Bhutan Himalaya. In M. Ohsawa (ed.), *Life Zone Ecology of Bhutan Himalaya* (pp. 1–71). Japan: Chiba University.

Pandit, B.H. & Thapa, G.B. (2004). Poverty and resource degradation under different common forest resource management systems in the mountains of Nepal. *Society and Natural Resources*, 17, 1–16.

Penjore, D. & Rapten, P. (2004). *Trends of Forestry Policy Concerning Local Participation in Bhutan* (pp. 21–27). Bhutan: The Centre for Bhutan Studies.

Phuntsho, S., Schmidt, K., Kuyakanon, R.S., & Temphel, K.J. (eds). (2011). *Community Forestry in Bhutan: Putting People at the Heart of Poverty Reduction.* Jakar and Thimphu, Bhutan: Ugyen Wangchuck Institute for Conservation and Environment and Social Forestry Division.

RGOB (Royal Government of Bhutan) (1969). *Forest and Nature Conservation Act of Bhutan.* Thimphu, Bhutan: Royal Government of Bhutan.

——— (1995). *Forest and Nature Conservation Act of Bhutan.* Thimphu, Bhutan: Ministry of Agriculture and Forests.

——— (2002a). *Conservation Management Plan: Jigme Singye Wangchuck National Park.* Thimphu, Bhutan: Department of Forest and Park Services.

——— (2002b). *Biodiversity Action Plan.* Royal Government of Bhutan: Ministry of Agriculture.

——— (2003). *Biodiversity Act of Bhutan.* Royal Government of Bhutan: Ministry of Agriculture.

——— (2004). *Bhutan Biological Conservation Complex.* Thimphu, Bhutan: Department of Forest and Nature Conservation Division.

——— (2006). *Forest and Nature Conservation Rules of Bhutan.* Thimphu, Bhutan: Ministry of Agriculture, Department of Forests, Royal Government of Bhutan.

——— (2007). *Land Act of Bhutan.* Thimphu, Bhutan: Royal Government of Bhutan.

——— (2008). *The Constitution of the Kingdom of Bhutan.* Thimphu, Bhutan: Royal Government of Bhutan.

——— (2010a). *Statistical Year Book of Bhutan.* Thimphu, Bhutan: National Statistics Bureau, Royal Government of Bhutan.

——— (2010b). *National Strategy for Community Forestry: The Way Ahead.* Thimphu, Bhutan: Ministry of Agriculture and Forests, Department of Forests and Park Services.

——— (2011). *Subsidized Timber and Non-wood Forest Produce Allotment Policy.* Thimphu, Bhutan: Ministry of Agriculture and Forests.

——— (2012). *Bhutan RNR Statistics.* Thimphu, Bhutan: Policy and Planning Division, Ministry of Agriculture and Forests.

Schweinfurth, U. (1982). Der innere Himalaya—Rueckzugsgebiet, Interferenzzone, Eigenentwicklung. *Erdkundliches Wissen*, 59, 15–24.

Seeland, K. (2000). National park policy and wildlife problems in Nepal and Bhutan. *Population and Environment*, 22(1), 43–62.

Temphel, K.J., Thinley, K., Wangchuk, T., & Moktan, M.R. (2005). *Assessment of Community Forestry Implementation in Bhutan* (PFMP Report No. 23). Thimphu, Bhutan, Participatory Forest Management Project.

Temphel, K.J. & Beukeboom, H.J.J. (2006). *Community Forestry Contributes to the National and Millennium Development Goals without Compromising the Forestry Policy.* Thimphu, Bhutan: Social Forestry Division, Ministry of Agriculture and Forests, Royal Government of Bhutan.

Wang, S.W., Lassoie, J.P., & Curtis, P.D. (2006a). Farmer perceptions of crop damage by wildlife in Jigme Singye Wangchuck National Park, Bhutan. *Wildlife Society Bulletin*, 34(2), 359–364.

——— (2006b). Farmers' attitudes towards conservation in Jigme Singye Wangchuck National Park, Bhutan. *Environmental Conservation*, 33(2), 1–9.

Wang, S.W. & Macdonald, D.W. (2006). Livestock predation by carnivores in Jigme Singye Wangchuck National Park, Bhutan. *Biological Conservation*, 129, 558–565.

Wangchuk, T. (2010). *Analysing Governance of Community Forestry in Bhutan: A Case Study of Punakha Dzongkhag* (MSc thesis). Graduate School, Faculty of Forestry, Kasesart University, Bangkok, Thailand.

Wangdi, R. & Tshering, N. (2006). *Is Community Forestry Making a Difference to Rural Communities? A Comparative Study of Three Community Forests in Mongar Dzongkhag.* Thimphu, Bhutan: Social Forestry Division, Ministry of Agriculture and Forests.

Xu, J., Chen, L., Lu, Y., & Fu, B. (2006). Local people's perceptions as decision support for protected area management in Wolong Biosphere Reserve, China. *Journal of Environmental Management*, 78, 362–372.

Zhou, D.Q. & Grumbine, R.E. (2011). National parks in China: Experiments with protecting nature and human livelihoods in Yunnan province, People's Republic of China. *Biological Conservation*, 144, 1314–1321.

# Southeast Asia

SECTION III

Southeast Asia

# 7

# Indonesia I: Review of Local Community Dimensions of Forest Policies

*Mustofa Agung Sardjono and Ndan Imang*

## Introduction: Community as an Integral Part of Forest Ecosystem

There are only very few countries in the world where forest covers more than two-thirds of its land territory. One of them is Indonesia, an archipelago country, which has more than 130 million hectares (Mha) of tropical 'dipterocarps' forests distributed mainly in four largest (of around 17,000) islands, namely Sumatra, Kalimantan, Sulawesi and Papua. It has to be noted that Indonesian forests are not only very large, but biologically and economically have been acknowledged as very rich (BAPPENAS, 2003a, 2003b; FWI/GFW, 2001; Sarsito, 2010). Although following legal and illegal exploitation, conversions and natural disasters (especially periodical forest fires), the annual deforestation and forest degradation rate during the last decade being in the range of 1.1–3.5 Mha (BAPPENAS, 2011; FWI/GFW, 2001; Ginting, 2000; Soemarwoto, Soemarwoto, & Brotoisworo, 1992; Sunderlain & Resosudarmo, 1997), the change in forest condition has not affected the forest areas according to the national official definition.

Indonesian forests have been classified into three categories based on their main function: production forests (66%), protection forests (21%) and conservation forests (13%). Production forests have been allocated for hundreds of commercial logging concessionaires since the beginning of 1970. Out of tens of marketable tree species, dipterocarps, for examle,

Meranti (*Shorea* spp), Kruing (*Dipterocarpus* spp) and Kapur (*Dryobalanops* spp), are the most valuable and dominant in local as well as international timber markets. Forestry once contributed significantly to the country's emerging revenue and became the second leading export commodity after oil and gas. During the golden age of forestry in the previous century, the ITTO (2001) reported that in the 1980s, forestry exports contributed USD 200 million/year and then increased to USD 2 billion/year in the 1990s. The contribution became USD 20 billion, or 10% of the total gross domestic product, before the monetary crisis in 1997. In the last decade, the total export value from forestry and timber industries has fluctuated, ranging from USD 7.3 to 9.7 billion (MoF, 2011).

In addition, protection forests, which are mostly located in hilly regions, according to the Indonesian Forestry Act No. 41/1999, are expected to play a primary role in regulating water balance and erosion control. Conservation forests, for example, national parks, nature sanctuaries and recreation forests, play a major role in protecting endemic or rare flora, fauna and fragile habitats from any disturbances for the welfare of human beings. One brief example reflecting the high diversity of Indonesian forests is that with only $\pm 200$–250 tree species per hectare of mixed dipterocarps, the forests of Kalimantan, typically referred to as Borneo (including Sarawak and Sabah/Malaysia, and Brunei), reflect the richest forest in Southeast Asia. Furthermore, various non-timber forest plants (e.g., 4,000 orchids, 25,000–30,000 of other flowering plants) and diverse fauna support the island as the world's biodiversity centre (BAPPENAS, 2003a, 2003b).

As it has been commonly found in other tropical countries, human beings (further referred to as local communities—see Kaufman, 1953 and also Sardjono, 2004a) and their cultures have become an integral part of the Indonesian forest ecosystem, in addition to biotic (flora, fauna, micro-organisms, etc.) and abiotic (soil, water, etc.) elements. Their existence in the forests predates the establishment of the country in 1945. The exact dates are unknown, but human presence existed in the Indonesian forests definitely hundreds or even thousands of years ago. Even at present, different community groups, both settlers and villagers, can be found living traditionally inside forest areas, in not only the production forest but also in the protection and conservation forests. Their physical and cultural existence depends on sustainable supply for their primary, secondary and tertiary consumption of different products and multiple services of the forest resources. Therefore, it is unquestioned that there are many communities who have developed and attempted to maintain conservation ethics, which is sometimes referred to as so-called local wisdom (Colfer et al., 1999; Moniaga, 1994; Sardjono, 2010; Sardjono & Samsoedin, 2001; Ukur, 1994; Widjono, 1998a, 1998b; Zakaria, 1994).

From the last available data of Dephut (2005), at least 48.8 of approximately 220 million people of Indonesia live in the areas surrounding the forests. In total, 10.2 million of them are poor, and 9.4 million depend on the forest and timber industries. However, Sardjono (2007; see also Perhutani, 2006) estimated a much higher number because in Java alone, 5,617 villages are located surrounding the forest areas, and approximately 21 million poor people have access to the forests. Local communities depend on many land use and production activities, such as shifting cultivation, collecting or hunting non-timber forest products and fishing. It has to be noted that these activities were initially of socio-cultural or subsistence economic orientations, but more recently they have also been oriented towards cash income (De Beer & McDermott, 1989; Sardjono & Inoue, 2007; Sardjono, Yasuhiro, Wijaya, & Kamaruddin, 2001; Sheil et al., 2002). This has emerged as many regions have higher accessibility, and some of the daily needs can be fulfilled from outside markets (instead of local natural resources).

The most interesting view concerning communities' forest-based income sources is that the production activities of the local community strive to adapt to the forest functions as declared by the government, such as production and conservation forests, although some of these people might have been there before the existence of the government. According to the research of Akhdiyat et al. (1998), in the case of villages within production forests, the structure of income sources of the local communities is dominated by shifting cultivation (36.5%), followed by timber industry labour (23.5%), rubber plantations (19%), non-timber forest products (10%), firewood (7%) and finally fruit gardens (5%). However, in villages in the conservation/recreation forest of West Kalimantan, shifting cultivation contributed only 6% of the total income, which was slightly higher than ecotourism services (5%). The top three income sources were non-timber forest products (58%), rubber plantations (21%) and fruit gardens (9%) (Fahruk, Sardjono, & Kuncoro, 2002).

Relevant to the above situation, many researchers have argued that local communities should be considered as keepers of the forests and consequently a part of the key stakeholders (Colfer et al., 1999; Poffenberger, 1990; Sardjono, 2004a). Sardjono (2007), therefore, underlined the importance of determining equitable benefits as one of the three principles of forest policies in Indonesia, in addition to maximum/progressive yield and sustaining resources.

Based on the above background, this chapter has the objective of understanding the forest policy dynamics from the beginning of the economic leap in the early 1970s until the last decade and the chances provided to local communities to be able to participate in forest utilisation and management activities. It has to be understood that in the Indonesian basic law of 1945, there is a mandate given to the government as the personification of the state, to regulate and to control forestry activities that may guarantee maximising benefits for mainly achieving community welfare. In addition, we will also try to analyse the relevance of community-oriented forest policies with prototype design guidelines for collaborative governance of natural resources as proposed by Inoue (2009). This analysis is considered important for determining whether community-oriented forest policies have also adopted theoretical concepts.

## Large-scale Timber Extraction Ignored Local Community Needs

It was mentioned in the previous chapter that the large-scale timber extraction era, involving granting hundreds of forest concessions in Indonesia, began in the 1970s. The decision of the Indonesian government to conduct intensive forest and other natural resource exploitation, especially in the outer large islands of Java, was made to increase the country's foreign exchange and to accelerate national development following monetary deficits of the succeeding older political governance regime (Ascher, 1999; Sardjono, 2004a, 2007). Despite the fact that the number of active concessionaires has decreased significantly following the lower timber stand volume, higher social conflicts and rapid development of other more attractive income sources, such as crop estates (especially oil palm) and mining industries (especially coal), forestry, which covers the majority of the total land territory is still expected to be one of the economic backbones of the country, not only from timber production but also from non-timber, ecological services as well as other types of forest land utilisation (as stated in the Government Regulation No. 6/2007 jo. No. 3/2008).

However, it should be noted that there was already a state political will to include local community welfare into the large-scale natural forest exploitation policy. In every forestry agreement, which has to be fulfilled by forest enterprises, the life and living of the local communities who are mostly forest dependants should be further ensured. Prioritising local manpower recruitment was even required when industrial tree plantation as well as timber industries were introduced. All efforts provided practically no optimum impacts, as there was no serious implementation by the companies; apart from very weak government control, the policies were not adapted to local characteristics (Sardjono, 2004a, 2004b, 2007), as stated in Table 7.1.

**Table 7.1:**
*Some factors affecting limited positive impacts of large-scale forestry policy to local community*

1. Natural forest exploitation (since the beginning of 1970s)
   - Most of the sub-activities of forest exploitation (tree felling using chainsaw or bulldozer) beyond community activities (shifting cultivation)
   - A lot of companies' regulations limit activities of local communities inside concession area, e.g., forbidden to collect forest products and to do shifting cultivation
   - Not all social obligations concerning local communities' social economy are implemented by timber companies (e.g., recruitment of local workers)

2. Timber/wood industries (since the mid of 1980s)
   - Almost all industries are located near big cities or far away from the local communities
   - Modern technologies used in timber industries offer limited opportunities only to those who have better education (especially young generation), to participate
   - Compared to its demand, the availability of working opportunities is too low, and therefore leads to hard competition
   - Generally, increasing urbanisation (in order to look for better jobs) leaves children, women and old people in villages

3. Industrial timber estate (since the end of 1980s)
   - Because of the negative perception of local communities' performance (less educated, lazy and strong bond to traditional culture), they are usually inferior in comparison with migrants in getting jobs in tree plantation
   - The development of industrial timber estate is only based on legal aspect (permit from central government)
   - The activities of industrial timber estate do not involve local institution and even frequently overlap (or are in contrary) with local interests
   - Wrong assimilation with local communities cause misperception among them that industrial timber estate causes destruction of potential natural forests

*Sources:* Sardjono (2004a, 2004b, 2007).

Since the beginning of the 1990s, the government has attempted to obligate timber concessionaires to share a small percentage of their profits to implement community development programmes. These programmes were not merely to increase community welfare but also were indirectly expected to create positive images among the community about the strong empathy of the companies and furthermore generate mutual cooperation. However, based on her observations, Wentzel (1997) was very pessimistic about the success of these programmes due to half-hearted efforts of many companies to implement such programmes and such charity programmes were far removed from the substance of community demands, such as land tenure acknowledgement and safe access to forest resources. Many community groups even have negative impressions because their traditional dependencies on forests would be broken through introduction of different non-forestry-based economic activities, and their local traditional values will be modernised under the concept of empowerment. Progress monitoring, which was conducted by Sardjono et al. (1999), revealed that based on the indicators of the three principles used for evaluation of programme performance, only administrative responsibilities were perfectly fulfilled by most timber concessionaires, and most aspects of the implementation process and programme achievements were failures. Unfortunately, many parties have argued that forestry administration is, based on much evidence, a source of corruption, collusion and nepotism (Ascher, 1999; Dauvergne, 2001; LATIN, 1998).

Instead of success, those situations worsened the relationship between concession holders (timber companies) and the surrounding communities in general, promoting disharmony and distrust. However, according to Sardjono (2004a), under the very strong (and even repressive) government of the New Order, these issues have remained uncovered for decades and proceeded through sequential processes (competitions, contraventions and conflicts), as it was discussed theoretically by Soekanto (1990). Just after the people's movement in the beginning of 1998 (what is referred to in Indonesian as 'reformasi'), after which the Soeharto regime collapsed, all of these social conflicts in almost all management units of forest concession in Indonesia came to light. Based on their observation of forest-related conflicts in Indonesia, Wulan, Yasmi, Purba, and Wollenberg (2004) reported that during 1997–2003, there were in total approximately 359 cases appearing in public media concerning conflicts that happened in all types of forests: production, protection and conservation forest areas.

Sardjono (2004b) stated that there were some social issues that have been considered as direct and indirect causes of the social conflicts between timber companies and local surrounding communities: (1) taking over forest areas/living spaces of the local community, (2) limitation of community activities in the exploited areas, (3) minimum communication between communities and companies, (4) labour recruitment of outsiders/migrants (not local community members, (5) minimum benefits of the existing companies to local communities, (6) encroachment on the traditional protected or sacred forests/places and (7) deforestation and its impacts on the rural agro-ecosystem.

## Social Forestry Schemes Focused More on National Economic Interests

Learning from the minimum positive impacts of the community development programmes and indeed the strong drive by national and international non-governmental organisations, as well as many academicians, for local communities to participate in management and obtain access to resources, the Indonesian government introduced community forestry schemes in the middle of the last decade of the 20th century through community forest (*Hutan Kemasyarakatan*/HKm in Indonesian), and then followed by community-based tree plantation (or *Hutan Tanaman Rakyat*/HTR) and village forests (or *Hutan Desa*/HD) schemes. These three main schemes complement private forests (*Hutan Rakyat*/HR), which were earlier acknowledged by the government, as the private forests are implemented outside the forest area (on private lands). Although the three schemes have been intentionally directed to empower local communities, they are indeed expected to be implemented under collaborative management with the support of local governments, facilitation of non-government organisations and especially private investors, because local communities have very limited capital (human resources, physical, financial and even social ones). However, based on more than 15 years of observation, the three schemes are only dynamic in legal context, and field implementation has demonstrated far too slow of progress. Table 7.2 discusses the information and data of the three official community forest schemes (HKm, HD and HTR).

From Table 7.2, it can be seen that although there is most likely 'political will' in the government to involve people or to provide access

**Table 7.2:**

*Legal dynamics and progress of community forests (HKm), village forests (HD) and community-based tree plantation (HTR) in Indonesia (until end of 2011)*

| Issues (1) | Community forests (HKm) (2) | Community-based forest plantation (HTR) (3) | Village forests (HD) (4) |
|---|---|---|---|
| 1. Commencement of the programme | 1995 | 2007 | 2008 |
| 2. Responsible institution | Directorate General of Watershed Management and Social Forestry, Ministry of Forestry | Directorate General of Forestry Business Management, Ministry of Forestry | Directorate General of Watershed Management and Social Forestry, Ministry of Forestry |
| 3. Legal basis | | | |
| a. Act | • FA No. 41/1999 (on Forestry) | – | • FA No. 41/1999 (on Forestry) |
| b. Government regulation | • No. 6/2007 jo. No. 3/2008 | • No. 6/2007 jo. No. 3/2008 | • No. 6/2007 jo. No. 3/2008 |
| c. Minister decree/ regulation | • No. 622/1995 (on guideline of HKm) • No. 677/1998 jo. No. 865/1999 (on HKm) • No. 31/2001 (on implementation of HKm) • No. P. 37/2007 jo. P. 18/2009 jo. P. 13/2010 jo P. 52/2011 (on HKm) | • No P. 23/2007 (on HTR) • No. P. 5/2008 (procedure of timber utilisation in HTR within plantation forests) • No. P. 9/2008 (on requirements to get revolved fund for HTR establishment) | • No. P 49/2008 jo. P 14/2010 jo. P 53/2011 (on HD) |
| d. Director general regulation | • P. 07/2009 jo P. 10/2010 (on procedure for HKm implementation) • P. 01/2010 (on guideline for identification and inventory of HKm/ HD areas) • P. 05/2010 (on technical guidance of forest area allocation for HKm/HTR/HD) | • P. 05/2010 (on technical uidance of forest area allocation for HKm/ HTR/HD) | • P. 01/2010 (on guideline for identification and inventory of HKm/ HD areas) • P. 05/2010 (on technical guidance of forest area allocation for HKm/ HTR/HD) • P. 11/2010 (on procedure for HD implementation) |

*(Table 7.2 Continued)*

*(Table 7.2 Continued)*

| Issues (1) | Community forests (HKm) (2) | Community-based forest plantation (HTR) (3) | Village forests (HD) (4) |
|---|---|---|---|
| 4. Target areas | • Production forests (no legal concession) <br> • Protection forests | Production forests (no legal concession) | • Production forests (no legal concession) <br> • Protection forests |
| 5. Duration | 35 years and can be extended every 5 years | 60 years and can be extended for max 35 years | 35 years and can be extended every 5 years |
| 6. Institution | Forest farmer groups | (a) Individuals <br> (b) Groups or <br> (c) Cooperation | Village institution |
| 7. Community rights | | | |
| a. Production forest | • Timber (only from re-forestation/ replanting) <br> • Non-timber forest products <br> • Ecological services <br> • Area utilisation | • Timber | • Timber <br> • Non-timber forest products <br> • Ecological services <br> • Area utilisation |
| b. Protection forests | • Non-timber forest products <br> • Ecological services <br> • Area utilisation | – | • Non-timber forest products <br> • Ecological services <br> • Area utilisation |
| 8. Achievements (until September 2011) | | | |
| a. Official target (until 2014) | 2,000,000 ha | 5,400,000 ha | 500,000 ha |
| b. Verified areas | 402,596 ha | – | 181,541 ha |
| c. Established areas | 170,820 ha | – | 65,234 ha |
| d. Permitted areas | 41,330 ha | 90,414 ha | 10,310 ha |
| e. Location | 24 Provinces | NA | 13 Provinces |

FMD/FMR, Forestry Minister Decision/Forestry Minister Regulation (*Permenhut*); DGR, directorate general regulation; jo (juncto = added); ha, hectares; NA, Data not available.
**Source:** Data modified from Silalahi and Santosa (2011).

to a local community in forest management, at the same time 'political commitments' for seriously implementing social or community forestry schemes are still far below expectations. Sardjono (2004a) stated that most government initiatives, especially related to forestry conflict resolution, are still at the 'appearance' rather than 'substance' levels.

Some factors affecting the slow progress of social/community forestry in Indonesia have been identified (e.g., Sardjono, 2011; Silalahi & Santosa, 2011), such as (1) limitation of official schemes (which means no recognition of local traditional forest management practices including customary forests); (2) difficulty finding forest areas where there are no concessionaires (or probably occupations); (3) complicated official procedures and long bureaucracy, especially for people with generally limited capital (a still unattractive future for the offered schemes) and (4) lack of harmonious relationships, or at least common understanding, between the central (esp. Ministry of Forestry) and local governments following reformation and regional autonomy (see Chapter 8).

From the analysis of the legal basis used for the implementation of social/community forestry in Indonesia, it can be hypothesised that one factor that may contribute to the unattractiveness of the schemes among local communities is their business orientation. There is a clear impression that all regulations related to community forestry (especially in the initial stage of their development, but it remains so currently) are directed to utilise local human resources to the optimum use of forest resources for economic interests, and, therefore, they cannot optimally meet the real multiple demands of local communities.

It is clearly understood that cash income is seen as the most important element of forest benefits, but money is not the only need of the people. In general, the community's interests in occupying, maintaining and surely managing the forests also involve social equity (i.e., land and resource tenure) as well as cultural identity (i.e., local wisdom and traditional knowledge). These interacting interests have been reflected through hundreds of forest-based local traditional resource management practices (including different forest-gardens—cum agro-forestry systems) that can be found in almost all local community groups in Indonesia (see Arifin et al., 2003; Suharjito, Khan, Djatmiko, Sirait, & Evelyna, 2000; Zakaria, 2004). Unfortunately, no single local practice or even their values and the different official schemes that were previously discussed are accommodated by the government regulation. Once again, it is not merely because traditional norms in local forest management practices do not fit with positive country laws but also their subsistence orientation is clearly not in line with national interests, especially to revitalise forestry sector for supporting the economic growth of the country.


## Forestry Decentralisation Highlighted Political Rather Than Community Issues


Political paradigm shifts including in forestry, from state- to community-based and from timber- to resource-oriented through the reformation movement at the end of the 20th century and then followed by implementation of decentralisation in the frame of regional autonomy (in 2001) brought high expectations that forests would be sustainably managed, especially for the benefit of the local communities. Wider space for the involvement of the local communities in forestry through the revision of Forestry Act No. 5/1667 with Act No. 41/1999 and also additional government regulations (including minister decrees) added to the optimism that political authority transferred from the central to local governments during decentralisation would prioritise community forestry in Indonesia and facilitate smooth growth.

However, since the beginning of the decentralisation era, most of the energy has been focused mainly on political rather than community issues, especially in relation to authority distributions between central and local governments (both province and district/city) and also between provinces and districts/cities and among districts/cities. In addition, the establishment of new districts or even provinces has been a common phenomenon and interpreted as concrete proof of political fights for power among regions (and most likely community groups) following the uncontrollable euphoria after political reformation. Although there are no data to validate the objective reasons of such a situation, based on the tendencies in the last decade, it can be considered that one reason for the greater amount of struggle for more authority and power, especially among local governments, is to obtain wider opportunities and indeed rights for commercial use of the existing natural resources. The massive exploitation of those resources is the easiest and fastest way for many regions to support the efforts for developing their economy and earning income, even though there is a common idea that the purpose of such exploitation is to increase local community welfare.

The above situation can be illustrated by the East Kalimantan case, one of Indonesia's resource-rich provinces. Data collected by the

Regional Forestry Council of East Kalimantan, which was cited by Sardjono (2010), revealed that at the end of 2007, there were already 427,850.50 ha of crop estate plantations (note mostly oil palm), with an average annual growth rate of 5.99% between 2003 and 2007. In addition to that, the province already had 25 large-scale coal mining sites with a total production of 97.3 million tons of coal, excluding hundreds or even thousands of small-scale companies with permits from the districts. Furthermore, in the forestry sector, of the total forest area of 14.7 Mha (or around 60% of the total provincial land territory), timber concessionaires occupied more than 8.2 Mha (where timber plantation reached a concession of approximately 1 Mha). Consequently, the gross regional domestic product (GRDP) of East Kalimantan in 2008 reached Rp. 315.2 billion, which was dominated by the mining and processing industries. However, though newer data are not available, from 2000 to 2005, the forestry sector still contributed Rp. 22.7 billion (or approximately USD 2.7 millions) to the GRDP.

The major reasons that forestry decentralisation in Indonesia has not worked smoothly as expected by many parties are as follows: (1) Under regional autonomy era, many local governments have opinions that even forests should be managed based on administrative boundaries, whereas the forestry sector retrains in its ecological principles that forests have to be managed as an ecosystem independent from administrative boundaries. (2) Based on Indonesian governmental structure and main departmental tasks/responsibilities, local governments (and indeed regional forestry services) are vertically subordinate more to the Ministry of Internal Affairs, and therefore, they do not seriously consider Forestry Ministry decrees or regulations bureaucratically binding. (3) Forests often cover major areas of provinces/districts/cities, and therefore, there have been perceptions that this sector limits non-forestry investments, which are very important for supporting regional economic development. Moreover, discussion or requesting permits from the Forestry Ministry for converting or lending forest areas is considered very difficult. Therefore, because the Forestry Ministry has been perceived as being inflexible and intolerant, many local governments have expressed similar responses to many programmes of the ministry (including community forestry). (4) According to Act No. 32/2004 (for Regional Government), forestry does not belong to the compulsory tasks of autonomous regions (unlike education or health), and therefore, some districts/cities do not have forestry services, or if they have forestry tasks, they have been integrated with other sectors, even though they have contradictory interests, such as combining forestry with the mining and energy sectors.

In relation to community forestry development, according to Sardjono (2005), observations concerning autonomy implementation indicate that social movements reached an anticlimax, becoming more fragile and finding themselves at crossroads. First, they have reached anticlimax over time as community paradigms have spread widely, and the spirit of collective actions to assist local people to obtain their authority over land and resource management has become weaker. Second, they have become more fragile because when local people gain access to land and resource management, this becomes the main cause of horizontal conflicts at the grass-roots level. Last but not least, the social movements are at the crossroads because in line with the autonomy, the focus of these movements is distributed throughout several administrative and structural levels. The whole phenomena mentioned above are affected by internal and external factors related to the social movements themselves. The internal factors stem from their disinclination to accept the logical consequences of land and resource tenure. In addition, the external factors involve the dynamics of politics, economy, socio-culture and ecology, parallel with the regional autonomy that requires adaptability of a social movement, particularly for land and resource tenure as important factors for community forestry development.

## Global Environmental Demands Bring Questions about Community Rights and Position

As one of the world's richest tropical forest countries, Indonesia in the Conference of the Parties (COP) 15, Copenhagen (Denmark, December 2009) underlined its commitment to reduce emissions by at least 26% (and even up to 41% with the support of international communities) by 2020. That target is indeed relatively higher than other countries, even higher than developed/industrial countries, such as England which proposed only 20%. To achieve its commitment, the Indonesian government, via the National Development Planning Board (BAPPENAS as it is called in Indonesia), has planned to reduce greenhouse emissions from 2010 to 2020 with emphasis on management of peat lands and waste, and to develop different key sectors such as forestry, agriculture, industry, transportation and energy. Furthermore, to ensure progress and achievement, it is also considered important to establish measurement, monitoring and verification programmes. Forestry through the REDD/REDD+ (Reduced Emission from Deforestation and Forest Degradation) scheme plays a very significant role related to the emissions target because it has been allocated by the government to contribute 14%, or more than half of the national commitment. The commitment of Indonesia and the focus on the forestry sector is surely reasonable because of (1) deforestation in developing countries has supplied 20% of global $CO_2$, and (2) the carbon that has been stocked in the forest ecosystem reaches ~4,500 Gt $CO_2$, which is one and a half times higher than the level in the atmosphere (~3,000 Gt $CO_2$).

Related to that commitment, shortly after the COP 13 in Bali (2007), the Indonesian Ministry of Forestry drafted some ministerial degrees related to the REDD. However, until the end of 2008, there were only three of them signed and issued, namely (a) Ministry of Forestry Regulation (in Indonesian *Permenhut*) No. P.68/Menhut-II/2008 (*on the Implementation of Demonstration of Reduction of Carbon Emission from Deforestation and Forest Degradation*); (b) Permenhut No. P.30/Menhut-II/2009 (*on the Procedure for Reduction of Emission from Deforestation and Forest Degradation/REDD*); and (c) Permenhut No. P.36/Menhut-II/2009 (*on the Procedure of Utilisation Permits for Carbon Sinks and Absorption in Production and Protection Forests*). All the above REDD regulations include a position and even roles for local communities, either primary/supplementary or direct/indirect. Indirect and supplementary involvements are their supports to other stakeholders' activities, while the implementation of different schemes of community forestry (i.e., HKm, HTR and HD) would enable their direct but primary roles. The relationship of all REDD+-related Forestry Ministry regulations and the position of different community forestry schemes (including HKm, HTR, HD and even HR) are shown in Figure 7.1.

However, many parties are still questioning the ultimate goals of involving local communities, whether it is to fulfil global requirements of the (international) buyers for ensuring financial compensation or because REDD+ is perceived as a source of additional complementary attractiveness for the forestry sector to involve the local community following threats from competition with other land uses such as tree-crop estates (especially oil palm) and mining (especially coal).

Those questions are definitely very logical, given that the implementation of REDD+ without any reforms to the existing forestry governance system in Indonesia could also produce negative perception in local communities or forest dependants for various reasons, that is, loss of access to the forests, suppression of traditional activities and/or displacement, and also horizontal social conflicts (e.g., Cortez & Stephen, 2009). In addition, a study by Silalahi and Santosa (2011) based on the readiness

**Figure 7.1:**
*Resume of substances of the existing REDD/REDD+ regulations issued by the Ministry of Forestry*

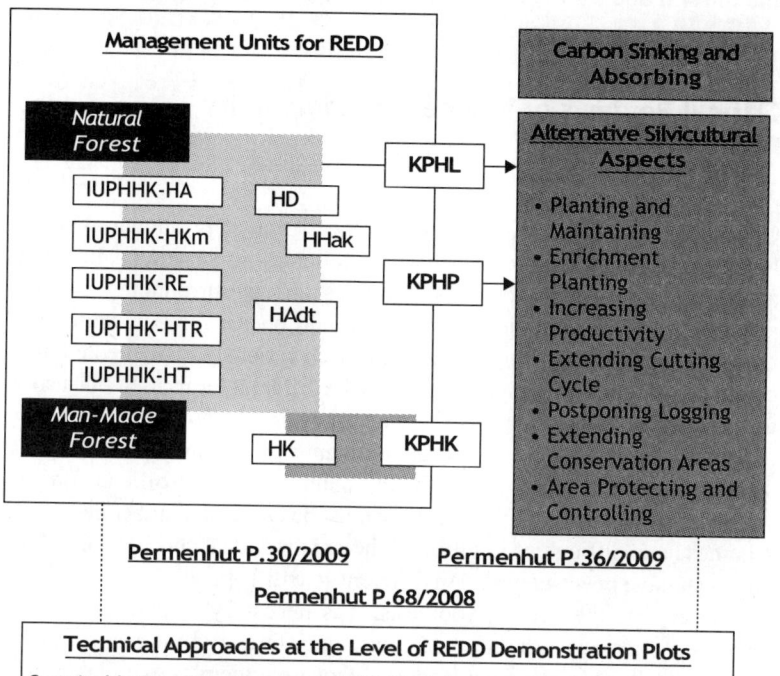

IUPHHK, *Izin Usaha Pemanfaatan Hasil Hutan Kayu* (timber utilisation permit); *HA, Hutan Alam* (Natural Forest); *HKm, Hutan Kemasyarakatan* (community forests); RE, *Restorasi Ekosistem* (ecosystem restoration); HTR, *Hutan Tanaman Rakyat* (community/small scale-based tree-plantations); HT, *Hutan Tanaman* (large-scale tree plantations); HD, *Hutan Desa* (village forests); HAdt, *Hutan Adat* (customary forests); HHak, *Hutan Hak* (forests under rights, e.g., private forests); HK, *Hutan Konservasi* (conservation forests); KPHP, *Kesatuan Pengelolaan Hutan Produksi* (production forest management); KPHL, *Kesatuan Pengelolaan Hutan Lindung* (protection forest management unit); KPHK, *Kesatuan Pengelolaan Hutan Konservasi* (conservation forest management unit).
**Source:** Sardjono and Inoue (2010).

of community forestry towards REDD+ concluded that despite several chances or opportunities, local communities are not ready to participate in the scheme for various reasons, for example, community forestry is in general small scale (and therefore is not profitable); the communities have low financial, human resource, institutional and technological capital; there are difficulties with long-term additions with no guarantee for drift; and the still unclear distribution of the REDD+ compensation.

## Critical Reviews of Indonesian Community Forestry Policies

Conceptually, community forestry in general is widely known as a collective term involving the response to challenges in sustaining functions and optimisation of the benefits of forests for the surrounding community groups, especially those who live in and on the forest resources. In reality, however, community forestry has been variously implemented from country to country, not only in relation to the different terms used but also in the substance of the policies used because they are absolutely dependent on the political interests of the ruling government as the main regulatory agent. This phenomenon is also valid for the main Indonesian official forestry schemes, HKm, HTR and HD. As they must be politically embedded, these schemes have difficulties in fulfilling the required ideal concepts or definitions that have been suggested in different scientific papers.

To examine the above hypothesis, it is necessary to perform critical reviews of Indonesian policies on community forestry based on nine prototype guidelines (*Kyouchi* in Japanese) that are principles of collaborative forest governance proposed by Inoue (2009). These are derived from and evolved out of the previously designed principles for common property rights. However, only three of them have been agreed upon and will be used in the following discussion. They are design guidelines 3, 4 and 5.

### Design Guideline 3 (Graduated Membership)

Conceptually, all schemes of community forestry in Indonesia (HKm, HTR and HD) actually acknowledge 'core memberships' of the local community as the main target of empowerment because they were considered inferior groups for more than two decades during forestry development since the beginning of the 1970s. However, especially for HKm and HTR schemes, which organisationally are separated from existing village structure, the real core members among local community are those who are directly involved or those who have position in the unit management, regardless of whether they consist of local community members or possibly outside but agreed-upon investors/operators. The unit managers are becoming first-class members with the strongest authority in steering forest or resource management, and other groups in the village (including village elites) would be only graduated members.

In the case of the HD scheme, and even if it is implemented among traditional communities, when local structure or traditional institutions have not been concretely formulated to confirm their position or their authority either in the organisation structures or in the determined statutes and rules of association, they will be by law powerless in every aspect of resource management (planning, organising, actuating and controlling). This has consequences, as once the HD concept is accepted and then implemented without any careful considerations of its implications, local existing valuable wisdom and local institutional roles related to resources used (as it is illustrated among Bahau Dayak in East Kalimantan shown by Simon Devung in Chapter 8) could be mostly (if not absolutely) ignored in the implementation of national policies-based community forest management schemes.

Therefore, it can be concluded that the initial steps of the designation and determination of organisations, governance as well as statutes and rules of association (including right and responsibilities of management unit especially in the case of HD to local villagers) are very important to avoid a situation where in a few forests users who belong to the management unit become more dominant in the resource utilisation and obtain more benefits than wider members of the local community. It has also to be emphasised, however, that all situations are possible because the areas used to implement the HKm, HTR and HD schemes, although partly occupied by individuals or groups of people, are classified as state forests.

## Design Guideline 4 (Commitment Principle)

As was mentioned in the previous principle (design guideline 3), unit management of HKm, HTR and even HD do not have responsibilities to the village authorities and/or villagers, if those are not deliberately included in the organisation structure, statutes and rules of association. The commitments of unit managers, especially related to maintenance of forest resource sustainability, are officially directed to the government (Ministry of Forestry) as the controller over the forests, although in fact, local traditional claims on the lands and resources used also exist. The government also has authorities to monitor, evaluate and consequently to give rewards and punishments for any performance displayed by the management unit. It can be assumed that small-scale management units such as HKm, HTR and HD are basically similar to the larger-scale ones (timber concessionaires), and therefore, in addition to ecological integrity, they also have to maintain (and to implement) social responsibilities to the local surrounding community.

Socially committed responsibilities in the frame of Indonesian sustainable forest management concepts consist of five elements: (1) determining (and avoiding any disturbances) traditionally used areas through participatory processes; (2) opening intensive communication with local communities prior to every activity implemented in their management unit; (3) protecting all traditionally sacred and conserved areas within their management units; (4) prioritising local surrounding community to recruit labour and staff, creating different income source opportunities and doing community empowerment programmes; and (5) emphasising more resolution and mediation processes for any social conflicts that occur with the local surrounding community rather than adjudication.

Those above social responsibilities of the management unit actually reflect that for certain cases or aspects, local (traditional) institutions (which are assumed representing all groups in the community members) are also in a better position to determine the resource management than the HKm, HTR and HD management units as utilisation rights holders only. The effectiveness of such traditional control depends on whether the local institution is still functioning internally and externally and whether the state or government is formally/informally acknowledging its existence. The customary *Nagari* system in West Sumatra, which is presented in Chapter 9 by Yonariza et al., has proven that such an existing traditional institution that obtains support from the government and also from community can reduce forest degradation (including through illegal felling), and at the same time, also provide a favourable atmosphere for implementation of officially introduced community-based forest management schemes, for example, HKm.

### Design Guideline 5 (Fair Benefit Distribution)

The benefit distribution of small-scale or community-based forest management systems in Indonesia is variously implemented but primarily depend on the management forms. HKm and HTR, which are under the control of groups of forest farmers, either in the form of cooperatives or other legal business institutions, would definitely be exclusive in benefit sharing (which means the benefits will be mostly enjoyed by members). When there are partnership systems with external investors applied, the benefit will be divided equally among participants or will be based on the agreements made between/among them. However, based on the experiences thus far, because members of farmer groups are dominantly local community members and to maintain good relationship with non-members, a unit management will also share a small part of the benefit (usually called as fee) for village development. Thus far, there has been no official standard fee standard established; therefore, it can vary from place to place.

In addition, HD is relatively more tolerant in terms of benefit sharing because based on the regulations, the management unit involved is institutionally established by village structure. Therefore, HD can possibly share more benefits or financial profit with the village or villagers (with a condition that operationally the forest management will be conducted by itself or there is/are no investor(s) involved in a partnership scheme form, which certainly also require(s) profit sharing). How a village will further distribute the benefits in the form of fee proportionally to individuals or to households or even to public facilities within the village depends on local community member discussion. As an example, based on a social forestry study in the Segah Area Berau District of East Kalimantan (Indonesia), Sutisna and Sardjono (2006) proposed that only 30% of the fee received from forest activities should be distributed equally to villagers, and approximately 50% should be used for public services such as operational funds for village administration, the village representative body, public facilities (especially education and public health), and women and youth empowerment programmes. They recommended that the remaining 20% of the fee have to be reinvested based on the forest management unit business plan.

All the prototype guidelines 3, 4 and 5 have been practically or explicitly stated or regulated in the community forestry policies (HKm, HTR or HD), but they can be found and implemented under villages' or involved actors' own initiatives. All issued regulations deal dominantly with administrative mechanisms and focus on the relationship between the government (as a regulatory agent) and the forest users as right holders. That means that creative initiatives are possible as there is no current official restriction of the technical forest governance at the community level. This situation should be seen as a great chance to ensure equality and fairness in the implementation of community forestry schemes in Indonesia. At the same time, it can provide an opportunity to determine more appropriate implementations that provide valuable inputs for future community forestry policy improvements.

## Conclusions

From the previous discussions, the following can be concluded. *First*, even after more than three decades of Indonesian forest policies development, the positions of local communities are still far from ideal in terms of empowerment and/or to increasing their welfare. Practically, the mutual interrelation between forests and surrounding communities has not been optimally formulated and implemented because of several possible reasons, among others: (1) different paradigms and focuses of direction between policy (referring more to state political interest) and reality (more based on sociological needs); (2) lack of complete understanding, especially among dominant policy makers about the community as integral element of forest ecosystem, given that their forestry knowledge is oriented around conventional temperate sciences; (3) participatory and transparent processes in many legal drafting and policy formulations (i.e., public consultation), even in the decentralisation era, do not work properly and therefore did not optimally absorb local community needs/interests; and (4) community forestry in Indonesia entered into the political discussion very late and most likely also suffered from poor timing (i.e., repressive regime, financial crisis, local social euphoria, political transition, resource scarcity and global demands).

*Second*, if we consider that community forestry in Indonesia demands necessarily ideal collaborative management concepts, because it not only deals with forest as common properties but also involves multiple stakeholders (government, local community, non-governmental organisation and possibly private enterprises), the existing policies based on their substance and implementation do not explicitly refer to the three prototype guidelines of collaborative forest governance proposed by Inoue (2009). In fact, the review based on those principles indicated that guideline 3 is practically difficult to implement because the Indonesian community forestry policies acknowledge only official forest management unit, even if they are not representative of the local community. In addition, the applicability of guideline 4 in the community forestry schemes is absolutely dependent on existing and functioning local institutions. Because it is not regulated in the government community forestry policy, strong local institution and committed management units would be the key aspects for fair benefit sharing of forest utilisation as reflected in guideline 5.

However, there are still chances to implement the collaborative governance guidelines at field or village levels, because officially, there is no restriction on developing better local forest governance initiatives. Empowered local actors and indeed other key stakeholders are therefore absolutely needed, especially when we consider that the future challenges of community forestry policies in Indonesia will be even more difficult mainly because of internal and external social changes (Sardjono, 2011). The internal factors include rapid demographic increases, including population density, shifts on perception and orientation (subsistence to semi- or fully commercial), weaker local institutions (including traditional norms, for example, conservation ethics) and social capital. External factors include many aspects, such as increase of resource users (competitors) especially the oil palm plantation and coal mining industries, the development of modern technologies, more complex networks and interrelations, and further drastic democratisation (at the grass-roots level as well).

# References

Akhdiyat, M., Sardjono, M.A., & Kuncoro, I. (1998). Analisis Kontribusi Hutan terhadap Pendapatan Masyarakat Desa di Sekitarnya. *Journal Pascasarjana Unmul*, 1(1), 44–68.

Arifin, H.S., Sardjono, M.A., Sundawati, L., Djogo, T., Wattimena, G.A., & Widianto. (2003). *Agroforestri di Indonesia*. Agroforestry Exercise Material. Bogor: World Agroforestry Center (ICRAF)/IPB/UNMUL/CIFOR.

Ascher, W. (1999). *Why Government Waste Natural Resources. Policy Failures in Developing Countries*. San Francisco: Institute for Contemporary Studies.

BAPPENAS (2003a). *Indonesian Biodiversity Strategy and Action Plan 2002–2020* (National Document). Jakarta: Badan Perencanaan Pembangunan Nasional.

———— (2003b). *Indonesian Biodiversity Strategy and Action Plan 2002–2020* (Regional Document). Jakarta: BadanPerencanaan Pembangunan Nasional.

———— (2011). *Strategi Nasional REDD+*. Jakarta: Badan Perencanaan Pembangunan Nasional.

Colfer, C.J.P., Prabhu, R., Guenter, M., McDougall, C., Porro, N.M., & Porro, R. (1999). *Siapa yang Perlu Dipertimbangkan? Menilai Kesejahteraan Manusiadalam Pengelolaan Hutan Lestari*. Bogor: CIFOR.

Cortez, R. & Stephen, P. (2009). *Introductory Course on Reducing Emissions from Deforestation and Forest Degradation (REDD): A Participant Resource Manual.* Jakarta: CCBA/Rainforest Alliance/the Nature Conservancy/WWF/Conservation International/GTZ.

Dauvergne, P. (2001). *Loggers and Degradation in the Asia Pacific. Corporations and Environmental Management.* Cambridge: Cambridge University Press.

De Beer, J.H. & McDermott, M.J. (1989). *The Economic Value of Non-Timber Forest Products in South East Asia.* Amsterdam: Netherland Committee for IUCN.

Dephut (2005). *Rencana Strategis Kementrian Negara/Lembaga Departemen Kehutanan Tahun 2005–2009.* Jakarta: Departemen Kehutananan Republik Indonesia.

Fahruk, E., Sardjono, M.A., & Kuncoro, I. (2002). Analisis Potensi dan Manfaat Kawasan Hutan Wisata Gunung Kelam Ditinjaudari Kontribusinyaterhadap Pendapatan Masyarakat Desa Sekitarnya. *Equator,* 1(1), 1–34.

FWI/GFW (2001). *Potret Keadaan Hutan Indonesia.* Bogor/Washington, DC: Forest Watch Indonesia and Global Forest Watch.

Ginting, L. (2000). *Debt for Nature Swaps. Konversi Utanguntuk Pembiayaan Konservasi. Siapa yang Sesungguhnya Diuntungkan?* Jakarta: Wahana Lingkungan Hidup Indonesia (WALHI).

Inoue, M. (2009). Design guidelines for collaborative governance (*Kyouchi*) on natural resources. In T. Murota (ed.), *Local Commons in Globalized Era* (pp 3–25). Tokyo, Japan: Minerva-Shobou (original article in Japanese).

ITTO (2001). *Mewujudkan Pengelolaan Hutan Lestari di Indonesia.* Laporan Misi Teknis ITTO untuk Indonesia. Jakarta: International Tropical Timber Organization.

Kaufman, H.F. (1953). Sociology of forestry. In W.A. Duerr & H.J. Vaux (eds), *Research in the Economics of Forestry.* Washington, DC: Charles Lathrop Pack Forestry Foundation.

LATIN (1998). *Kahutanan Indonesia Pasca Soeharto: Reformasi Tanpa Perubahan.* Bogor: Pustaka LATIN.

MoF Indonesia (2011). *Kehutanan Masyarakat dan Tantangan Legalitas Kayu.* Jakarta: Indonesian Ministry of Forestry.

Moniaga, S. (1994). Pengetahuan Masyarakat Dayak sebagai Alternatifdalam Penanganan Permasalahan Kerusakan Sumberdaya Alam di Kalimantan: Sebuah Kebutuhan Mendesak. In P. Florus, S. Djuweng, J. Bamba, & N. Andasaputra (eds), *Kebudayaan Dayak. Aktualisasidan Transformasi.* Pontianak: Grasindo - LP3S - Institute of Dayakology Research and Development.

Perhutani (2006). *Strategi Pengelolaan Hutan Jati Perum Perhutani Saat Ini dan Yang Akan Datang.* Paper presented at Discussion on 'Mencari Jawaban tentang Forest Governance di Jawa'. Jakarta, 15–16 June 2006.

Poffenberger, M. (ed.). (1990). *Keepers of the Forests: Land Management Alternatives in Southeast Asia.* Berkeley (USA)/Gland (Switzerland): WG-CIFM.

Sardjono, M.A. (2004a). *Mosaik Sosiologis Kehutanan. Masyarakat Lokal, Politik dan Kelestarian Hutan.* Yogyakarta: Debut Press.

——— (2004b, December). *Role and Responsibility of Local Communities in Forestry Politics in Indonesia.* Paper presented at Forest Governance Workshop in Tokyo University, Tokyo, Samarinda, Center for Social Forestry (CSF).

——— (2007). *Social Forestry Roles in Participatory Rural Development (Indonesian Context). Narrative Materials for Intensive Lecture* (14/07/2007). Tokyo: The University of Tokyo.

Sardjono, M.A. (2010, November). *Tropical Rain Forest as Global Literature through Conserving Local Wisdoms of the Traditional Communities in Kalimantan.* Paper prepared for the WISDOM 2010 World Conference on Science, Education and Culture and Colloquium in Honour of Dr Ann Dunham Soetoro – Gajah Mada University, Yogyakarta, Samarinda, Center for Social Forestry (CSF).

———— (2011, September). *Tantangan Perubahan Sosialdalam Kehutanan Masyarakat di Indonesia.* Materials presented in the FKKM National Meeting, Bogor, Samarinda, Center for Social Forestry.

Sardjono, M.A. & Inoue, M. (2007, August). *Why Do Local Communities Shift Their Orientation? Exploring Important Social Values of Tropical Rain Forests in East Kalimantan (Indonesia).* Paper presented at the 16th Indonesian Scientific Conference 2007 in Japan, Samarinda/Tokyo, Center for Social Forestry (CSF)/ Laboratory of Global Agricultural Sciences.

———— (2010). *Analysis of Indonesian Forestry Initiatives towards Mechanism Development for Reducing Emission from Deforestation and Forest Degradation (REDD+).* Samarinda/Tokyo: Center for Social Forestry (CSF)/Laboratory of Global Forest Environmental Studies.

Sardjono, M.A. & Samsoedin, I. (2001). Traditional knowledge and practice of biodiversity conservation among Benuaq Dayak community in East Kalimantan. In C.J.P. Colfer & Y. Byron (eds), *People Managing Forests: The Links between Human Wellbeing and Sustainability.* Washington, DC: RFF Press Book.

Sardjono, M.A., Yasuhiro, Y., Wijaya, A., & Kamaruddin, I. (2001). *Social Structure and Production Activities of the Community Surrounding Forest Concessionaires in Sangkulirang East Kutai District (Indonesia).* Samarinda/Morioka/Jakarta: UNMUL-YayasanBiOMA-Tohoku Research Center-PT Sumalindo Lestari Jaya.

Sarsito, A. (2010). *An Overview: Forestry in Indonesia.* Jakarta: Ministry of Forestry.

Sheil, D., Puri, R.K., Basuki, I., Van Heist, M., Syaefuddin, Rukmiyati, Sardjono, M.A., Samsoedin, I., Sidiyasa, K., Chrisandini, Permana, E., Angi, E.M. Gatzweiler, F., Johson, B., & Wijaya, A. (2002). *Exploring Biological Diversity, Environment and Local People's Perspectives in Forest Landscapes. Method for a Multidisciplinary Landscape Assessment.* Bogor: CIFOR.

Silalahi, M. & Santosa, A. (2011). *Kajian Kebijakan Kehutanan Masyarakat dan Kesiapan Menghadapi REDD+.* Materials presented in the National Meeting of FKKM, Bogor 6-8 September 2011. Bogor: Forum Komunikasi Kehutanan Masyarakat.

Soekanto, S. (1990). *Sosiologi Suatu Pengantar.* Jakarta: Rajawali Press.

Soemarwoto, O., Soemarwoto, I., & Brotoisworo, E., (1992). *Pembangunan Terlanjutkan Kehutanan.Menjawab Tantangan Gerakan Anti KayuTropik.* Jakarta: Departemen Kehutanandan PPSDAL Universitas Padjadjaran.

Suharjito, D., Khan, A., Djatmiko, W.A., Sirait, M.T., & Evelyna, S. (2000). *Karakteristik Pengelolaan Hutan Berbasiskan Masyarakat.* Yogyakarta: Aditya Media.

Sunderlain, W.D. & Resosudarmo, I.A.P. (1977). *Laju dan Penyebab Deforestasi di Indonesia. Penelaahan Kerancuan dan Penyelesaiannya.* Bogor: CIFOR.

Sutisna, M. & Sardjono, M.A. (2006). *Proposal Pembangunan Perhutanan Sosialberupa Kebun Meranti Masyarakat di Kawasan Budidaya Non-Kehutanan PT. SLJ IV Kecamatan Segah Kabupaten Berau* (Report of Study funded by TNC). Samarinda: The Nature Conservancy (unpublished).

Ukur, F. (1994). Makna Religi dari Alam Sekitar dalam Kebudayaan Dayak. In P. Florus, S. Djuweng, J. Bamba, & N. Andasaputra (eds), *Kebudayaan Dayak. Aktualisasidan*

*Transformasi.* Pontianak: Grasindo- LP3S -Institute of Dayakology Research and Development.

Wentzel, S. (1997). *PMDH Wrap Up in the Form of Policy Options.* Samarinda: GTZ-SFMP.

Widjono, A.M.Z.R.H. (1998a). *Masyarakat Dayak Menatap Hari Esok.* Jakarta: Gresindo.

——— (1998b). *Menata Kembali Hubungan Negara dengan Masyarakat Adat.* Samarinda: Jaringan Pembelaan Hak-Hak Masyarakat Adat LBBPJ.

Wulan, Y.C., Yasmi, Y., Purba, C., & Wollenberg, E. (2004). *Analisa Konflik Sektor Kehutanan di Indonesia 1997–2003.* Bogor: CIFOR.

Zakaria, R.Y. (1994). *Hutan dan Kesejahteraan Masyarakat Lokal.* Jakarta: Wahana Lingkungan Hidup Indonesia (WALHI).

# 8

# Indonesia II: Customary Land Tenure in East Kalimantan

*G. Simon Devung*

## Introduction

As in other forest-rich countries in Asia, the involvement of local people in forest management in Indonesia has long been articulated by academics, non-government organisations and the government. A number of policies and initiatives have also been designed by the Indonesian government to accommodate local community interests in using and managing local forests. More can be done, however, to strengthen the rights of local people under customary law and resolve competing interests over local forest lands for more effective involvement of and commitment by the local people in collaborative forest management programmes.

The primary legal basis accommodating local community interests in current Indonesian forestry policy is Act No. 41/1999 on forestry. In this act, three community-based forestry schemes are stipulated as choices for the community: *Hutan Kemasyarakatan—HKm* (Community Forest), *Hutan Desa—HD* (Village Forest) and *Hutan Rakyat—HR* (Private Forest). The *HKm* scheme is offered to forest farmer groups for 35 years with the possibility of further extension at 5-year intervals, given the successful operation. The *HD* scheme is offered to village authorities under the same terms. The target areas for both *HKm* and *HD* are production forests and protection forests, with specified use and management rights for each target area and scheme (Sardjono & Imang, 2011). *HR*, on the other hand, is a scheme for individuals possessing forested lands in the Non-Forest Use Area (*Kawasan Non Budidaya Kehutanan—KBNK*). Based on government regulations (*Peraturan Pemerintah—PP* No. 6/2007), there is also a new scheme called *Hutan Tanaman Rakyat—HTR* (Community-based Forest Plantation), offered to individual forest farmers, forest farmer groups or community cooperatives for 65 years with the possibility of further extension for 35 years. The target areas for this scheme are the production forests, where activities are focused on planting and harvesting timber (Sardjono & Imang, 2011). In addition to these four community-oriented forestry management schemes, one more scheme is mentioned in Act No. 41/1999 on forestry, namely *Hutan Adat—HA* (Customary Forest), which is defined as 'state forests located in the traditional jurisdiction areas' that are 'handed over to the *adat* community to manage' (Anonymous, 1999). Unfortunately, as of yet, there have been no supporting regulations issued for its implementation.

Thus, from the formal legal perspective, there are multiple choices available for stakeholders to develop schemes for collaborative forest management. For the purpose of this chapter, a broad comprehension of the characteristics of collaborative forest management, as well as of the forest uses and tenure situations, is needed. In general, collaborative forest management is defined as a working partnership between the key stakeholders in the management of a given forest. This definition includes the perspective of a variety of partnerships in different tenure situations and implies a need to manage complex social, institutional and silvicultural issues (Carter & Gronow, 2005). Two prominent variables imperative to include in the design and development of collaborative forest management schemes are therefore 'which local forest lands will be managed?' and 'who are the holders of the forest land tenure rights?' as these rights' holders will be key stakeholders to be involved in the collaborative forest management scheme.

Identifying those two variables at the local level in current Indonesian policy and socio-cultural contexts remains confusing at times. The origins of this confusion lie largely in diverse interpretations of what and where the forests are, and disagreement and even conflict over who controls and should manage the forest lands (Hermosilla & Fay, 2005). Ambiguities concerning 'what and where the forests are' originate from the confusion between the definitions of 'forest' (*hutan*) and 'forest area' (*kawasan hutan*) as stipulated in the Basic Forestry Law (No. 41/1999) compared to those used by the local community according to their customary classifications.

This chapter focuses on the discussion on 'what forests to manage and which local people need to be involved' in collaborative forest management schemes based on local customary land tenure arrangement characteristics. Specifically, the discussion revolves around the characteristics of different types of local forest lands as the objects of local customary land tenure rights, and various subjects holding different rights over different types of local forest lands; these are the two prominent variables influencing the probability of the success of collaborative forest management programmes, and therefore need to be accounted for in forest governance policies.

In the current Basic Forestry Law (No. 41/1999), which replaced the first Basic Forestry Law (No. 5/1967), a 'forest' (*hutan*) is defined as 'A unit of ecosystem in the form of lands comprising biological resources, dominated by trees in their natural forms and environment, which cannot be separated each other'. This legal definition is typically ecological and scientific (Lund, 2011) and very conceptual in nature. It is therefore unfamiliar to local community members and is operationally difficult to understand. The local community members identify a 'forest' (*hutan*) from its vegetation type (Abdoellah, Lahjie, Wangsadidjaja, Hadikusumah, Iskandar, & Sukmananto, 1993; Goenner, 2001) with a variety of vernacular names, depending on the local sub-ethnic groups in the area. Within most sub-ethnic groups, the vernacular names for 'forest' also clearly denote whether the forest is a 'primary forest' (*hutan perawan*) that has never been farmed, or an 'old secondary forest' (*hutan sekunder tua*) that has been farmed (*bekas garapan*) and has reached its more or less climax vegetation type, which is very similar to that of the 'primary forest' (Devung, 2008; Lahajir, 2001; Sindju, 2003). Such local identification is not entirely compatible with the legal definition of 'forest' (*hutan*) as stated in the Basic Forestry Law (No. 41/1999).

Confusion also arises from the definition of 'forest area' (*kawasan hutan*) in the Basic Forestry Law (No. 41/1999), which states that the 'forest area' (*kawasan hutan*) is 'A certain area, which is designated and or stipulated by the government to be retained as permanent forest'. This legal definition of forest area (*kawasan hutan*) neglects the fact that within the so-called designated forest area, there have been patches of both 'old secondary forest' (*hutan sekunder tua*) and 'young secondary forest' (*hutan sekunder muda*), which are part of the local community's rotating cultivation system. As mentioned by a number of researchers, the 'government-classified state forest areas' have in fact been used by the local communities for a variety of purposes, including their livelihoods, inclusive settlements, rice fields and orchards from generation to generation (Colchester et al., 2006; Devung, 2003; Nanang & Inoue, 2000; Sirait, Prasodjo, Podger, Plavelle, & Fox, 1994).

Further complications emerge from the interpretation of the 'state forest' (*Hutan Negara*), stated in the Basic Forestry Law (No. 41/1999) as 'A forest located on lands bearing no ownership rights'. This definition practically nullifies local people's customary rights over their lands within the forest area (*kawasan hutan*), which, as defined in the Basic Forestry Law (No. 41/1999), are areas with 'no rights attached' (Colchester et al., 2006), although that actually means 'no statutory rights' as defined by the Basic Agrarian Law (No. 5/1960) (Anonymous, 1976) or the Basic Forestry Law (No. 41/1999). In fact, however, those lands have 'bundles of rights' held by a variety of tenure right holders according to the local customary land tenure, which tend to be ignored by government officials issuing large scale timber extraction licenses and/or by the forest concessionaires holding the licenses (Sardjono & Imang, 2011).

According to land tenure specialists, customary land tenure is a traditional system of land holding derived from the aspirations and operations of people's traditions and customs, based on the accepted practices and principles underlying those practices. This type of land tenure hinges on the special relationship between man and land, and it exists in many variations (Berger, 2006; Chikhwenda, 2002). In the local customary land tenure system in East Kalimantan, 'customary land' (*tanah adat*) covers the sense of territory, a particular area within the territory, and pieces of land occupied, farmed or cultivated by households or kinship groups (Devung, 2008). This of course includes areas with primary forest (*hutan perawan*) and secondary forest (*hutan sekunder/bekas garapan*) within the territory, which are ignored in the policy formulations of both the Spatial Use Planning (*Rencana Tata Ruang*) and the Basic Forestry Law (*Undang Undang Pokok Kehutanan*).

In the Spatial Use Planning (*Rencana Tata Ruang*) arrangements, in spite of ample types of planned space areas mentioned in the Spatial Management Law (*UU No. 24/1992 tentang Penataan Ruang*, which was then superseded by *UU No. 26/2007 tentang Penataan Ruang*), in the public discourse and implementation at the provincial and district levels, those types of planned space area (five types in *UU No. 24/1992* and 10 types in *UU No. 26/2007*) are popularly simplified into two broad categories: the 'Forest Use Area' (*Kawasan Budidaya Kehutanan—KBK*), which is treated practically as the 'State Forest Area' (*Kawasan*

*Hutan Negara*) and the 'Non-Forest Use Area' (*Kawasan Budidaya Non Kehutanan—KBNK*) or 'Area for Other Uses' (*Areal Penggunaan Lain—APL*), as seen in Figure 8.1.

Research on land use and remote sensing data, however, have revealed that significant areas of the so-called forest use area (*Kawasan Budidaya Kehutanan—KBK*) or forest zone (*Kawasan Hutan*) are in fact community-planted agro-forests (fruits, resins and timber trees), agricultural lands or grasslands (Hermosilla & Fay, 2005). On the other hand, the so-called non-forest use area (*Kawasan Budidaya Non Kehutanan— KBNK*) or area for other uses (*Areal Penggunaan Lain—APL*) actually contain large tracts of primary forest (*hutan primer/hutan perawan*) and old secondary forest (*hutan sekunder tua/bekas garapan*) that are traditionally used by the local communities for a variety of uses relating to their subsistence livelihood, such as hunting, harvesting fruits and edible plants, gathering building materials, and collecting commercial forest products (Colfer, 1995; Devung, 2003; Nanang & Inoue, 2000).

The following section will briefly elaborate on the types of local forest lands and the various subjects holding different rights over different types of the forest lands, be they in the government-designated 'forest

**Figure 8.1:**
*East Kalimantan RTRWP spatial plan 2005 and* Huang Tring *customary area position*

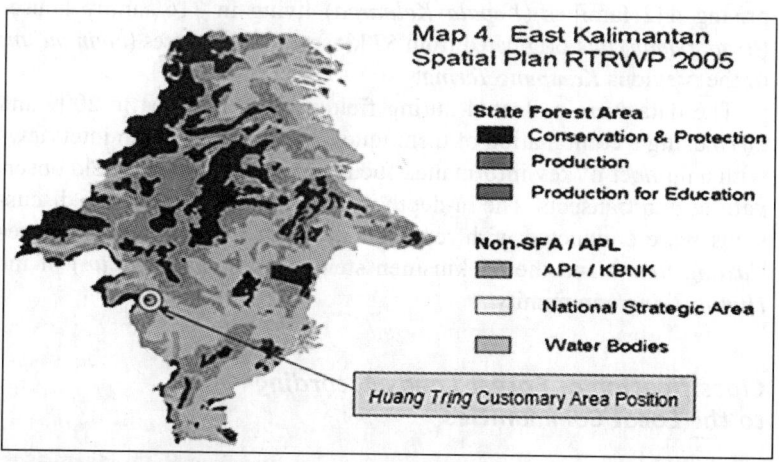

**Source:** ADB TA 4687-INO Natural Resource Management in a Decentralised Framework Final Report 2007, Map 4, p. 72.
(Disclaimer: This image has been redrawn by the author (G. Simon Devung) and is not to scale. It does not represent any authentic national or international boundaries and is used for illustrative purposes only.)

use area' (*Kawasan Budidaya Kehutanan—KBK*) or in the 'non-forest use area' (*Kawasan Budidaya Non Kehutanan—KBNK*), as perceived by the East Kalimantan's indigenous local communities from their emic point of views based on their local tradition.

## A Case Study of Local Forest Land Classification and Tenure Rights

The case was taken from the *Bahau Huang Tring* community in West Kutai district of East Kalimantan, located as shown in Figure 8.1.

The *Bahau Huang Tring* community has been settled in the *Huang Tring* customary area (*Wilayah Adat Huang Tring*) for at least 2.5 ages since the arrival of the first settlers in the mid-18th century (Devung, 2011; Devung, Ding, Avun, & Bayau, 1986). Until 1955, there was only one administrative village (*desa administratif*) in the area, called *Kampung Tering* (Tering village). Today there are five administrative villages: *Tering Lama, Tering Baru, Tering Seberang, Jelemuq* and *Pur-worejo*, as seen in Figure 8.2. *Tering Lama* is the new name for the previous *Kampung Tering*, the original village of the *Bahau Huang Tring* community. In 2010, the village had a population of 1,445 people, comprising 442 families (*Kepala Keluarga*) living in 316 family houses (*luvang amin*) that originated from 57 kinsmen stem houses (*amin pu'un*) of the previous *Kampung Tering*.

The data were collected during fieldwork in the area in 2009 and 2010 using a combination of techniques, including in-depth interviews with a number of key informants, focus group discussions, field observations and transects. The in-depth interviews and focus group discussions were conducted with representatives of the 316 family houses (*luvang amin*) and the 57 kinsmen stem houses (*amin pu'un*) of the *Huang Tring* community.

### Classification of Forest Lands According to the Local Communities

In principle, the classification of forest lands by the local communities in East Kalimantan is closely related to their shifting cultivation system with its fallow cycle, vegetation succession and forest regeneration (Abdoellah

**Figure 8.2:**
Huang Tr<u>i</u>ng *customary area*

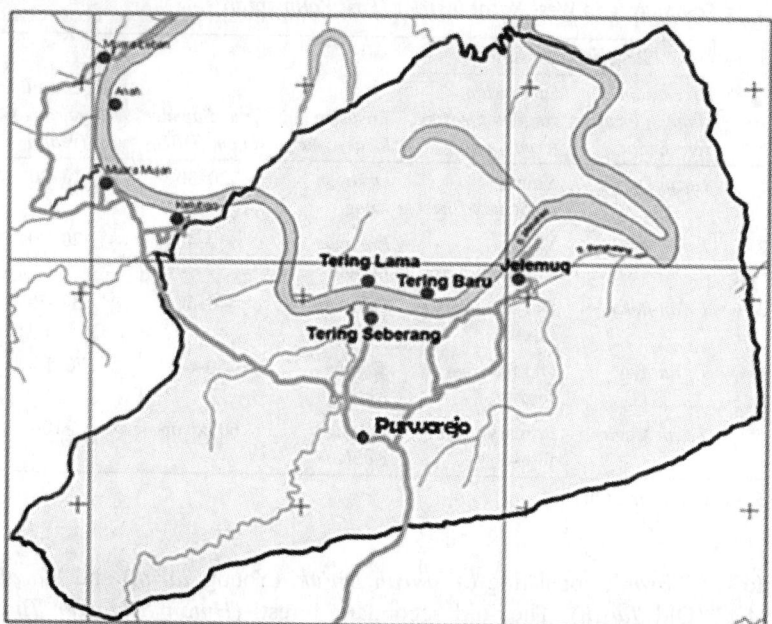

*Source:* Tering Sub-District Map 2008 and Bakorsurtanal Base Map 1992.
(Disclaimer: This image has been redrawn by the author (G. Simon Devung) and is not to scale. It does not represent any authentic national or international boundaries and is used for illustrative purposes only.)

et al., 1993; Sindju, 2003). The general classifications of forests as 'primary/virgin forest' (*hutan primer/hutan perawan*) and 'secondary forest' (*hutan sekunder/bekas garapan*) is identified by the local communities in distinctively different ways using a variety of local vernacular terms as previously mentioned (Abdoellah et al., 1993; Devung, 2011; Gunawan et al., 1999; Nanang & Inoue, 2000; Sardjono, 2004; Sindju, 2003). In the *Bahau Huang Tring* community in West Kutai District, the sequence of the vegetation succession and forest regeneration from the 'young secondary forest (*tarah du'uk*)' to the 'old growth forest' (*tu'an aya'*) and 'primary forest' (*tu'an megan*) and their forestry term analogues are listed in Table 8.1.

As shown in Table 8.1, in the local forest land classification, the forestry term 'young secondary forest' (*Hutan Sekunder Muda*) is equivalent

**Table 8.1:**
*Vegetation succession and forest regeneration according to the* Bahau Huang Tring *Community in West Kutai District, East Kalimantan*

| No. | Succession stage in local vernacular | Succession stage in forestry terminology | Tree size in local name | Tree diameter in cm (DBH) | Regeneration age range (years) |
|---|---|---|---|---|---|
| | *Regenaration stage* | | | | |
| 1 | *Tarah Du'uk* | Young secondary forest | *Dekayan Kung* | 20–30 | 10–20 |
| 2 | *Tarah Aya'* | Young secondary forest | *Dekayan Lu'ung* | 30–40 | 20–30 |
| 3 | *Tu'an Bekan* | Old secondary forest | *Dekayan Keliham* | 40–50 | 30–70 |
| 4 | *Tu'an Aya'* | Old frowth forest | *Tapah Keliham* | 50–60 | 70–100 |
| 5 | *Tu'an Megan* | Primary/virgin forest | *Lekang Keliham* | 60 cm up | >100 |

**Source:** Devung (2011).

to the '*Tarah*', consisting of '*Tarah Du'uk*' (Young *Tarah*) and '*Tarah Aya*' (Old *Tarah*). The 'old secondary forest' (*Hutan Sekunder Tua*) is identified as '*Tu'an Bekan*' denoting that the forest land used to be farmed by someone who is still recognised, and old enough that the vegetation structures and species are very similar to those of the 'old growth forest' and 'primary forest'. The 'old growth forest' (*Hutan Klimaks*) is identified as '*Tu'an Aya*', denoting a mature forest that was farmed by someone who has been forgotten; its vegetation structures and species are very similar to those of the 'primary forest'. The 'primary forest' (*Hutan Primer/Hutan Perawan*) is identified as '*Tu'an Megan*', which means 'virgin forest with ample resources'.

## Customary Land Tenure and Forest Land Rights

From a theoretical perspective, the definition of 'land', including the 'forest land', is quite broad. FAO and UNEP experts, for example, asserted that 'land is not regarded simply in terms of soils and surface topography, but encompasses such features as underlying superficial deposits, climate and water resources, and also the plant and animal communities which have developed as a result of the interaction of these physical conditions' (Kutter & Neely, 1999). Therefore, rights over forest lands should be seen in emic terms from each forest land classification as described earlier because the subjects bearing the rights over each type of forest land are distinctively different.

Relating to the rights over the forest lands and forest resources, some experts have noted that rights over forest lands and resources are in fact 'bundles of rights' that depict 'multiple rights' over the land and its resources borne by certain individuals or groups of different people (Coleman, 2010; Munro-Faure, Groppo, Hererra, Lindsay, Mathieu, & Palmer, 2002; Ostrom, 2003; Schlager & Ostrom, 1992). The synthesis of the multiple rights for East Kalimantan comprises eight types (Devung, 2011):

1. Access rights
2. Withdrawal rights
3. Use rights
4. Control rights
5. Management rights
6. Transfer rights
7. Residuary rights
8. Ownership rights

From the perspective of the subjects bearing the rights (Crewett, Bogale, & Korf, 2008; Myers, Hetz, & Roth, 2007; Platteau, 1995; Tenaw, Zahidul, & Parviainen, 2009), however, they might be identified as (Devung, 2011):

1. Individual rights
2. Collective rights
3. Corporate rights
4. Communal rights
5. State rights
6. Open access

To analyse the contents of the bundles of rights, it is worth examining the main elements embedded in a right, i.e., the objects of the rights, the subject holding the rights, the nature of the rights, the types of rights and the authority regulating the rights (Dietz, 2005; Fauzi, 1998). The following discussion will revolve around these entities in terms of the customary land tenure and forest land rights in East Kalimantan, focusing on the *Bahau Huang Tring* community in West Kutai District.

## Rights over Primary and Old Growth Forest Lands

In most indigenous local communities in East Kalimantan, like the *Bahau Huang Tring* community, the prime rights over primary and old growth forest lands are *communal* in nature. The rights are embedded in the rights of the local people as a community (*huang ji' ukung*) to manage and use their customary territory that they have occupied and controlled from generation to generation, including the primary and old growth forest lands therein. As mentioned above, the Primary Forest Lands are identified as '*Tu'an Megan*', which means 'virgin forests having abundant resources', because they have never been farmed or cultivated and therefore have not yet been under the exclusive rights of an individual, household or kinship group within the community. Old growth forest lands are identified as '*Tu'an Aya*', which denotes mature forests that were farmed by someone in the unknown past, and which are therefore no longer under the exclusive rights of an individual, household or kinship group within the community. The prime rights over the primary and old growth forest lands are therefore *communal* (Devung, 2011).

An exception is observed, however, within the *Benua'* and *Bentian* Dayak communities in West Kutai District, where certain primary and old growth forest lands are sometimes controlled collectively by kinship groups whose ancestors had the privilege rights in the past as the guardians of the area (Devung, 2008). Among other communities in East Kalimantan, there is also a tradition of allocating a specific area of primary and old growth forest lands within their customary/ village territory as a reserved forest area that is used only on occasions of need for the whole community's common interests. Two examples of communities of this type include the *Tana' Ulen* among the *Kenyah* Dayak communities and the *Tana' Mawa'* among the *Bahau* Dayak (Eghenter, 2000; Nanang & Inoue, 2000).

In addition to the rights of the local people as a community to manage and use the primary and old growth forest lands within their customary territory, every community member also has individual rights to use certain resources or forest products in the primary and old growth forest lands. For example, every community member can go hunting, take wood for building materials, gather forest fruits and vegetables, and collect marketable forest products and other life necessities in the primary and old growth forest lands within their customary territory (Devung, 2011), except in the primary and old growth forest lands belonging to the above-mentioned reserved forest area (*Tana' Ulen* or *Tana' Mawa'*).

**Table 8.2:**
*Bundle of rights over primary and old growth forest lands*

| No. | Objects of the rights | Subjects holding the rights | Nature of the rights | Types of the rights | The authority regulating the rights |
|---|---|---|---|---|---|
| 1 | Primary and old growth forest lands | Local people as a community | Communal | • Access rights<br>• Withdrawal rights<br>• Use rights<br>• Control rights<br>• Management rights<br>• Transfer rights<br>• Residuary rights<br>• Ownership rights | • Customary and village authority |
| | | Local community members as individuals | Individual | • Access rights<br>• Withdrawal rights<br>• Use rights | • Customary and village authority |
| | | Other parties outside the community members | Individual | • Access rights<br>• Limited withdrawal rights<br>• Limited use rights | • Customary and village authority |

*Source:* Devung (2011).

Because the rights over the primary and old growth forest lands are embedded in the rights of the local people as a community to manage, control and to use their customary territory, according to the customary law, the local people as a community and as individuals have the right to prohibit other outside parties from using resources or forest products in the primary and old growth forest lands within the customary territory, if they lack a permit from the customary authority. Other outside parties have only limited withdrawal and use rights for access (Devung, 2011). Table 8.2 shows an example of the bundle of rights over the primary and old forest lands in the *Bahau Huang Tring* community as seen from the perspective of the objects of the rights, the subject holding the rights, the nature of the rights, the types of rights and the authority regulating the rights.

### Rights over Old Secondary Forest Lands

The rights over old secondary forests (*Tu'an Bekan*) are *collective*, and controlled by kinship groups originating from the same stem house (*huang ji' amin pu'un*). The 'old secondary forests' (*Tu'an Bekan*) have usually reached a vegetation succession of 30–70 years, which is equivalent to the age of 2–3 descending generations. Typically, some of the kinsmen have moved from the stem house to their own houses and still share the rights over the old secondary forest lands that are owned collectively by all the kinsmen of the stem house (*huang ji' amin pu'un*) (Devung, 2011).

Trees and other forest products in the old secondary forests can be taken and used by any household belonging to the stem house provided that other kinsmen are notified. Other parties outside the stem house can also take and use a limited number of trees or other forest products in the old secondary forest lands provided that they have a permit from the stem house elders (Devung, 2011). Table 8.3 describes the details of the bundle of rights over old secondary forests in the *Bahau Huang Tring* community.

**Table 8.3:**
*Bundle of rights over old secondary forest lands*

| No. | Objects of the rights | Subjects holding the rights | Nature of the rights | Types of the rights | The authority regulating the rights |
|---|---|---|---|---|---|
| 2 | Old secondary forest lands | Kinsmen of the stem house | Collective | • Use rights<br>• Control rights<br>• Management rights<br>• Transfer rights<br>• Residuary rights<br>• Ownership rights | • Stem house elders |
| | | Families living in the stem house | Collective | • Access rights<br>• Withdrawal rights<br>• Use rights<br>• Control rights<br>• Management rights | • Stem house elders |
| | | Families living outside the stem house | Collective | • Access rights<br>• Withdrawal rights<br>• Use rights | • Stem house elders |
| | | Other parties outside the kinsmen of the stem house | Individual | • Access rights<br>• Limited withdrawal rights<br>• Limited use rights | • Stem house elders |

**Source:** Devung (2011).

## Rights over Young Secondary Forest Lands

The rights over young secondary forests are also *collective* and are typically controlled by households (*luvang amin*) that may consist of a single family or several families living in the same house (*amin*) (Table 8.4). The young secondary forests (*Tarah*) comprise the '*Tarah Du'uk*' and '*Tarah Aya*', the fallow forest lands of 10–30 years of age, farmed or cultivated by household members who generally still live in the same house (*huang ji' luvang amin*) (Devung, 2011).

The young secondary forests are household land reserves that may be opened for rice fields (*uma'*), orchards (*lepu'un*) or gardens (*lida'*) and are also used for gathering or collecting a variety of minor forest products needed by household members. Other parties outside the household can also take and use a limited number of the forest products in the young secondary forest lands provided that they have a permit from the household elders (Devung, 2011).

Until 1955, the local community retained full control over all the forest lands in their customary area, according to the autonomous governance (*zelf bertuur*) system under the Dutch and the *Kutai* Sultanate (*Swapraja*), under which forests were classified as either primary and old growth forests or old and young secondary forests. After the first Indonesian General Election (*Pemilu* 1955), however, the central

**Table 8.4:**
*Bundle of rights over young secondary forest lands*

| No. | Objects of the rights | Subjects holding the rights | Nature of the rights | Types of the rights | The authority regulating the rights |
|-----|-----------------------|------------------------------|----------------------|---------------------|--------------------------------------|
| 3 | Young secondary forest lands | Household members | Collective | • Access rights <br> • Withdrawal rights <br> • Use rights <br> • Control rights <br> • Management rights <br> • Transfer rights <br> • Residuary rights <br> • Ownership rights | • Household elders |
| | | Other parties outside the household members | Individual | • Access rights <br> • Limited withdrawal rights <br> • Limited use rights | • Household elders |

**Source:** Devung (2011).

**Table 8.5:**
*Shift of control over forest lands in the* Bahau Huang Tṛing *customary area from 1955 to 2010*

| No. | Type of forest lands | Total forest areas under local community control (ha) 1955 | Shift of control over the forest areas generated by government land use policy (ha) | | | | |
|---|---|---|---|---|---|---|---|
| | | | 1955–1965 | 1965–1980 | 1980–1990 | 1990–2000 | 2000–2010 |
| 1 | Primary and old growth forests | 4,816.08 | 2,317.72 | 1,200.76 | 1,200.76 | 1,200.76 | 774.31 |
| 2 | Old secondary forests | 1,631.76 | 1,554.45 | 1,550.26 | 1,543.68 | 1,525.48 | 1,513.70 |
| 3 | Young secondary forests | 3,435.23 | 3,290.05 | 3,155.31 | 2,983.58 | 2,916.36 | 2,742.66 |
| | Total forest areas | 11,838.07 | 7,162.22 | 5,906.33 | 5,728.02 | 5,642.60 | 5,030.67 |
| | Total forest areas shift of control | 0 | 4,675.85 | 5,931.74 | 6,110.05 | 6,195.47 | 6,807,40 |
| | % of shift of control over the forest areas | 0 | 39.50 | 50.11 | 51.61 | 52.34 | 57.50 |

*Source:* Devung (2011).

government gradually increased its control over the use and management of local forest lands. This is clearly shown by the changing government arrangements and policies on the establishment of new administrative villages, the placement of transmigration settlements, and the operation of forest concessions and sub-district development facilities within the customary area from 1955 to 2010. The government had taken control of 57.5% of the forest area from local communities by 2010, as shown in Table 8.5; the cumulative decrease of each type of forest area under the control of the *Huang Tṛing* community from 1955 to 2010 is shown in Table 8.6 and Figure 8.3.

## Notes for Policy Development

As it has been widely accepted that collaborative forest governance needs to be organised through collaboration among various stakeholders who have a range of interests in local forest use and management through a

**Table 8.6:**
*Cumulative decrease of each forest types in the* Bahau Huang Tring *community control from 1995 to 2010*

| No. | Forest types | Total forest areas in the community control in 1955 (ha) | Total forest areas in the community control in 2010 (ha) | Total forest areas in the community control decrease (ha) | Total forest areas in the community control decrease percentage |
|---|---|---|---|---|---|
| 1 | Primary/old growth forests | 4,816.07 | 774.31 | 4,041.76 | 83.92 |
| 2 | Old secondary forests | 1,631.75 | 1,513.70 | 118.05 | 7.23 |
| 3 | Young secondary forests | 3,435.23 | 2,742.66 | 692.57 | 20.16 |
|  | Total | 9,883.05 | 5,030.67 | 4,852.38 | 49.10 |

*Source:* Devung (2011).

**Figure 8.3:**
*Comparison of the forest areas under local community control in the* Bahau Huang Tring *customary area in 1955 and 2010*

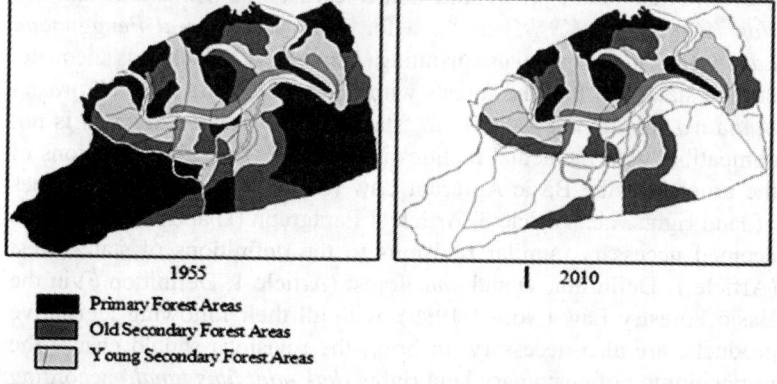

1955          2010

■ Primary Forest Areas
▨ Old Secondary Forest Areas
▧ Young Secondary Forest Areas

*Source:* Devung (2011).
(Disclaimer: This image has been redrawn by the author (G. Simon Devung) and is not to scale. It does not represent any authentic national or international boundaries and is used for illustrative purposes only.)

consensual 'principle of involvement' (Inoue, 2003), it is imperative to identify relevant stakeholders who must be involved in the collaboration (Colfer, 1995; Purnomo, Mendoza, Prabhu, & Yasmi, 2005), particularly the local communities whose livelihoods depend on the forests, and who in most cases also have traditional rights over the land (Carter & Gronow, 2005; Isozaki, 2003; Purnomo et al., 2005; Tribowo & Haryanto, 2001). The existence of numerous interest groups within the local communities sometimes makes it difficult to identify the prime stakeholders. In designing and developing a collaborative forest management scheme, it is therefore highly advisable to first discern which forest lands to manage, and then trace the holders of the tenure rights, as these groups and/or individuals will be key stakeholders in the proposed collaborative forest management scheme.

As seen in the case study, forest lands are generally identified in four main categories. Each category has a bundle of rights, with different subjects holding different rights and an authority regulating the rights as described above. It is important to account for these realities in the design and development of a collaborative forest management scheme, especially in (1) choosing which forest lands to manage and for what purpose; and (2) determining which relevant stakeholders to involve and what their expected roles will be.

Consequently, at the policy level, there needs to be a substantial reconsideration of the existing legislation related to forest land uses and tenure. For example, the determination of the 'forest use area' (*Kawasan Budidaya Kehutanan—KBK*) and 'non-forest use area' (*Kawasan Budidaya Non Kehutanan—KBNK*) or 'area for other uses' (*Areal Penggunaan Lain—APL*) in the spatial use planning (*Rencana Tata Ruang*) is adequate, but the statement that the forests within the 'forest use area' (*Kawasan Budidaya Kehutanan—KBK*) are 'state forests' (*Hutan Negara*) is not compatible with the actual realities in the field. Therefore, revisions of the articles of the Basic Agrarian Law (No. 5/1960) relating to types of land rights, i.e., Article 3, Article 4 Paragraph (I) and Article 16, are deemed necessary; similar revisions to the definitions of state forest (Article 1, Definition 4) and *adat* forest (Article 1, Definition 6) in the Basic Forestry Law (No. 41/1999), with all their following legislative products, are also necessary. In brief, the revisions should ensure the re-positioning of customary land rights (*hak adat atas tanah*) according to customary land law (*hukum adat tentang tanah*), which according to the Basic Agrarian Law (No. 5/1960) itself is the primary basis for the establishment of the national agrarian law (Opinion Consideration No. a). In that way, the overlap in tenure systems (statutory and customary forest land rights on the same piece of land) could be avoided and settled through a harmonious integration of customary land rights into the national agrarian law.

As seen from the variety of the forest types and the characteristics of their tenure right structures at the local level, there must be a distinction between *communal* and *collective* forest management schemes in the design and development of collaborative forest management schemes. A communal forest management scheme refers to the management of the forest by people as a community or fundamental group (a group of people living in the same locality such as a hamlet, village or territory). Contrastingly, a collective forest management scheme refers to the management of the forest by people as a collective unit, including functional groups within a community in a single locality (such as kinsmen, families or households living in the same village) or within communities across localities (such as kinsmen and families with members living in two or more different villages). The *Hutan Desa* (Village Forest) and the *Hutan Adat* (customary forest) described in the Basic Forestry Law (No. 41/1999) are ideally compatible for *communal* forest management schemes to manage the primary and old growth forest lands because, as seen in Table 8.2, at the community level, the prime subjects holding the rights over the primary and old growth forest lands are the local people as a community (both in terms of the village as their locality and the *adat* as their social binder). Therefore, some adjustments are needed in the regulations, as the schemes should cover not only the primary and old growth forest lands in the 'forest use area' (*Kawasan Budidaya Kehutanan—KBK*) as they are now but also include the primary and old growth forest lands in the 'non-forest use area' (*Kawasan Budidaya Non Kehutanan—KBNK*) or 'area for other uses' (*Areal Penggunaan Lain— APL*) in the design. In accordance with those notions, adjustments are needed in the policy concerning the eligible applicants for the schemes. For the *Hutan Desa* (Village Forest), the target groups should be village institutions (*Lembaga Desa*) representing the village community—not necessarily newly created village institutions as stipulated in current regulations, but already existing village institutions such as the village community institutions (*Lembaga Masyarakat Desa—LMD*) or the village cooperative units (*Koperasi Unit Desa—KUD*). For the *Hutan Adat* (customary forest), the logical target groups should be the customary institutions (*Lembaga Adat*) both at the village and sub-district levels (*Lembaga Adat Desa dan Lembaga Adat Kecamatan*), representing the *adat* communities (*Masyarakat Adat*) at the village level and cross-localities at sub-district levels.

On the other hand, the *HR* (People Forest), which is attributed to private forest lands, and so far regarded as a type of individual forest management (neither communal nor collective management) is actually more compatible with the *collective* forest management scheme for managing old and young secondary forest lands because, as seen in Tables 8.3 and 8.4, at the community level, the prime subjects holding the rights over the old secondary forest lands are kinsmen groups, while the prime subjects holding the rights over the young secondary forest lands are household members collectively. Consequently, some adjustments are needed in the regulations, which should cover not only old and young secondary forest lands in the 'non-forest use area' (*Kawasan Budidaya Non Kehutanan—KBNK*) or 'area for other uses' (*Areal Penggunaan Lain—APL*) as they are now but also old and young secondary forest lands in the 'forest use area' (*Kawasan Budidaya Kehutanan—KBK*) in the scheme. Adjustments are also needed in the policy concerning the eligible applicants for the schemes. For the *HR* (People Forest) scheme, the target groups should also cover the kinsmen groups who have prime rights over the old secondary forest lands and household groups, and who collectively have prime rights over the young secondary forest lands. Local tenure arrangements will be effective even though the *HR* is not changed.

From the perspective of local customary land tenure, the existing *HKm* (Community Forest) and *Hutan Tanaman Rakyat* (People Planted Forest) schemes are only fit for *collective* forest management schemes that manage forest lands beyond the village or *adat* territories, in those areas that are purely state forest lands. Adjustments are therefore needed in the existing policies and regulations pertaining to the designation of the areas allocated for the two schemes and in the policy concerning eligible applicants.

As a final note, adjustments are also needed in the design and development of collaborative forest management schemes related to the dynamics and changes in some elements of the customary land rights at the local level (Cramb & Wills, 1998; Jessup, 1992). In that way, a sound basis and framework for the re-positioning of customary land rights in the Indonesian legislative system, from the national macro-level down to the village micro-level, could be ensured. It ultimately depends, however, on the political will of the Indonesian government to fairly accommodate customary land rights in an adequate and secure legal basis and framework, which the government remains reluctant to do, as government authorities feel that changes will effectively reduce state control over resources that are important capital for the country or the region where the resources exist (Simorangkir & Sardjono, 2006).

## Discussion on the Design Guidelines

The success of the design and development of collaborative forest management schemes will depend very much on the treatment of the issues mentioned earlier both at the policy level and at the field implementation level. A number of experts have striven to develop and introduce design principles for common property resources, such as the community forests (McKean, 1999; Ostrom, 1990, 2005; Stern et al., 2002). Based on the design principles, Inoue (2011a) proposed prototype design guidelines for collaborative forest governance comprising nine design guidelines: (1) degree of local autonomy; (2) clearly defined resource boundaries; (3) graduated membership; (4) a commitment principle; (5) fair benefit distribution; (6) a two-storied monitoring system; (7) two-storied sanctions; (8) a nested conflict management mechanism and (9) trust building.

Following the design principles, at the policy level, appropriate arrangements should be designed to allow a sufficient degree of local autonomy in the use and management of local forest lands and the resources therein. The overlap of tenure systems (statutory and customary forest land rights on the same piece of land) caused by the government's unilateral determination that the forests within the 'forest use area' (*Kawasan Budidaya Kehutanan—KBK*) are all classified as 'state forest' (*Hutan Negara*) has practically reduced the degree of local autonomy in the use and management of local forest lands and their resources, as well as mixing up the local tenure systems that have been effective for decades among the community.

Case study data reveal that the total forest area under local community control in the *Bahau Huang Tring* Customary Area in 1955 was 9,883.05 ha. Up to 2010, the number had cumulatively decreased by 49.10% as shown in Table 8.6 due to the above-mentioned tenure overlap, most notably since the issuance of the Basic Forestry Law No. 5/1967 and Government Regulation No. 22/1967 on the concession license, fees and royalty. With the enactment of Law No. 1/1967 on foreign investment, ample opportunities for the massive, government-controlled exploitation of the forests were opened. The central government granted permits and allocations of forest concession areas, not only without due consideration for the forests' ecological functions but also with little regard for the social, cultural and economic functions of the forests for the communities who have been living in them for centuries (Resosudarmo &

Dermawan, 2002). By such arrangements, the primary forest areas under local community control in the *Bahau Huang Tring* Customary Area in 2010 had shrunk to 774.31 ha from the previous total primary forest areas of 4,816.08 ha in 1955, a decrease of 83.92%, described in Table 8.6, and visually depicted in Figure 8.3. This case shows that the degree of local autonomy in the use and management of local forest lands and their resources, as recommended in design guideline 1 (*degree of local autonomy*), is quite low because control is vested in the authority of the central government. Therefore, if there are no substantial changes in forest policy arrangements, we cannot expect the implementation of healthy, community-based, collaborative forest management in the near future.

At the field implementation level, sufficient knowledge of the types of forest lands and the local customary land tenure, along with familiarity with the existing statutory regulations, would be very helpful in the application of the design guidelines. For example, understanding the types of forest lands within the forest area (in terms of their bio-structure) will help to clearly define the resource boundaries of the primary, old growth, old secondary and young secondary forests in the area, allowing the application of design guideline 2 (*clearly defined resource boundary*). Appropriate identification of the holders of the tenure rights and the authority regulating the rights over each type of forest land (the tenurial/socio-structure) will help to determine the local people who will act as 'core members' and those parties who will act as the second and third class members in the collaboration, as per design guideline 3 (*graduated membership*). The involvement of appropriate core, second and third class members from among the local people and outsiders in the collaboration will help in developing and maintaining the commitment of the parties involved in the collaborative forest management scheme, as per design guideline 4 (*commitment principle*); this, in turn, will then help to guarantee the fair distribution of management and forest use benefits among the parties involved in the collaborative forest management scheme, as per design guideline 5 (*fair benefit distribution*).

As the case of *Bahau Huang Tring* shows, there is a low degree of local autonomy in the use and management of local forest lands and their resources because the control over the forest lands lies with the central government. According to current policy arrangements, local governments at the provincial and district levels, let alone the village governments, have little authority in the programmes designed for the forest area (*Kawasan Budidaya Kehutanan/KBK*). Under such condition, we can expect little from the implementation of the principles asserted in design guidelines 6–8.

In design guideline 6 (*two-storied monitoring system*), the core members of Collaborative Forest Governance are expected to monitor other members' rule compliance; the local government monitors whether the rule itself is appropriate for sustainable and scientific forest management. Similarly, in design guideline 7 (*two-storied sanctions*) the core members have responsibility, which is supported by the local government; and in design guideline 8 (*nested conflict management mechanism*) there is an assertion of informal conflict resolution in the community, with informal intercession by the local government supported by formal mechanisms at the local and national levels (Inoue, 2011b). As the local government has little prescribed authority in the programmes designed for the forest area (*Kawasan Budidaya Kehutanan/KBK*), we can doubt the implementation of the *two-storied monitoring system, two-storied sanctions and nested conflict management mechanism* as recommended in design guidelines 6–8, respectively.

For the final principle, design guideline 9 (*trust building*), there is an assertion that for cooperation with outsiders, forming, maintaining and strengthening social capital is essential (Inoue, 2011b). Social capital refers to the institutions, relationships and norms that shape the quality and quantity of a society's social interactions; it is not just the sum of the institutions that underpin a society, but the glue that holds them together (The World Bank, 1999). The central thesis of social capital theory is 'relationships matter', as do social networks. A sense of belonging and the concrete experience of social networks (and the relationships of trust and tolerance that accompany them) can bring great benefits to people (Field, 2003). In the case of the *Bahau Huang Tring*, there have been webs of social networks pertaining to the use and management of the local forest lands and resources at the community level. Every type of forest land has its local tenure rights structure showing the subjects holding the rights, the nature of the rights, the types of rights and the authority regulating the rights.

As noted in the case study, however, various inappropriate legislative products and incompatible government policies reveal that the knowledge relating to the variety of forest types and their tenure rights structures at the community level is still insufficient among the government authorities. This is reflected in the overlap of the designated forest concession areas and the local communities' lands, generating horizontal and vertical conflicts at the local level. These conditions do not support design guideline 9 because conflict decays the trust, sense of belonging, social interactions and social networks that are needed for healthy collaborative forest governance and collaborative forest management schemes.

## Conclusion and Suggestions

As shown earlier, the ideal characteristics of the prototype design guidelines for collaborative forest governance as proposed by Inoue (2011) have yet to be fully met in the arrangements of existing forest policy and initiatives to accommodate local community interests in using and managing local forests. To appropriately implement the above-mentioned design guidelines in Indonesia, adequate attention should be given to the variety of bio-structures and tenurial socio-structures of the local forest lands in the outer islands such as Kalimantan and Papua, the characteristics of which are somewhat different from those commonly encountered in Java. Adjustments in government regulations are therefore needed, particularly in the Basic Agrarian Law and Basic Forestry Law, with their following implementation rules and regulations.

Ultimately, all national laws, government policies, rules and regulations dealing with forest rights need to be revisited and revised to accommodate as far as possible the variability of forest land characteristics throughout Indonesia in terms of both the bio-structures and the tenurial socio-structures of the forest lands. This will allow the recognition of the rights of the local communities over their territories and the natural resources on them, as well as the support of the community-based natural resource management that fosters local social resources in the implementation of the Collaborative Forest Management. The revision of national laws, government policies, rules and regulations should be followed by wider forest resource management schemes aimed at the maintenance and enrichment of the biological resources of the existing forest lands and their intertwined social aspects of tenure rights and use.

Embedded in the scheme is the need to identify the existing forest lands, their legal designation/land use status (*KBK* or *KBNK*), the local communities settling in or nearby, their tenure rights over the forest lands, and the uses of the forest lands by the local communities and/ or other parties. Sufficient knowledge about the characteristics of forest resources, as well as the community use and rights over the forest lands, will help in developing a proper strategy and policy for the Collaborative Forest Governance within the reality that most forest lands are not formally registered. The first priority is the treatment of agricultural lands in the *KBK* that have been used by local communities for subsistence agriculture for decades; another priority is the strategy and policy for the primary and old growth forest lands in the *KBNK* that are still used by the local communities for fulfilling a variety of their subsistence needs.

In practice, the schemes should also be able to remove various existing obstacles for community-based forest management and the maintenance of forest resources, as noted by forestry experts, i.e., (1) government mega projects undertaken at the expense of forest lands; (2) the common trend of plundering the forest frontier rather than developing existing open land; (3) the simplistic and rigid assumption of state ownership over all forests within the *KBK* and the state control of any unregistered land in the *KBNK* and (4) poor coordination among government departments and overlap in forest land use plans (Elson, 2011).

# References

Abdoellah, O., Lahjie, A.B., Wangsadidjaja, S.S., Hadikusumah, H., Iskandar, J., & Sukmananto, B. (1993). *Communities and Forest Management in East Kalimantan* (Research Network Report No. 3). Berkeley: Center for Southeast Asia Studies, University of California.

Anonymous (1976). *Basic agrarian law (Undang-Undang Pokok Agraria – UUPA)*. Translation Land—494. Jakarta: Directorate General of Agrarian Affairs of the Department of Home Affairs.

———— (1999). *The Law of the Republic of Indonesia Number 41 Year 1999 on Forestry*. Official Translation. Jakarta: Bureau of Laws and Regulations I, State Secretariat of the Republic of Indonesia.

Berger, T. (2006). *Slum Areas and Insecure Tenure in Urban Sub-Saharan Africa: A Conceptual Review of African Best Practices* (Master Thesis). Upsala Universitet. Retrieved from http://www.uu.diva-ortal.org/smash/get/diva2:131022/FULL-TEXT01.

Carter, J. & Gronow, J. (2005). *Recent Experience in Collaborative Forest Management* (CIFOR Occasional Paper No. 43). Bogor: Center for International Forestry Research.

Chikhwenda, E.J.W. (2002). *Transactive Land Tenure System in the Face of Globalization in Malawi*. Paper for the Commons in the Age of Globalization Conference, Chichiri, Blantyre, University of Malawi Polytechnic.

Colchester, M. (2006). *Forest Peoples, Customary Use and State Forests: The Case for Reform*. Forest Peoples Programme Paper for IASCP 11th Biennial Conference, Bali, 19–23 June 2006.

Coleman, E.A. (2010). *Common Property Rights, Adaptive Capacity, and Response to Forest Disturbance.* Miami: Department of Political Science, Florida State University.

Colfer, C.J.P. (1995). *Who Counts Most in Sustainable Forest Management.* Bogor: Center for International Forestry Research.

Cramb, R.A. & Wills, I.R. (1998). Private property, common property and collective choice: The evolution of Iban Land Tenure Institutions. *Borneo Research Bulletin,* 29, 57–70.

Crewett, W., Bogale, A., & Korf, B. (2008). *Land Tenure in Ethiopia: Continuity and Change, Shifting Rulers, and the Quest for State Control.* Washington, DC: CGIAR, Systemwide Program on Collective Action and Property Rights (CARPRI).

Devung, S.G. (2003). Anthropological aspects of the people of Kayan Mentarang National Park, East Kalimantan. In A. Mardiastuti & T. Soehartono (eds), *Joint Biodiversity Expedition in Kayan Mentarang National Park Report.* Jakarta: Ministry of Forestry-WWF Indonesia-ITTO.

——— (2008). *Laporan Penelitian: Upaya Penguatan Tanah Adat di Kabupaten Kutai Barat.* Sendawar: Dinas Kebudayaan dan Pariwisata Kabupaten Kutai Barat.

——— (2011). *Dampak Perubahan Peruntukan Lahan Hutan Terhadap Hak Tenurial Adat dan Sumber Penghidupan Masyarakat Lokal di Wilayah Adat Tering, Kabupaten Kutai Barat.* Disertasi, Program Studi Doktor Ilmu Kehutanan. Samarinda: Program Pascasarjana Universitas Mulawarman.

Devung, G.S., Ding, L.H., Avun, W., & Bayau, Y. (1986). *Naskah Rekaman Upacara Tradisional Lalii' Ugaal Suku Dayak Bahau Hwang Triing, Kalimantan Timur.* Samarinda: Depdikbud, Proyek Inventarisasi dan Dokumentasi Kebudayaan Daerah.

Dietz, T. (2005). *Pengakuan Hak atas Sumberdaya Alam: Kontur Geografi Lingkungan Politik.* Yogyakarta: Insist Press.

Eghenter, C. (2000). What is Tana Ulen Good for? Considerations on indigenous forest management, conservation, and research in the interior of Indonesian Borneo. *Human Ecology,* 28(3), 331–357.

Elson, D. (2011). *An Economic Case for Tenure Reform in Indonesia's Forests.* Paper prepared for Right and Resources 5th Anniversary, July 2011, Trevaylor Consulting (www.trevaylor.org).

Fauzi, N. (1998). *Pengakuan Sistem Penguasaan Tanah Masyarakat Adat: Suatu Agenda NGO Indonesia.* Paper INFID—Forum Indonesia, Jakarta, 2 April 1998.

Field, J. (2003). *Social Capital.* London: Routledge.

Goenner, C. (2001). *Pengelolaan Sumberdaya di Sebuah Desa Dayak Benuaq: Strategi, Dinamika dan Prospek.* Eschborn: Deutsche Gesellschaft fur Technische Zusammenarbeit (GTZ) Gmbh.

Hermosilla, A.C. & Fay, C. (2005). *Strengthening Forest Management in Indonesia through Land Tenure Reform: Issues and Framework for Action.* Bogor: World Agroforestry Center.

Inoue, M. (2003, April 4). Diverse management of Indonesian forests: A new governor gives locals a greater say in their resources. *International Herald Tribune and the Asahi Shimbun.* Retrieved from http://www.asahi.com/english/asianet/hatsu/eng_hatsu030404b.html.

——— (2011a). *Prototype Design Guidelines for 'Collaborative Governance' of Natural Resource.* Presented at 13th Biennial Conference of the International Association for the Study of the Commons, Hyderabad, India, 12 January.

Inoue, M. (2011b). *Summary of the Design Guidelines: Instruction for the Authors of Multi-level Forest Governance in Asia—Recognising Diversity*. Tokyo: GSALS University of Tokyo.

Isozaki, H. (2003). Sustainable and participatory forest management: Legal and administrative supporting measures, and final recommendations. In M. Inoue & H. Isozaki (eds), *People and Forest: Policy and Local Reality in Southeast Asia, the Russian Far East and Japan* (pp. 73–89). Dordrecht, Netherlands: Kluwer Academic Publishers.

Jessup, T.C. (1992). Persistence and change in the Practice of Shifting Cultivation in the Apo Kayan, East Kalimantan, Indonesia. In J.J. Fox (ed.), *The Heritage of Traditional Agriculture among the Western Austronesians*. Canberra: The Department of Anthropology, Australian National University.

Kutter, A. & Neely, C.L. (eds). (1999). *The Future of Our Land: Facing the Challenge— Guidelines for Integrated Planning for Sustainable Management of Land Resources*. Rome: FAO and UNEP.

Lahajir, J. (2001). *Etnologi Perladangan Orang Dayak—Tunjung Linggang*. Yogyakarta: Galang Printika.

Lund, H.G. (2011). *Definitions of Forest, Seforestation, Afforestation, and Reforestation*. Gainesville, VA: Forest Information Service.

McKean, M.A. (1999). Common property: What it is, what it is good for, and what makes it work? In C. Gibson, M.A. Mckean, & E. Ostrom (eds), *Forest Resources and Institutions*. Rome: FAO.

Munro-Faure, P., Groppo, P., Hererra, A., Lindsay, J., Mathieu, P., & Palmer, P. (2002). *Land Tenure and Rural Development*. FAO Land Tenure Studies 3. Rome: FAO Publishing Management Service, Information Division.

Myers, G., Hetz, P.E., & Roth, M. (2007). *Land Tenure and Property Rights*. Volume I Framework, USAID. Burlington: ARD Inc.

Nanang, M. & Inoue, M. (2000). Local forest management in Indonesia: A contradiction between national forest policy and reality. *International Review for Environmental Strategies*, 1(1), 175–191.

Ostrom, E. (1990). *Governing the Commons*. Cambridge: Cambridge University Press.

——— (2003). How types of goods and property rights jointly affect collective actions. *Journal of Theoretical Politics*, 15(3), 239–270.

——— (2005). *Understanding Institutional Diversity*. Princeton: Princeton University Press.

Platteau, J.-P. (1995). *Reforming Land Rights in Sub-Saharan Africa*. Geneva: United Nations Research Institute for Social Development.

Purnomo, H., Mendoza, G.A., Prabhu, R., & Yasmi, Y. (2005). Developing multi-stakeholder forest management scenarios: A multi-agent system simulation approach applied in Indonesia. *Forest Policy and Economics*, 7, 475–491.

Resosudarmo, I.A.P. & Dermawan, A. (2002). Regional autonomy and its preliminary impacts on forestry sector. In B.P. Resosudarmo et al. (eds), *Indonesia's Sustainable Development in a Decentralization Era*. Jakarta: Indonesian Regional Science Association (IRSA).

Sardjono, M.A. (2004). *Mosaik Sosiologis Kehutanan: Masyarakat Lokal, Politik dan Kelestarian Sumberdaya*. Jogjakarta: Debut Press.

Sardjono, M.A. & Imang, N. (2011). *Local Community Dimension in the Last Half Century of Indonesian Forest Policy*. Paper for the International Workshop on Variations in Policies for Governance and Management of Community Forestry—Bridging Multi-Level Outcomes in Asia. Faculty of Agriculture, University of Tokyo, 13 October 2011.

Schlager, E. & Ostrom, E. (1992). Property rights regimes and natural resources: A conceptual analysis. *Land Economics*, 68(3), 249–262.

Simorangkir, D. & Sardjono, M.A. (2006). *Implications of Forest Utilization, Conversion Policy and Tenure Dynamics on Resource Management and Poverty Reduction.* Roma: Food and Agriculture Organization (FAO).

Sindju, H.B. (2003). Making a Swidden: Social and technological aspects of Leppo' Ke agricultural practices. In C. Eghenter et al. (eds), *Social Science Research and Conservation Management in the Interior of Borneo*. Bogor: Center for International Forestry Research.

Sirait, M., Prasodjo, S., Podger, N., Plavelle, A., & Fox, J. (1994). *Mapping Customary Land in East Kalimantan, Indonesia: A Tool for Forest Management.* Honolulu: East-West Center.

Stern, P., Dietz, T. Ostrom, E., & Stonich. S. (2002). Knowledge and questions after 15 years of research. In E. Ostrom, T. Dietz, N. Dolsak, P. Stern, S. Stonich, & E. Weber (eds), *The Drama of the Commons*. Washington, DC: National Academy Press.

Tenaw, S., K.M. Zahidul Islam, K.M., & Parviainen, T. (2009). *Effects of Land Tenure and Property Rights on Agricultural Productivity in Ethiopia, Namibia and Bangladesh.* Discussion Paper No. 33. Helsinki: Department of Economics and Management University of Helsinki.

The World Bank (1999). *What Is Social Capital?* Retrieved from http://www.worldbank. org/poverty/scapital/whatsc.htm.

Tribowo, D. & Haryanto (2001). *Disappearing Diversity: An Overview on Indonesia's Degrading Forest and Its Biodiversity.* Paper for the Workshop on Integration of Biodiversity in National Forestry Planning Programme, CIFOR Bogor, 13–16 August 2001.

# 9

# Indonesia III: Characteristics of Forest Management Policy in West Sumatra

*Yonariza, Ganesh P. Shivakoti, Mahdi, Tri Martial and Yolamalinda*

## Introduction

### Background

The synergy among national, regional and local forest management policies is highly important in countries such as Indonesia for the following reasons. Government-declared forest areas amount to nearly 70% of total land area; these forest areas straddle islands and provinces with varying proportions of forest in each province. These forest areas are categorised into conservation forest, production forest and protection forest. Under Forestry Law No. 41/1999 and following the decentralisation policy, forest management authority resides in the national, provincial and district levels of government where conservation forests are under the jurisdiction of the national government. Production and protection forests are regulated under either provincial or district governments. The composition of a forest by management category also varies widely in each province; certain provinces are dominated by conservation forest, while others are dominated by production forest. Presently, no effective unit oversees forest management activities at the field level in all types of forest management category. Without synergy among the levels of government, forest resources become open access. Hence policies developed at the community level are highly important, yet knowledge of local management policies is incomplete and scattered.

The government of Indonesia recognises the importance of community involvement in forest management and continues to improve community forestry policy; community forestry has evolved through at least four milestones. In the early 1980s, the Indonesian Ministry of Forestry (MoF) introduced a social forestry programme, known as the taungya system (*sistem tumpangsari*), in a state-owned teak forest plantation (Perum Perhutani) in Java. In 1995, the government issued MoF decree No. 622/kpts-ll/1995 in which the concept of community forestry was no longer limited to the technical and production aspects of forest management but also supports the development of a community's capacity and right to manage forest resources as a long-term objective. In 2003, the ministry created the social forestry programme. To implement this programme, regulation No. 1/Menhut-II/2004 was issued in 2004. This regulation concerns the empowerment of the people living within and around the forest in the implementation of social forestry. By the end of 2004, the MoF has declared five priority policies. The policy of empowering the economy of communities within and surrounding the forests is pertinent to enacting community forestry projects (Hindra, 2007). This policy is consistent with forestry law where understanding the needs of the community is a key to success in forest management. Therefore, forest management practices concerned with only timber production without considering the community should be reoriented to forest resource management that empowers the community.

The empowerment of the community, however, is limited by the absence of secure community property rights on forest lands and by inadequate participation in forest management, which contribute to the Indonesian forestry crisis (Contreras-Hermosilla & Fay, 2005). To resolve this crisis, the MoF has introduced policies and programmes using community-based forest management (CBFM) under a variety of names and approaches. Community forestry, called *hutan kemasyarakatan* in some ministerial decrees, has been a centrepiece of MoF policy from the early post-Suharto era through the wave of policy decentralisation (Safitri, 2006).

More than half of West Sumatra province of Indonesia has been declared a forest area, and due to its topographic conditions, that is, mountainous areas, most of the forest is designated conservation and protection forest. These types of forests provide little direct benefit to local governments and communities in terms of revenue; therefore, the active forest management for silviculture is narrowly applied. The province is home to the Minangkabau people, the largest contemporary matrilineal society; the society is organised into a complete shape called a *nagari*, and each *nagari* has its own territory with clear boundary. Most of the areas declared by the central government as forest areas overlap with the *nagari* territories. As an autonomous social unit governing resources within its territory, there is wide variation of local forest management policy within each *nagari* depending on the type of forest found in the territory. Some *nagari*s are dominated by conservation forest while others are dominated by protection forest; few *nagari*s fall under the production forest category. There are more than 500 *nagari*s in the West Sumatra; therefore, there is considerable variation in forest management policies at the *nagari* level. In addition, the location of the *nagari* in relation to the forest varies considerably. Based on statistical data, 7.78% of *nagari*s are located inside the forest area, and 49.68% of villages are bordered by forest, which means that more than half of *nagari*s have high levels of interaction with forest land (Table 9.1). Hence, the *nagari* plays an important role in forest management. In local understanding, almost 60% of *nagari*s have forest area within their territories.

Local policies significantly influence forest management quality. Yonariza and Webb (2007) found that communities regulating forest management, illegal timber cutting can be curtailed; in the absence of such regulation, illegal timber cutting continues. Therefore, it is important to analyse the variation of local forest policy among communities (*nagari*) in West Sumatra and analyse the factors influencing local forest management policy.

This chapter raises the following questions: (1) How conducive is current government policy to CBFM? (2) How do local policies of

**Table 9.1:**
*Village area and its proximity to forest land*

| Village location | Area (ha) | Per cent |
|---|---|---|
| Within forest area | 321,582 | 7.78 |
| In direct border with forest | 2,054,524 | 49.68 |
| Outside forest land | 1,759,357 | 42.54 |
| Total | 4,135,463 | 100.00 |

**Source:** Provincial Statistical Bureau (2006).

forest management vary in West Sumatra? (3) How are the concepts of graduated membership, commitment and fair benefits of collaborative management applied to communities in the West Sumatran context? (4) What are the implications of local forest management policy variation for future CBFM in Indonesia?

## Objective

This chapter aims to (1) present recent CBFM policy in Indonesia; (2) discuss variation of local forest management policy among *nagaris* in the West Sumatran context; (3) discuss the application of prototype common pool resources co-management as elaborated by Inoue (2011) and (4) discuss the implications of local forest management policy variation for future forest management in Indonesia.

## Significance

Sustainably managing forests using a common property regime remains challenging. Additional efforts to improve the productive capacity of forest common pool resource (CPR) systems, especially in the context of the current debates on the effects of climate change and the implementation of new programmes such as REDD[+], are needed. It remains to be seen whether polycentric policy approaches for governance and management of forest CPRs, which are environmentally sustainable and gender-balanced, are effective (Inoue & Shivakoti, introductory chapter). This chapter focuses on the proponents of common property resources management systems who argue that devolving natural resource management to the local community is a way to solve forest problems, i.e., deforestation, enhancing the role played by communities.

Following this introductory section, the chapter describes the methodology employed in writing this chapter. It then introduces current CBFM policy in Indonesia. It is followed by the presentation of forest management in West Sumatra and a brief review of the socio-economic and resource characteristics of the communities studied. The following section describes the characteristics of local forest management policy and the reasons for policy variation. It then discusses the application of forest management in the context of principles designed for West

Sumatra. Lastly, the chapter discusses the implications of local forest management policy variation for the future of forest management in the province.

## Methodology

This chapter is based on the recent dissertation and thesis work by the authors, which were all carried out in West Sumatra, Western Indonesia. Yonariza (2007) explored how the decentralisation policies of the government have affected the protected areas and livelihoods at the local level; he found that co-management practices between local people and state agencies exist. Mahdi (2008) investigated the institutions and livelihoods of people in the upland and lowland areas of the sub-watershed. It focused on recording the local institutional dynamism resulting from the changing context of the last 10 years in the upland and lowland areas of the watershed. Tri Martial (2010) analysed the forms and the roles of local institutions with respect to land tenure and tree tenure in agro-forestry and forest systems in West Sumatra. Malinda (2008) studied forest management and illegal logging and identified the existing illegal logging practices; the socio-economic characteristics of the community and the major factors contributing to illegal logging. These studies address local policies and institutions covering all types of forest management categories, that is, conservation forest, protection forest and production forest.

These studies share a concern with CBFM, and they include comparisons among communities in West Sumatra. However, the studies focus on different areas of forest management: Yonariza and Yolamalinda studied conservation forests, Mahdi addressed degraded production forests, and Tri Martial included conservation, protection and production forests. Regardless of the type of the forest being studied, assessment of local policy interventions in forest management depends on the benefits and costs accrued. Hence, whether local people become involved in either conservation or depletion depends on the direct benefits and costs they bear. The variation in forest management policy among *nagari*s with regard to these involvement opportunities becomes important. The *nagari*s included in these studies, the landscape type, the forest characteristics and the local forest management policies found in *nagari* are presented in Table 9.2.

**Table 9.2:**
*Characteristics of nagari covered in this chapter*

| *Authors* | Nagari | *Topography* | *Forest characteristics* | *Existence of local forest policy* |
|---|---|---|---|---|
| Yonariza | Anduring | Steep slope, requires people participation in protecting the forest | Conservation forest | Weak |
| | Koto Sani | Steep slope, requires people participation in protecting the forest | Conservation forest | Strong after decentralisation |
| Tri Martial | Koto Malintang | Steep slope, requires people participation in protecting the forest | Conservation forest Agro-forestry | Strong ever since |
| | Paru | Mountainous village | Protection forest | Strong ever since |
| | Paninggahan | Steep slope, requires people participation in protecting the forest | Conservation forest and protection forest | Strong ever since |
| Mahdi | *Nagari Selayo Tanang Bukik Sileh* | Dominated by steep slope mountainous areas | Depleted production forest cover reduced sharply whiledry land agriculture and settlement increased between the period of 1996 and 2006 | Weak ever since |
| | Nagari Dilam | Slightly sloppy, steep slope mountainous areas | Production forest, almost converted to agricultural land, only a small portion of forest, both primary and secondary, can be seen | Weak |
| Yolamalinda | Lubuk Gadang | Steep slope | Conservation forest, strong presence of state forest agency | None |

*Source:* Authors.

## Decentralisation and Community-based Forest Management

It is well documented that under the new order regime, all Indonesian forests were declared state forests and the outer island forests were opened to large-scale timber extraction to generate the much needed revenue. This centralised control of Indonesia's forestry sector followed the period 1967–1970 when policy was relatively relaxed and allowed district authorities and village communities to engage in small-scale logging activities. However, these activities were marginalised in favour of multinational corporations linked to the central government. The centralisation of natural resource management has decreased community participation and local autonomy (Casson et al., 2002). This trend was reversed under the decentralisation era at the turn of the 21st century. CBFM was widely promoted.

Following the fall of Suharto, regulation reform has been especially concerned with the adoption of CBFM that considers the tenure rights of indigenous people and local communities to forest lands and resources (Colchester, 2002). Regulatory changes during the political reform largely known as *reformasi* included allowing communities residing in or near forest areas to be actively involved in forest exploitation, to play a role in forest management and to carry out low impact extraction activities, primarily of non-timber forest products (NTFPs).

The fall of Suharto also triggered illegal logging. This was also the case in West Sumatra where some *nagari*s did not employ policies to curtail illegal logging; Yolamalinda (2007) observes that in South Solok respondents prefer permissive forestry rules for the villagers who still depend on the forest to earn their living. This dependence means that it is important for authorities to consider what the local community needs when making rules and regulations. However, according to the respondents the authorities never take this local need into consideration. Respondents also state that the government should be clear with the rules and take action coherently; villagers do not understand the rules and they become disappointed in governments when the government agencies do not act coherently. Officials confiscate illegal timber belonging to villagers and then release that illegal timber to the authorities. Corruption among the authorities is no secret among villagers and others, but no system could avoid it.

Yolamalinda (2007) reports that the socio-economic characteristics of the communities in the study area suggest that these communities do not have alternatives to the forests for their livelihoods. Most of the land area, approximately 60% of the total area, is declared forest land. Meanwhile, the primary occupation in these communities is farming, which also drives people towards the forest. The agricultural production of farmers in this area is insufficient to meet their daily needs and they must find additional sources of income. Forest and timber products are favoured because the wages earned are comparable with the wages obtained from farming.

A study by Yonariza (2007) conducted in villages surrounding the Barisan I Nature Reserve found that the presence of local regulations negatively affects rural household participation in illegal timber felling. What are the conditions that produce this outcome? There are four possible answers: the proportion of arable land, the homogeneity of the population, the history of the settlement and the type of benefits received from the forest/protected areas. Under the current forestry law (Law No. 41/1999), the state recognises forests managed by customary community rights known as *Hutan Adat,* which literally means state forest that falls under *adat* territory; the management of these forests devolves to the *adat* community (CIFOR, 2002). How is post-decentralisation forest management practised by local communities?

## Overview of Forest Management in West Sumatra

West Sumatra is one of the important conservation areas in Indonesia; more than 60% of its forest area is classified as protected area. Total forest area in the province is 2,600,286 ha while the total province land area is 4,178,830 ha; therefore, 62.23% of the land is considered forest area (West Sumatra Provincial Forestry Service, 2002). A total of 1,756,708 ha, or 67.56%, of the forest area is protected, consisting of preservation/conservation and protection forest areas (Table 9.3). Due to its topographic conditions and mountainous areas, most of the forest is considered conservation forest and protection forest. These types of forest provide little direct benefit to local governments and communities in terms of revenue; therefore, the active forest management required for silviculture is not applicable. The province is known for its matrilineal culture and its traditional social unit, the *nagari*. There are more than 530 *nagari*s in the province (Kato, 1978); there is currently considerable variety in forest management policies at this level.

**Table 9.3:**
*Distribution of forest by function in West Sumatra Province, 2002-2012*

| S. no. | Forest function | 2002[1] | | 2012[2] | |
|---|---|---|---|---|---|
| | | *Area (ha)* | *%* | *Area (ha)* | *%* |
| 1 | Preservation and conservation | 846,175 | 32.54 | 846,200 | 34.11 |
| 2 | Protection forest | 910,533 | 35.02 | 792,114 | 31.93 |
| 3 | Limited production forest | 246,383 | 9.48 | 246,400 | 9.93 |
| 4 | Permanent production forest | 407,849 | 15.68 | 407,800 | 16.44 |
| 5 | Convertible production forest | 189,346 | 7.28 | 188,257 | 7.59 |
| | Total | 2,600,286 | 100.00 | 2,480,771 | 100.00 |

**Sources:** [1] Monografi Kehutanan *(Forest in Figure)* West Sumatra Provincial Forestry Service (2002).
[2] Forestry Statistics of Indonesia (2011). Ministry of Forestry, Jakarta (2012).

In this province, land and forest have been clearly demarcated by the *nagari*s. As a socio-political unit, each *nagari* possesses the characteristics of a democratic society including the division of power among legislative, judicative and executive bodies. As an economic entity, a *nagari* controls resources such as land, forest and water. Before 1970, there were more than 531 *nagari*s; although the central government abolished the *nagari* as the lowest administrative unit, favouring the Javanese *desa* system, the Minangkabau were able to maintain the *nagari* as a customary community. After the collapse of the Indonesian 'new order' regime, the Minangkabau began to revitalise the *nagari* government system in 2001. All the land in the province, including the forests, has been divided into *nagari*. Each *nagari* has either natural or man-made boundaries. The natural boundaries include watershed boundaries that imply that the particular *nagari* has an interest in protecting the watershed and forest contained therein.

The *nagari*s traditionally divide the forest into several categories based on forest conditions and function, which are similar to the government's forest classification by function. According to Rochmayanto, Sasmita, and Jannetta (2006), the forest types according to the Minangkabau *adat* are:

1. *rimbo tuo* (old forest), forest for taking NTFPs such as *manau*, *gaharu*,
2. *rimbo besar* (big forest), forest of wild animals,
3. *rimbo* raya (great forest), forest for harvesting fruit,

4. *rimbo dalam* (inner forest), forest for water and conservation,
5. *rimbo luas* (vast forest), forest for taking timber for building houses,
6. *rimbo* lepas (free forest), forest for settlement expansion,
7. *rimbo ana,* forest under the management of the *penghulu/rajo* (matrilineal clan leader/king), and
8. *rimbo piatu,* forest for the ancient spirits. According to Rochmayanto et al. (2006), the Minangkabau people also traditionally divided the forest by type.

Scholars also recognise forest zoning systems among Minangkabau people that consider the ecological functions of a forest (Firmansyah, 2007; Warman, 2007). The zones are:

1. *Hutan larangan* (forbidden forest), a zero growth area in which it is forbidden to change the ecosystem of the zone but the utilisation of non-timber products is allowed;
2. *Hutan simpanan* (reserve forest), a reserve area that is allocated as a reserve for future generations;
3. *Hutan olahan* (production forest), a forest area that is managed intensively by the people. This type of forest area is used mostly for agro-forestry (*parak*[1]).

Forest land can also be classified according to land ownership. According to LBH Padang (2005), Rochmayanto et al. (2006) and Firmansyah (2007), *ulayat* land (the communal ancestral land of a village or matriclan) based on Minangkabau custom consists of:

1. *Ulayat rajo,* managed by the *penghulu* (matrilineal clan leader) from several *nagaris,* which are commonly located far away from the *nagari.*
2. *Ulayat nagari,* managed by the whole *penghulu* in a *nagari.*
3. *Ulayat suku,* managed by the whole clan under *penghulu pucuk/ andiko* leadership.
4. *Ulayat kaum,* managed by a member of a sub-clan member through the matrilineal hereditary system under the *penghulu* or *datuk* of a particular clan.

---

[1] *Parak* is a local terminology for agro-forestry. This system is common in the Minangkabau area. See Michon (1986).

5. *Ulayat paruik*, managed by a genealogic clan member (*paruik*).
6. *Ulayat keluarga inti* (nuclear family), managed by a family that originates either from an *ulayat paruik* or from private property.

In this chapter, we explore the variation in local forest management policies among the Minangkabau *nagari* of West Sumatra. Prior to this analysis, we present the socio-economic and resource characteristics of the communities (*nagaris*) studied.

## Socio-economic and Resources Characteristics of the Communities

To provide the background information on variation in local forest management policy, it is necessary to present some of the socio-economic and resource characteristics of the studied communities. Among the Minangkabau, land, including forest area, is classified into *nagari*, clan and extended family land. *Nagari* land is common to all *nagari* members and each member can access its resources by customary law; members can benefit from the land but not own it. Clan land is the common property of a clan; members can benefit from the land, including the forests but they do not own the land privately. *Kaum* land, or extended family land, is usually land owned by a group of descendants from a great grand-mother. The distribution of land by ownership varies among the *nagari*; in *nagari*, such as Paninggahan, Koto Malintang, Koto Sani, Salayo Tanang and Anduring, *nagari* land or *ulayat nagari* is dominant. Forest ownership in other *nagari* tends to be *kaum* land ownership, a smaller unit of social organisation. The membership of each group is different, namely in the identity of the core member, who is the descendant of an original settler or the descendant of a late comer.

Forest areas are found on the steep slopes and mountainous areas of these *nagaris*. Under the government forest management categories, these forests are conservation or protection forests. Only a small part of the forest, as in *Nagari Paru* and *Nagari Dilam*, is production forest category.

The main source of livelihoods varies among the *nagaris*. In Koto Malintang, Paninggahan, Koto Sani and Anduring, which are located on irrigated paddy, water is important and the forest is protected to conserve water. In *Nagari Paru*, villagers rely on rubber cultivation and the forest is protected for the expansion of this rubber cultivation. In

*Nagari Salayo Tanang* and *Nagari Dilam*, the population relies on dry land farming and upland cash crops, and the forest land is used as farm land extension. In *nagaris* where irrigated paddy fields are dominant, the forest remains well protected, while in other *nagaris*, the forest has been cleared for upland farming. There is considerable competition between agricultural and forest land use. Furthermore, in *nagaris* where irrigated *sawah* is dominant, main products collected from the forest are NTFPs, while in other *nagaris*, the villagers extract timber. As demonstrated in the next section, different forest characteristics and alternative means of obtaining a livelihood cause forest management policy differentiation among the *nagaris*.

## Local Forest Management Policy

Following Inoue (2011), '... by policies it is meant both legislative (national, district) and non-legislative (local) institutions'. What are the forest management policies among the studied *nagaris*? The forest is an important source of livelihood for local people, directly or indirectly. In addition to providing environmental services such as water, the forest provides NTFP. As such, the forest has been part of their customary territories and is an important ancestral property. Since the establishment of the *nagaris*, the forest has been an important element within the institution. Each *nagari* manages the forest by designating it for different purposes that are managed differently. Under the new order regime, however, much of this forest area has been declared state forest.

   Our case studies reveal clear variation in forest management policy by *nagari*. In Nagari Paru, 4,000 ha of forest is considered *nagari*-protected area (*rimbo larangan*). By 1992, *nagari* regulations were formally issued to curtail illegal logging within the *nagari*. In *Nagari Paninggahan*, the conservation forest is protected by the *nagari* communities from illegal logging. In *Nagari Koto Malintang*, there is *de facto* co-management of the conservation forest by the *nagari* and the state forest agency. Locals are allowed to collect NTFP. But, there is no clear forest management policy in *Nagari Selayo Tanang*. In *Nagari Dilam*, rules concerning the harvesting of forest products have been recently implemented, which the *nagari* government monitors and evaluates. In *Nagari Koto Sani*, there is also co-management between the state forest agency and the local community. *Nagari* regulations are formally issued to curtail illegal logging. In *Nagari Anduring*, no specific policy was identified.

It can be concluded that forest management policy exists primarily for forest protection. The protection forests are sources of water for many *nagari*s whose livelihoods depend on irrigated paddies. Nevertheless, in *nagari*s such as *Pakan Rabaa*, where Kerinci Seblat National Park is located, no local forest policy has been developed. The lack of local policy is caused by the predominance of government agencies. Likewise, in villages where villagers do not depend on paddies for their livelihoods, no forest policy seems to have developed. Hence, the importance of the forest for ecosystem services determines local forest policy interventions. In *nagari*s where the forest is an important source of water, illegal logging poses a threat to the water supply; therefore, the *nagari*s devise regulations to curtail this illegal logging. These policies have been in effect since the revitalisation of the *nagari* government system in 2001 after being abandoned from 1983 to 2001 when the forests became open access resources that were accessed by people from different parts of province for timber.

In *Nagari Koto Malintang*, local institutions accommodate both conservation and livelihoods by developing a complex system of agro-forestry. Among the crop cultivated in the agroforestry system are durian (*Durio zibenheis*), bayur (*Pterspermum javanicum*), surian (*Toona sinensis*), cinnamon (*Cinamomum burmanii*) and nutmeg (*Myristica fragrans*). These perennial tree crops protect water sources and provide NTFP. However, there must be a more specific reason for the observed policy variation.

## Reasons for Policy Variation

The *nagari*s are close, related by Minangkabau culture; therefore, one needs to identify another source of the observed policy variation. The following reasons may explain the policy variation: semi-autonomous social fields, policy incentives or decentralisation of forest policy.

### Semi-autonomous Social Fields: Room for Local Policy Variation

The *nagari*s of West Sumatra are considered semi-autonomous social fields; they are capable of self-regulating but existing within a larger system. As such, variation in local forest management policy is very possible. Current decentralisation trends in Indonesian policy make it more obvious that these semi-autonomous social fields are capable of issuing regulations for their community members.

The *nagari* is the lowest-level political unit of the matrilineal Minangkabau ethnic group. Each *nagari* is composed of several neighbouring hamlets from a clan (*suku*). A clan is composed of several lineages (*kaum*) (Mahdi et al., 2009). Each *nagari* has its own rules and laws because the *nagari*s are independent institutions. The *nagari* has a democratic, autonomous and informal structure, with the clan and hamlet leaders placed on top (Naim, 1984). In West Sumatra province, Indonesia, the *nagari* was the form of village organisation from colonial and post-colonial times until 1983 when it was replaced with a system based on smaller administrative villages, *desa*. The *nagari* functioned as an informal institution from 1983 until provincial regulation no. 10/2000 was enacted (Benda-Beckmann & Benda-Beckmann, 2001). This regulation triggered the revitalisation of the *nagari* role in forest management to some extent.

## Nagari Revitalisation

Soon after the implementation of Law 22/1999, there was a movement to revitalise indigenous villages. People who no longer used the terms for '*desa*' in their territories re-established indigenous names such as '*nagari*' in West Sumatra, '*kampong*' in West Kalimantan and '*pekon*' in Lampung. Local forestry norms have been reproduced and revitalised and even written into local laws such as village or kampong regulations (*Peraturan Desa or Peraturan Kampung*) and community charters (*Piagam Kesepakatan Masyarakat*). All of these changes are directed towards forest protection and enhancing community participation in forest management. Similarly, district governments have enacted innumerable forestry regulations; most of these regulations aim to legalise forest exploitation (Colcester, 2002).

Along with the restoration of the *nagari*, these laws provide an opportunity for the people of West Sumatra to write and to formalise their customary laws regarding forest land and forest resources. *Nagari Sungai Kamuyang* of the 50 *Kota* district, for instance, issued *nagari* regulations in 2003 regarding the utilisation of *nagari*-owned land, including forest resources (Mahdi, 2008).

McCarthy (2005) asserts that in Indonesia,

... with the recent eruption of local struggles over resources and now with the new decentralisation reforms, there is renewed interest in the role of customary *adat* institutional arrangements in village government, land tenure and forest management, the *adat* arrangements are constantly renegotiated. *Adat* customary orders are tied to local notions of identity and associated notions of appropriateness, and as such constitute patterns of social ordering associated with both implicit deeply held social norms and more explicit rules. This article concludes that, while the State and *adat* regimes often compete to control the direction of social change, they also constantly make accommodations and, in some respects, need to be considered as mutually adjusting, intertwined orders. (p. 57)

This situation contributes to local variation in forest management policy.

## Incentives

Incentives are the main instruments for developing community forest management institutions. Studies of institutions aim at understanding the incentives that motivate human behaviour at a particular time and space and the impact of the behaviour on natural resources. Categories of incentive include (1) incentives in relation to characteristics of resources, (2) incentives related to community characteristics and (3) incentives related to the characteristics of the rules (Djogo, Sunaryo, Suharjito, & Sirait, 2003). The institutional arrangements determine access to present and future benefits; the arrangements also affect the distribution of farm income (Dorner, 1971). The community involved in protecting the forest, for example, must gain some benefits out of their involvement.

According to the UNEP (2001, p. 34), 'Forests will be protected when the people conclude that forest conservation is more beneficial (e.g., generates higher incomes or has ecological or social values) than their clearance'. This quotation summarises the argument for economic policies that as long as there is a greater incentive to cut forests than there is to preserve them, deforestation is likely to continue (Sherbiniin, 2002). This pattern is clearly observed among the *nagaris* studied; the *nagaris* were involved in forest management when benefits to the community are likely to accrue. The most prominent benefit from community involvement is obtaining environmental services such as water and land slide protection. Community members may receive different benefits from these environmental services; however, because water is the most useful, especially for irrigation, the landowners benefit the most. The landowners are mostly the descendants of the upper class community members whose ancestors controlled a greater portion of the land.

## Discussion

This part of the chapter discusses two aspects: design principles and their implications for the future of forest management in Indonesia.

### Design Principles

Community linkages with outsiders in forest management are obvious in the West Sumatran context; co-management occurs in many parts of the forest where the local community takes an active role in cooperation with state agencies. In this context, design principles of collaborative management as described by Inoue (2011) (see Chapter 1) apply to West Sumatran cases. There are nine principles: the degree of local autonomy, the existence of clearly defined resource boundaries, the presence of graduated membership, the use of the commitment principle, the presence of a fair benefit distribution, the use of a two-tiered monitoring system and sanctions, the development of nested conflict management mechanisms, and the continuation of trust-building by strengthening social capital. This chapter only discusses the application of design principles 3 (graduated membership), 4 (the commitment principle) and 5 (fair benefits).

As elaborated by Inoue (2011), design principle 3 (graduated membership) is 'based on "open-minded localism", some of the local people act as "core members" (first class members), who have the strongest authority, cooperating with other graduated members who have relatively weaker authority (second class and third class members)'. Each *nagari* in West Sumatra has its own history. Historically, each *nagari* was first established by the families who first cleared the forest for settlement. These families tended to choose the best forestland for settlement. Following this initial settlement, boundaries were created. Other groups of people, called late comers, could be invited to join the initial settlers. The late comers established their settlement sites on second-class land, for a pattern that continued for subsequent settlers. This pattern of settlement created stratification in the citizenship of the *nagari*s. The first settlers tend to be the core members and the late comers become second- or third-class members. Furthermore, local policy is governed by the *adat* functionaries where the original settlers have the right to become *adat* functionaries and the descendants of these original settlers benefit from the ecosystem services.

The first settlers to clear the forest become powerful in determining the rights of second-class members to the land and its resources. The first settlers, or first-class members, enjoy the environmental benefits of the forest, such as water for irrigation, because the best agricultural land is used for paddy fields and benefits those core members the most. The descendants of the first settlers enjoy titles as community leaders; they inherit the *panghuluship*. The second- or third-class members may only cultivate dry land or become shareholders of a paddy field. The second-class members benefit from forest products such as non-timber products. In the creation of forest policy, the interests of first-class members are predominant. So, the graduate principle is inherent to the creation of the community, or *nagari*, itself.

*Design guideline 4 (commitment principle):* Inoue (2012) states that 'this principle recognises the authority to make decisions in a capacity that corresponds to their degree of commitment to forest use and management. Decision-making is not equal, but should be fair and just' (p. 8). In the *nagari*s, it is the *adat* council or council of community elders that has the authority to make a decision. The council consists of representatives from community sub-groups such as the representatives of a clan. Being the representative of a sub-group, the clan representative must ensure that the decision made benefits the members of the sub-group. These representatives are role models in forest management and provide an example of good forest management to the other community members. Commoners and the descendants of late comers are mostly excluded from the policy process. This exclusion may cause policies to be ineffective.

*Design guideline 5 (fair benefit distribution):* According to Inoue (2011), benefits are not necessarily distributed equally, but the distribution is fair in accordance with cost bearing. In West Sumatran *nagari*s, the most important benefit derived from the forest is water. The opportunity cost of using forest land should include environmental services. The main benefits of conservation forests are ecosystem services, mainly water; these services may be enjoyed mainly by descendants of the first settlers who cultivate the paddy fields.

## Implications of Recent Co-management Forest Programmes in Indonesia

CBFM had actually existed long before colonial times. However, this native practice has been disturbed by larger political changes. Despite government efforts to decentralised forest management to the local level, the government has strengthened its role in forest management through the forest management unit (FMU). With the FMU, the government intends to intensify its role in forest management by recruiting staff and deploying them to forest areas so that their presence is felt on a daily basis. The FMU aims to overcome the previous problem of the under-management of state forests. Some people claim that recent changes in forest management in Indonesia, including adopting participatory approaches, allow the government to reduce budgets for forest rehabilitation. The FMU is responsible for supervising CBFM within its jurisdiction. The government of Indonesia has also adopted participatory forest management where local people are considered the primary stakeholders. The FMU should adopt this participatory approach.

The FMU should coordinate all stakeholders and different types of forest rights. Because the FMU may cover an administrative unit, it recognises both traditional and state granted use rights within its jurisdiction and integrates those various rights into a management system. Following Government Regulation No. 6/2007, the government has played a role in empowering communities by the provision of legal status.

Traditional or indigenous management practice should be integrated through the participatory planning process (Hindra, 2007). This regulation seriously takes into account the level of community involvement under a co-management system. The FMU is the future of Indonesian forest management and it needs to consider the guiding principles presented by Inoue, especially graduated membership, commitment and fair benefits.

## Concluding Remarks

Forests should remain common property, where the benefits of these forests extend beyond the official boundaries. It is a necessity to manage the forest in a collaborative way to ensure benefits for all people. Unfortunately, communities living near forest areas only develop policy if they derive direct benefits from the forest, such as forest products and environmental services. These communities also resist strong intervention by outsiders; therefore, the negotiation of a win–win solution is crucial.

At the same time, existing studies neither elaborate the interrelationship of resources, such as water and land, as mediated by institutional arrangements nor consider the implications of the management of forest resources in an integrated manner regarding poverty reduction. The focus of forest management in the future must be on integrating the global benefits of the forest into local livelihoods to strengthen collaborative forest management.

# References

Benda-Beckmann, F. & Benda-Beckman, K. (2001). *Recreating the Nagari: Decentralization in West Sumatra.* Max Planck Institute for Social Anthropology Working Papers. *Working Paper No. 31.* Halle: Max Planck Institute for Social Anthropology.

Casson, A.A. & Krystof, O.O. (2002). From new order to regional autonomy shifting dynamics of "illegal" logging in kalimantan, Indonesia. *World Development,* 30(February).

CIFOR (2002). *Hutan Adat* (customary right forest). Warta Kebijakan. CIFOR No. 3 February 2002.

Colchester, M. (2002). *Bridging the Gap: Challenges to Community Forestry Networking in Indonesia.* Learning from International Community Forestry Networks: Indonesia Country Study.

Contreras-Hermosilla, A. & Fay, C. (2005). *Strengthening Forest Management in Indonesia through Land Tenure Reform: Issues and Framework for Action.* Bogor: CIFOR.

Djogo, T., Suharjito, D., & Sirait, M. (2003). Kelembagaan dan Kebijakan dalam Pengembangan Agroforestri. *Bahan Ajar Agroforestri 8.* ICRAF. www.worldagroforestrycentre.org.

Dorner, P. (1971). Land tenure institutions. In Melvin G. Blase Blase (ed.), *Institutions in Agricultural Development.* Ames Iowa: The Iowa State University Press.

Firmansyah, N. (2007). Eksistensi pengelolaan hutan oleh masyarakat nagari (catatan kecil pengalaman advokasi di Nagari Kambang, Nagari Guguk Malalo dan Nagari Simanau) [existence Existence of forest management by nagari community, a note from advocacy work in three nagari. Chapter 7 in Kurniawarman Warman et al Potret Pengelolaan Hutan di Nagari [portrait of nagari forest management]. Padang: HuMA and Qibar.

Hindra, B. (2007). *Community Forestry in Indonesia.* Paper presented at Asia Pacific Tropical Forest Investment Forum Bangkok, 6–8 August 2007.

Inoue, M. (2011). *Prototype Design Guidelines for Collaborative Governance of Natural Resources.* Tokyo: Department of Global Agricultural Science, Graduate School of Agriculture and Life Science, the University of Tokyo.

Kato, T. (1978). Change and continuity in the Minangkabau matrilineal system. *Indonesia*, 25, 1–16.

Mahdi (2008). *Local Responses to Changing Contexts of Natural Resources Management: Case Study at Lembang Sub-watershed of West Sumatra, Indonesia* (Dissertation). Bangkok, Asian Institute of Technology (AIT), Thailand.

Mahdi, Shivakoti, G.P., & Schmidt-Vogt, D. (2009). Livelihood change and livelihood sustainability in the uplands of Lembang sub-watershed, West Sumatra, Indonesia, in a changing natural resources management context. *Environmental Management*, 43(1), 84–99.

Martial, T. (2010). *Penataan kelembagaan penguasaan tanah dan pohon pada sistem agroforestri di sumatera barat* (Institutional Arrangement of Land And Tree Tenure in Agroforestry System In West Sumatera) (Dissertation). Padang: Program Doktor Program Pascasarjana, Universitas Andalas.

McCarthy, J.F. (2005). Between *adat* and state: Institutional arrangements on Sumatra's forest frontier. *Human Ecology*, 33(1), 57–82.

Michon, G., Mary, F., & Bompard, J. (1986). Multistoried agroforestry garden system in West Sumatra, Indonesia. *Agroforestry Systems*, 4, 315–338.

Naim, M. (1984). *Merantau pola migrasi suku* Minangkabau (merantau: migration pattern among Minangkabau ethnic). Yogyakarta: Gajah Mada University Press.

Provincial Forestry Service, West Sumatra Province (2012). *Forest Statistics 2012*. Padang: Provincial Forestry Service.

Provincial Statistical Bureau (2006). West Sumatra in figure 2006. Padang.

Rochmayanto, Y., Sasmita, T., & Jannetta (2006). Perspektif hutan ulayat dalam budaya Minangkabau Minangkabau (Studi Kasus Jorong Koto Malintang, Kab. Agam) [Indigenous forest management perspective Among Minangkabau]. Jakarta:

Safitri, M.A. (2006). *Change without Reform Community Forestry in Decentralizing Indonesia*. Paper presented at the 11th IASCP Conference, Bali, 19–23 June 2006.

Sherbinin, A. (2002). A guide to land-use and land-cover change (LUCC): A collaborative effort of SEDAC and the IGBP/IHDP LUCC Project.

Tucker, C.M. (2010). Learning on governance in forest ecosystems: Lessons from recent research. *International Journal of the Commons*, 4(2), 687–706.

UNEP (2001). *An Assessment of the Status of the World's Remaining Closed Forest*. Kenya: UNEP/DEWA/TR01-2. Division of Early Warning and Assessemt (DEWA) United Nations Environment Programme (UNEP) Nairobi.

Warman, K. (2007). Kajian Hukum Tentang Peluang dan Kendala Bagi Kebijakan Daerah Dalam Penguatan Tenurial Adat (Legal study on prospect and constraint provincial policy in strengthening traditional forest tenure). In Perkumpulan Qibar Potret Pengelolaan Hutan di Nagari. Padang: Perkumpulan Qbar.

West Sumatra Provincial Forestry Service (2002). Forest Statistics 2002. Padang, Provincial Forestry Service.

Yolamalinda (2007). *Forest Management and Illegal Logging in West Sumatra, Indonesia* (MS Thesis). Graduate School, Andalas University.

Yonariza & Webb, E.L. (2007). Rural household participation in illegal timber felling in a protected area of West Sumatra, Indonesia. *Environmental Conservation*, 34(1), 73–82.

Yonariza (2007). *Protected Area and Local Livelihood, Study of People Forest Interaction in Barisan I Nature Reserve, West Sumatra, Indonesia* (Dissertation). Asian Institute of Technology (AIT), Bangkok, Thailand.

# 10

# Malaysia: Governance and Community Participation in Forestry

*A. Ainuddin Nuruddin*

## Introduction

Malaysia has been endowed with rich biodiversity and has been rec-ognised as one of the top 12 mega-diverse countries in the world (Mittermeier, Gil, & Mittermeier, 1997). Malaysia has 12,500 species of flowering plants (NBDP, 1998) that contribute immensely to the eco-nomic development of the country.

Malaysia has a land area of approximately 32.83 million hectares (Mha); 13.16 Mha cover the Peninsular Malaysia, and 19.67 Mha cover Sabah and Sarawak, located in the island of Borneo. Of the total land area, 18.25 Mha are forested and 14.52 Mha of those (79.6%) are categorised as permanent reserved forest (MTC, 2009). Permanent reserved forests are areas gazetted under the National Forestry Act of 1984 and can be further designated into various types: 11.82 Mha have been designated as production forest, 2.65 Mha as protection forest and 0.05 Mha as national parks and wildlife sanctuaries (Figure 10.1). Forests designated as production forests are managed using sustainable forest management (SFM) principles and harvested commercially for timber production.

Forests in Malaysia provide timber resources for the wood-based sector, which was the fifth largest industry in the country and earned

**Figure 10.1:**
*Types and sizes of forested lands in Malaysia*

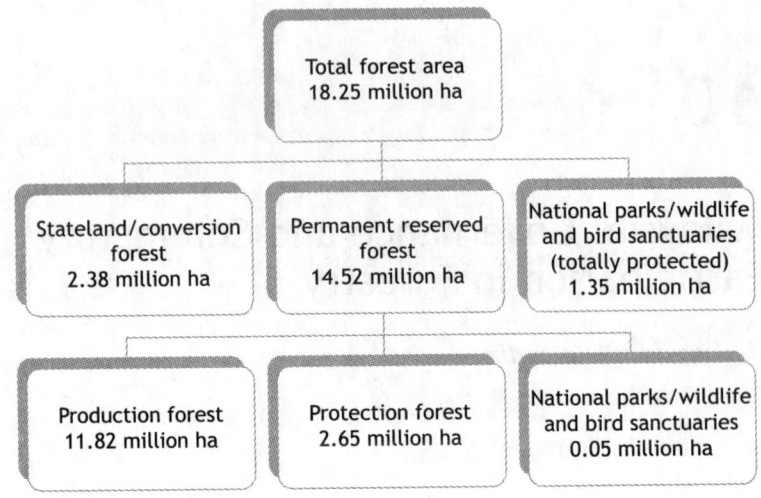

*Source:* MTC (2009).

a total of RM22.5 billion in 2008 (MIDA, 2010). It is expected that the wood-based sector will continue to be an important contributor to the Malaysian economy, and this sector is predicted to expand its earnings to RM53 billion annually by 2020 (MTIB, 2009). To achieve this target, meet an increase in demand by the wood-based industry, improve production efficiency and increase the quality of wood-based products, many factors, such as continuous and sustainable raw materials, are needed (Ainuddin, Shahwahid, & Awang Noor, 2011).

The purpose of this chapter is to discuss the history and development of forest management in Malaysia and the various levels of involvement of local communities in forest management. The level of involvement of local communities was evaluated based on Inoue's (2009) nine design guidelines. Briefly, Inoue's nine design guidelines, or *kyouchi* principles, are derived from common pool resource concepts and comprised the following nine principles: degree of local autonomy (design guideline 1), clearly defined resource boundary (design guideline 2), graduated membership (design guideline 3), commitment principle (design guideline 4), fair benefit distribution (design guideline 5), two-storied monitoring system (design guideline 6), two-storied sanctions (design guideline 7), nested conflict management mechanisms (design guideline 8) and trust building (design guideline 9) (Inoue, 2011).

# History

One of the oldest forest products commercially extracted from the Malaysian forest is agarwood, which was mainly for the Middle East markets. Arab traders sailed their ships from Arabian ports to many ports in this region, especially Malacca, in search of agarwood in exchange for products from Arabic regions. Ibn Battuta, a famous Muslim traveller from Morocco, chronicled his travels in this region, and he noted that agarwood was exchanged with products brought from the Arab world (Ibn Battuta, 2012). This barter trade has been going on for many centuries and has led to interactions between the locals and traders, sometimes resulting in intermarriages. Additionally, the increase in trade has also increased the prominence of Malacca as an international port where many traders from different parts of the region come to barter their products. The importance of Malacca was recognised by the West, and after several attempts, in 1511, Malacca was conquered by the Portuguese. However, in 1641, Malacca was conquered by the Dutch and was ceded to the British in the Anglo-Dutch Treaty of 1824. Malacca was first ruled by the British East India Company, was later a Crown Colony and subsequently combined with other states in the Peninsular Malaysia to form the Federation of Malaya. In 1963, Sabah and Sarawak joined the Federation of Malaya to form Malaysia.

# Forest Policy and Governing

In the Malaysian constitution, land is under the jurisdiction of the state governments, and they are empowered to enact laws and regulations for forestry. However, the federal government has the authority to conduct research and training and to provide technical assistance to the states. Under the terms of the constitution, Sabah and Sarawak can conduct forest research activities independently.

To coordinate and adopt a common approach to forestry, the National Forestry Council (NFC) was established on 20 December 1971 by the National Land Council (NLC). NFC is chaired by the deputy prime minister, and its membership includes the chief ministers of the 13 states. It serves as an avenue for the federal and state governments to discuss and harmonise forestry policy, administration and management. All decisions of the NFC have to be endorsed by the NLC (Thang, 1987), and the implementation of these decisions is undertaken by the state governments.

The earliest forest policy was the Interim Forest Policy adopted after the end of the Emergency in July 1960, when forests that had been closed due to the communist insurgency were opened for agricultural developments (Wyatt-Smith, 1963). This policy called for the permanent reservation of land for the present and future inhabitants of the country. This policy categorised two types of reserves: protective and productive. Protective reserves are forests for safeguarding the water supply and preventing flooding and soil erosion, while productive reserves were for supplying, in perpetuity, all forms of forest products.

The Interim Forest Policy was adopted as the National Forestry Policy in 1978, and forestry laws were streamlined and strengthened in the areas of forest management planning and forest renewal operations (MTC, 2007). In 1993, the National Forest Policy was further amended to incorporate SFM as recommended in the United Nations Conference on Environment and Development (UNCED). The two objectives of the amended National Forest Policy are to conserve and manage the nation's forests based on the principles of sustainable management and to protect the environment, as well as to conserve biological diversity and genetic resources and to enhance research and education. The amended policy also includes conservation of biological diversity, sustainability and recognition of the role of local communities in forest developments (MTC, 2007).

The National Forest Policy was strengthened with the passing of the National Forestry Act (Act 313) by the Malaysian parliament in 1984. This act was promulgated to ensure effective forest administration, utilisation, management, harvesting and reforestation in accordance with the concept and principles of SFM (Thang, 1987). It provides a legal basis for the enforcement of forest-related activities. This act is complemented by the Wood-based Industries Act (1984), which regulates the development of the wood-based industries. The National Forestry Act was amended in 1993 to strengthen its effectiveness in dealing with illegal logging and forest encroachment by increasing the fines and the jail terms for the offenders.

## Forest Management

When the British colonised Malaysia, they exploited the forest for timbers and other products, such as gutta-percha. Trees of *Palaquium gutta* were destructively felled for their latex, and silviculture operations were confined to the establishment of *P. gutta* and para rubber (*Hevea brasiliensis*) plantations (Smith, 1959) and the enrichment planting of *Neobalanocarpus heimii* (Thang, 1987). The next phase in Malaysian forest management was the introduction of silviculture treatments, where species whose crowns interfered with the valuable timber species were removed, and all bertam (*Eugeissona tristis*) and climbers within a fixed distance were cut (Wyatt-Smith, 1963).

The increased use of forest resources for poles and railway sleepers led to different silviculture treatments. The silviculture treatments were improved by implementing a series of silvicultural treatments, known as 'commercial regeneration fellings', which involved the gradual removal, in several stages, of inferior species before the felling of the useful species was permitted by commercial operations; this approach was permitted when there was a market for firewoods and poles that was sufficient to generate a profit to the Forestry Department (Thang, 1987). This system was interrupted by World War II, when large areas of forest were cleared for food production.

After the second world war, there was an increase in the demand for raw materials. The use of heavy machinery increased the amount of timber extracted, and the Forestry Department perceived the need for different silvicultural systems. The Malayan Uniform System was introduced with the aim of converting virgin lowland dipterocarp forests to even-aged forests that would be stocked by commercial tree species so that it could be harvested in a single felling. This system was successful for the lowland dipterocarp forest but was unsuccessful for the hill dipterocarp forest.

The Selective Forest Management System was introduced to further upgrade the management of forest resources and input from the new knowledge gained from the conducted research. This system calls for the rotation of forest harvesting between 25 and 30 years ensuring economic viable stock that can be harvested on the same stand after 25–30 years. This rotation is undertaken by setting a diameter cutting limit for dipterocarp and non-dipterocarp trees of no less than 50 cm and 45 cm dbh, respectively. The residual trees left unlogged are able to grow after 25–30 years to provide timber for the country. In this system, it is recommended that a minimum of 32 trees per hectare are maintained in the plots. Furthermore, during harvesting, directional felling is performed to ensure minimal damage to the stand, and forest roads for logging and skid trails are constructed with minimal impact to the environment. The establishment of buffer strips along streams and rivers also minimises soil erosion during the road construction and harvesting activities. Surveying the area after logging and after five years will determine the types of silviculture treatment best for ensuring good regeneration of the area.

In 1992, after UNCED in Rio de Janeiro, at which several recommendations on SFM were made, Malaysia initiated SFM following the recommendations of UNCED to upgrade its management of forest resources to international practices. SFM as defined by ITTO (1992) is the process of managing forests to achieve one or more clearly specified objectives of management with regard to the production and continuous flow of desired forest products and services. This management must also be performed without undue reduction of the forests' inherent values and future productivity and without undue desirable effects on the physical and social environment. Based on this definition, forests are managed not only for their products but also for their services, such as biodiversity, conservation and environmental and social functions. In the context of Malaysian forestry, SFM was implemented with three main pillars that are economically viable, environmentally sound and socially acceptable. SFM must be economically viable to bring benefits to the current stakeholders and the future generations of Malaysia, and the forests under SFM must remain healthy and be able to support their organisms while at the same time maintaining their productivity and renewal capability. SFM should also be socially acceptable; its activities should be socially sustainable by conforming to social norms.

To guide, implement and monitor SFM, a set of criteria and indicators was developed based on the ITTO criteria and indicators for SFM. This led to the formulation of the Malaysian Criteria and Indicators, which serves as a guideline for forest management activities and also acts as a standard for forest certification. An independent organisation called the Malaysian Timber Certification Council (MTCC) was established in 1998 to further develop and operate the certification process named the Malaysian Timber Certification Scheme. To date, nine forest management units (FMUs) with a total of 4.65 Mha of forests have been certified.

The centralisation of forest management by state governments restricts the role played by the local community near the boundary of the forest and the indigenous people living in the forest. This is due to the perceived view that the local community and indigenous people living in a forest lack the skills of managing that forest (Gill, Ross, & Panya, 2009); therefore, the perceived view is that a third party can take advantage of the local community and indigenous people to extract the forest resources leading to forest degradation. However, there is a current trend of decentralisation of forest governance (Agrawal, Chhatre, & Hardin, 2008), which allows local communities to play a bigger role in the management of forests. Gill, Ross, and Panya (2008) proposed co-management of the Southeast Pahang Peat Swamp Forest between the Jakun (ethnic indigenous people) and the State Forest Department to maintain the ecological integrity of the forest and to meet the subsistence needs of the local people. The decentralisation of forest governance will also help to reduce the conflict between the stakeholders.

Another issue that is related to SFM is the customary land rights of the indigenous forest-dependent community. These communities claim portions of the forest, which occasionally overlap with the forest concessions that are given to timber companies. This overlap causes conflicts and land disputes, including the blocking of the logging road by the indigenous community, between the local communities and the timber concessions. A number of disputes have been taken to court to be resolved by the judicial process.

## People and Forest

The forest is an important source of food, non-timber products and livelihood, and the indigenous people living in the forest and also local communities living near the forest are very dependent on the products the forest offers. Many studies have been conducted to show the dependence of the Orang Asli (aborigine people of Peninsular Malaysia) on the forest.

Shahwahid and Nik Mustapha (1991) studied the aborigines living near the Tasik Bera (Lake Bera) in Pahang and estimated the benefits generated from the utilisation of the forest products in the Tasik Bera ecosystem. They interviewed 43 families (23% of the total population) and solicited their consumption patterns of forest products. The respondents were requested to specify the quantity of each plant product collected from the forest during a period of a week or a month, the number of times they collect the plant product during the course of a year and, when possible, the market values of the plant products. The study found out that the mean imputed economic value per household was RM2,105 per annum with 35% of the families surveyed obtaining an economic value from the forest of less than RM3,772.5 per annum and 5% obtaining RM6,605 per annum. The collection of various rattan species contributed to 68% of the total economic value, followed by resins collected from *Dipterocarpus kerrii* (11%), mats and baskets made from *Pandanus* sp. (10%), and the remaining economic value was from the selling of aromatic wood from *antidesmaefolia* and *Aquilaria malaccensis*.

Another study on the role of timber and non-timber products to the Orang Asli was conducted by Rusli et al. (1997). The main aim of that study was to estimate the quantity of timber and non-timber forest products collected by the Orang Asli as well as the revenue that could have been generated by collecting those products. Interviews were held with each of the household heads of two Orang Asli communities residing at Sungai Rasau Luar and Sungai Rasau Dalam in November and December, 1996, near the Air Hitam Reserved Forest, Puchong. Data on price estimates of the various products were obtained by surveying market outlets in the vicinity as well as in the city of Kuala Lumpur. The household heads were also asked about the prices of some of the products when their prices could not be determined from the markets. The results showed that the Orang Asli communities are more dependent on the forest reserve for food and fruits compared to other purposes of the forest. The revenue generated by plant species was about seven times more than that of animal species. The greatest source of revenue came from housing construction, followed by handicraft making and fruits. However, these communities are currently less dependent on the forest compared to in the past, and the forests are providing a lower number of useable number plants and animals compared to in the past. Furthermore, the Orang Asli communities are now economically better off than before and they can depend more on the markets than the forest for their daily necessities. Other studies, such as those performed by Norini and Ahmed (2007) and Rosta et al. (2010), also found similar trend; the Orang Asli are obtaining a lower amount of products from the forest, especially after some of the forest is converted to agricultural land.

Other than the indigenous people living in the forest, local communities living near the fringes of the forest also depend on the forest. Most of the people living in these communities are poor and practice rudimentary cultivation methods to meet their daily needs. Therefore, they depend on the forest as a supplementary resource of food. This also causes encroachment into the forest areas and leads to the degradation of the forests. Consequently, in the long term, this will cause further damage and economic loss to the local communities.

To address this issue and prevent further degradation of the forest, the Sabah Forest Department (SFD) successfully developed a Joint Forest

Management (JFM) approach for managing the degraded forest. This approach, also called participatory forestry, is a collaborative effort by both the government agency and the local community to conserve the biodiversity in the remaining natural forest and the biodiversity of the degraded areas, usually using an agro-forestry system (UNDP, 2008). Most of the management, protection and utilisation of forest resources are performed collectively with the Forestry Department, which acts as a facilitator.

This approach was implemented in the Kelawat Forest Reserve, which had been degraded due to hill paddy cultivation by the local people. In 1992, a decision was made that the SFD and local communities would jointly manage the forest reserve. This effort was successful, resulting in planting of more than 20,000 mixed indigenous species of trees and silviculture treatments to allow the trees to grow well. Now, many projects are being conducted using the JFM approach to address the issues of forest degradation and the socio-economics of the local communities.

The forest is not only important as a source of food and income but also for ecological balance. An analysis by Ainuddin et al. (2007) of stakeholders of the Air Hitam Reserved Forest showed that the stakeholders ranked the ecological role of the forest as the most important issue. They viewed that this forest served an important function as a buffer against development and improved the regional climate of the area. They recognised the need to maintain and conserve the forest so that it would continue its role as an ecological balancer against housing areas that were built near the periphery of the forest.

The multi-dependence of the local community on the forest shows that the forest is not only necessary for the production of timber but has many other uses. Hence forest management should incorporate these needs. Innovative forest management is needed for this special case.

## Case Study 1

The Matang Mangrove Forest (MMF) is one of Malaysia's mangrove forests and one of its dynamic and productive coastal ecosystems (Figure 10.2). The MMF is situated between latitude 4° 152 N–5° 12 N and longitude 100° 22 E–100° 452 E (Azahar & Nik Mohd Shah, 2003). It covers an area of 40,496 ha and is divided into two main parts: 30% of the forest

**Figure 10.2:**
*Matang mangrove forest reserve is known for its good mangrove forest management system*

*Source:* Author.

is located in the mainland of Perak State, Peninsular Malaysia and 70% is located in several islands near the mainland.

The MMF is an important fish breeding area in the west coast of Peninsular Malaysia (Chong, 2005). Mangrove leaves decompose and form detritus, a source of food for micro- and macro-fauna. These animals form potential food sources for fishes and shrimps. The mangrove forest provides refugia for these fishes and shrimps as mangrove root structures, such as pneumatophores, provide protection to fishes and shrimps (Chong, 2007).

The presence of mudflats, which are rich with invertebrates, attracts many types of birds to the MMF. The MMF also attracts migratory birds migrating during the winter from Asia to Australia. There have been 61 species of birds recorded (Noramly, 2005), including 2 threatened bird species, in the MMF.

The MMF also contributes to the socio-economic well-being of the communities in the vicinity of the forest. It is estimated that RM280 million was generated by coastal and deep fishing, fish aquaculture, cockle rearing and the production of charcoal activities. Lim and Mohd Parid (2002) estimated that 92% of the population near the vicinity of the forest depended on the mangrove forest resources. Most of the population worked as fisherman, aquaculture operators, and charcoal kiln owners and operators. However, the community was experiencing a low fish harvest, insufficient space for aquaculture projects, a lack of local labour and river water pollution. The existence of illegal methods of netting also contributed to the decrease in fish and shrimp harvests.

The MMF is one of the best managed mangroves and has been managed since 1904 when the first management plan for the forest was developed. In 1930, a 10-year working plan was formulated, and this planning document is reformulated every 10 years. The MMF is gazetted as a reserved forest under the National Forestry Act. This gazette protects the forest from other competing land use and allows the forest to be managed sustainably. The MMF is managed by the Perak State Forestry Department, since, under the Malaysian Federal Constitution, forest resources are under the jurisdiction of the state government. Operationally, the management is performed by the LarutMatang Forest District under the leadership of the district forest officer.

The MMF is a source of poles and trees for the production of charcoal. In the working plan for 2000–2009, about 830 ha were allocated for harvesting yearly, and 116 ha were allocated for charcoal production. It was estimated that the production of poles and charcoal was RM30 million per year (Azahar & Nik Mohd Shah, 2003).

In preparation for the next working plan, a stakeholder analysis was conducted (Awang et al., 2010). Stakeholders of the MMF were interviewed, and several focus groups were conducted (Figure 10.3). The results of the meetings, interviews and focus meetings showed that the local community received benefits from the MMF employment opportunities arising from the forest-based industry and fishing activities. They also appreciated the good environmental quality provided by the MMF. The local community has shown high interest in the management of the MMF because their employment depends on the timber resources of the MMF and many of them work in the extraction of mangrove poles and charcoal industry. This community has the potential to be included in the monitoring of the MMF from illegal logging; they feel responsible for safeguarding the forest because they depend on its resources.

Other stakeholders of the MMF are the mangrove pole permit holders and the charcoal kiln operators. They are very dependent on the MMF for

**Figure 10.3:**
*Interviewing local community on dependency to MMF*

*Source:* Author.

their operations and work closely with the District Forest Department. The permits given by the District Forest Department allow a certain amount of pole wood or mangrove stems to be harvested and extracted from the compartment allocated. The allowable amount ensures that the forest resources are capable of regeneration and that the MMF is sustainable. However, these stakeholders desire more areas for harvesting due to the demand for their products.

Government agencies play important roles in the management of the MMF, and they formulate the policies and legislation that relates to the management of the MMF. The level of involvement of different government agencies varies. For example, the Department of Urban and Rural Planning focuses more on regional planning and how the MMF fits into this planning. Other agencies, such as the Department of Environment, deal with the monitoring of the water quality within the MMF. Furthermore, the Department of Drainage and Irrigation deals with the monitoring and maintenance of the rivers and streams within the MMF. There are other agencies which primarily monitor the resources within the MMF. The Forestry Department deals with the tree resources, while the Department of Wildlife and Parks monitors wildlife within the MMF.

Furthermore, the Department of Fisheries is more focused on the aquatic resources, with a particular emphasis on fishes and shrimps. The issues resulting from the overlapping jurisdictions of these agencies can be overcome by good coordination and cooperation between the concerned agencies. Non-governmental organisations also play an important role in the conservation of the MMF. They help in increasing the awareness of the importance of the wildlife in the MMF, especially birds. They also develop programmes to involve the local community in the conservation effort of the mangrove and the wildlife of the MMF.

Overall, most of the stakeholders desire better management of the MMF so that it is protected and its resources can be used sustainably. It is recommended that the District Forest Department organise regular meetings with the stakeholders so that the views of the stakeholders are included in the formulation of management plans and activities.

## Case Study 2

The Mangkuwagu Forest Reserve (MFR) is one of the forest reserves within FMU 17 and is managed by the SFD, Malaysia. This reserve is located approximately 210 km from Sandakan, the largest city on the east of Sabah. The size of the MFR is 8,325 ha, and it is classified as Sabah's Class II Productive Forest Reserve (UNDP, 2008). However, 76.8% of the MFR is classified under stratum 4 (four commercial trees per ha) due to intensive logging and forest fires in 1983 and 1988 (SFD, 2006). Additionally, approximately 1,846 ha of the MFR have been cleared for agricultural activities by the local community living within the MFR and those living near the MFR boundary. This land use conflict was due to land tenure issues resulting from the fact that people were living in the forest long before it was gazetted into forest reserve. These villagers feel that they have the right to use the land, and they cleared it for agricultural activities, which resulted in encroachment and forest degradation (Figure 10.4).

There are four villages within and in the vicinity of the MFR comprising 265 families with 1,610 individuals (UNDP, 2008). In 2006, when this project started, all four villages lacked basic amenities and services, such as paved roads, piped water, electricity and a sewage treatment facility. Most of the villagers had low income that depended on traditional non-timber forest products and the swidden agriculture involving

**Figure 10.4:**
*Degraded forest of Mangkuwagu Forest Reserve: The lower slope of the forest was dominated with shrubs indicating forest degradation*

*Source:* Author.

planting of subsistence crops, such as tapioca and upland rice. Some villagers obtained cash income from employment at oil palm and rubber plantations but the majority obtained low cash income and can be categorised as hardcore poor (SFD, 2008).

The lack of employment opportunities has driven the local communities to cut the trees, clear the forest and plant subsistence crops for food production. To stop further depletion of the MFR resources, discussions were initiated by the SFD, and the villagers led a rehabilitation project using the agro-forestry approach (Figure 10.5). This project is a win–win strategy for both the SFD and local communities that is aimed at creating a healthy, productive and continuous source of timber through rehabilitation planting and at increasing the socio-economic level of the local communities residing within and adjacent to MFR (SFD, 2007). In this project, rubber is planted and integrated with short-term cash crops, such as ground nuts, yams, sweet potato, ginger and maize.

The project was initiated by meetings, consultations and engagements with the SFD staff and villagers. During the consultations, the SFD staff and villagers determined the management approach that caters to the needs of the people and uses the technical knowledge of the SFD staff (Figure 10.6). This approach is people-centred and was based on

**Figure 10.5:**
*Allocated land was planted with rubber trees and interspersed hill paddy: The hill paddy will provide some income for the local community*

**Source:** Author.

a previous project in the Kelawat Forest Reserve in the 1990s. A joint management approach was conceived for the project. The main thrust of this approach is to empower the local communities with the knowledge and technical understanding that will enable them to modify their way of living (SDF, 2007).

For this project, 340 families from four villagers were selected to participate, and each family was allocated 1 ha of land for planting rubber. The selections were based on a set of criteria, such as income, residency, age, family members, land ownership and education, and the first 85 villagers were selected based on the highest score garnered. In 2008, approximately 117.23 ha out of 306.5 ha were planted with rubber, and in 2011, the planted areas increased to 254.99 ha. This represents 83% of the land allocated for this project. The reason for this shortfall was an insufficient supply of seedlings and bad road conditions. The villagers carried

**Figure 10.6:**
*Relationship and interaction between technical and villager input in project implementation in MFR*

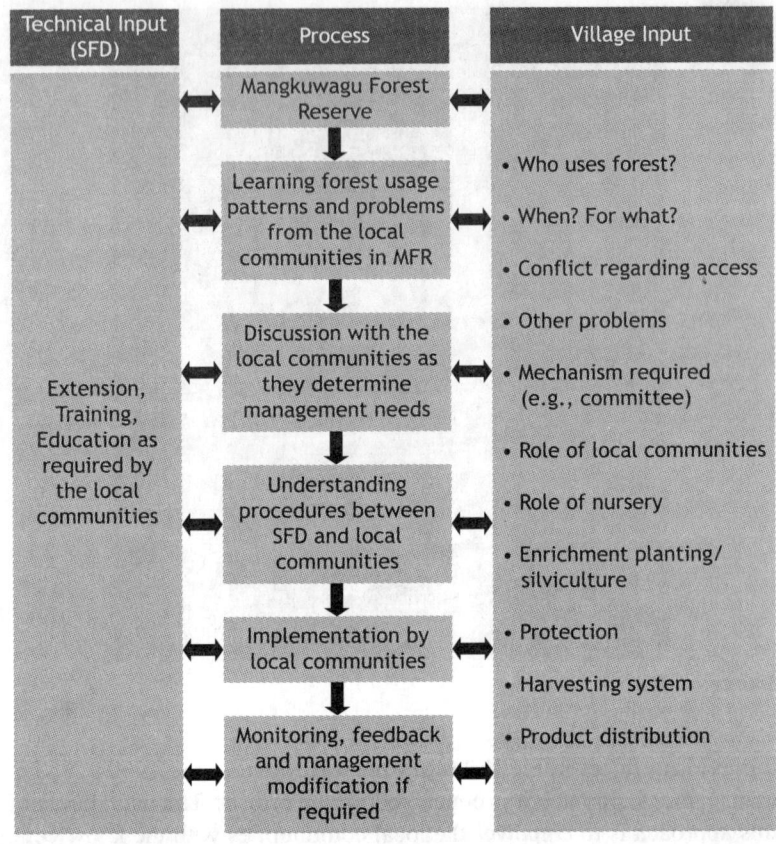

*Source:* SFD (2007).

out weeding and fertiliser input to the plants and were paid for these activities. This cash income for the participants who are poor served as an incentive for the participants to be involved in this project.

In MFR, a committee known as the Forest Management and Certification Committee was formed to ensure the participation of the local communities in the forest certification process. This committee consisted of 10 participants from each village and will represent the four villages in any social dispute issues with regards to certification requirements based on the Forest Stewardship Council and MTCC standards (Inoue, 2009;

SFD, 2007). The committee members were trained on communication and negotiating skills by the WWF to enhance their involvement in the social participatory process.

Even though this project is still ongoing, the once barren land in 2006 has become green with planted trees and in two years, the rubber trees will be ready to be harvested for latex. This will further increase the income of the participating farmers. Shifting cultivation activities were also reduced due to the focus of the farmers' activities on the rehabilitated land. The many courses and training programmes organised by the SFD helped to improve the organisational and communication skills of the villagers (SFD, 2011).

## Implications to Design Principles

The two case studies, Matang Forest Reserve and MFR, showed two different levels of involvement of local communities in forest management. When evaluated based on Inoue's (2009) first design principle, the local community near Matang Forest Reserve had less autonomy for the management of forest resources. The use and exploitation of the forest resources, such as mangrove poles, is regulated by the District Forest Department by the issuance of permits and licenses. This has been accepted by the local community because they believe that this approach can ensure the sustainability of the mangrove resources. Furthermore, they are aware that the mangrove forest is an important habitat for fish and other marine resources, which are important resources for their livelihood and are important economically for the local community. In the MFR, the local community had greater autonomy in managing the land allocated to them by the SFD. Even though they were required to plant rubber trees, they were given the autonomy to also plant short-term cash crops in the allocated area. This gave them a sense of ownership of the land and motivated them to take care of the land and the rubber trees that they planted. The members of that local community are also planning to apply for long-term ownership of the land, but they have to go through a lengthy process to be awarded a temporary occupational license.

In reference to Inoue's second design principle on clearly defined resource boundaries, both the Matang Forest Reserve and MFR boundaries are clearly demarcated, and the respective state forestry departments have jurisdiction of those forest reserves. In the case of the Matang

Forest Reserve, all the local communities are living outside the forest reserve and illegal land encroachment was rare. On the other hand, in the MFR, some local communities are living in the forest reserve and this causes conflicting land use and forest degradation. The project described above was aimed at providing a win–win solution to that problem.

There exists a graduated membership as described in Inoue's third design principle for both Matang Forest Reserve and MFR projects. In the Matang Forest Reserve project, the local communities are led by the village headman and supported by the village development and security committee, whose members are officially elected by the state government. They also act as the eyes and ears of the government. The village headman and its village development and security committee have strong authority and act as core members, while the other villagers are graduated members with weaker authority. In the Matang Forest Reserve project, policies that need to be conveyed to the villagers will be conveyed by the head villager and the village committee. The village committee has the responsibility to report any illegal encroachment and illegal felling to the authority. This is similar to the MFR project: the village headman and the village development and security committee of the local communities have strong authority and act as core members, while the other villagers are graduated members with weaker authority. However, in the MFR project, the participants are also consulted and are part of the decision-making process. In the JFM Committee, the core members and the SFD interact with each other to discuss ways to jointly manage the forest and execute the planned activities. This committee also looks at the problems and the solutions for the problems in the project.

With regard to the fourth design principle, which recognises the authority of community members to make decisions in a capacity that corresponds to their degree of commitment to forest use and management, both the Matang Forest Reserve and MFR studies revealed that the local communities recognised the state forestry authorities to be responsible for the management of the forest resources. This is because forestry is clearly stipulated by the state enactment to be the jurisdiction of the State Forest Department or, in the case of Sabah, the state ordinances. There is a clear line of authority and commitment by the state forestry departments, and this has been accepted and acknowledged by the local communities. Lately, there is a trend of local communities wanting to have more input in the management of forest resources, and the state forestry departments have responded to this by having meetings and engagements with the local communities and other stakeholders.

The formation of a joint management committee for the MFR project enhances the process of monitoring the project and also provides a feedback mechanism to the SFD. This is in line with the sixth design guideline, which states that core members monitor whether other participants abide by the conditions and also that the SFD can monitor the progress of the project in meeting its objectives. The SFD can also ensure the activities are in accordance with the scientific and technical requirements of SFM.

## Conclusions and Recommendations

Malaysia has a land area of approximately 32.83 Mha and has been blessed with abundant biodiversity and forest resources. Malaysia's successful tropical forest management originated from the late 19th century work of British colonial officers and has evolved to the present SFM. The SFM is further strengthened by the international standard certification process. The dependence of the indigenous people and local community is recognised, and adjustments are being made for them to play a greater role in the decision making in forest management. Numerous projects have demonstrated that joint management of forest resources between the major stakeholders leads to mutual benefits.

The two case studies presented two different approaches. In the Matang Forest Reserve, most of the forest management activities were initiated by the district forestry officers with little local community input. On the other hand, the second study the MFR showed that the local community is actively involved in the rehabilitation of the degraded forest by working together with the state forestry officers. When framed against the design guideline principles, the MFR project fits more closely to Inoue's design principles because it significantly involves the local community in the management of forest resources.

The results of the involvement of the local community in the MFR management project indicated the importance of the local community in rehabilitating degraded forest and the ability to design a project that can improve the local community members' livelihood through concordant agro-forestry projects. This approach can be included in forest governance of forest reserves that have many villages near their boundaries and a high dependency of the local community on the forest resources. The role of stakeholders, such as local community members, in the forest governance can be further enhanced and consolidated through changes to forest management policies.

Majority of the forest areas in Malaysia are owned by the state and governed through state-enacted legislation. These legislations are in harmony with the federal National Forestry Act (Act 313), which guides forest management and reforestation activities in accordance with the principles of SFM. In complying with the principles of the SFM, efforts are being taken to include indigenous and local communities into forest management decision making. However, this will take time since all the stakeholders need to understand each other's needs and further negotiations are needed to ensure these needs can be achieved.

In order to accelerate the involvement of the local community in forest governance, a review of the current National Forest Policy is needed. Current policy should highlight the importance of the role played by the local community in safeguarding the forest and should incorporate the local community in the forest-governing process. Examples from different parts of the world have shown that inclusion of the local community in forest governing process results in benefits for parties involved by not only increasing the livelihood of the local community but also improving the forest management and providing stability to the environment.

# References

Ainuddin, N.A, Shahwahid, M.O., & Awang Noor, A.G. (2011). Mapping issues in Malaysian wood-based industry using cognitive mapping approach. *The Malaysian Forester*, 74(1), 37–44.

Ainuddin, N.A., Hjortso, C.N.P., Norini, H., Khamuruddin, M.N., Awang Nor, A.G., & Ismariah, A. (2007). Introducing stakeholder analysis in Malaysian Forestry: The case of Ayer Hitam Forest Reserve. *Pertanika Journal of Tropical Agriculture*, 30(2), 131–139.

Agrawal, A., Chhatre, A., & Hardin, R. (2008). Changing governance of the world's forests. *Science*, 320, 1460–1462.

Awang Nor, A.G., Abdullah, M., Ahmad Ainuddin, N., Khamuruddin, M.N., Kamziah, A.K., Ismail Adnan, A.M., & Yip, H.W. (2010). Laporan akhir penyediaan pelan kerja Hutan Paya Laut Matang, Perak 2010–2019 (in Malay) Jilid 1.

Azahar, M., & dan Nik Mohd Shah, N.M. (2003). *A Working Plan for the Matang Mangrove Forest Reserve, Perak (5th Revision): The Third 10-year Period (2000–2009)*. Ipoh: Jabatan Perhutanan Negeri Perak.

Chong, T. (2005). Fifteen years of mangrove fisheries research in Matang: What have we learnt? In M.I. Shaharuddin, M. Azahar, U. Razani, A.B. Kamaruzzaman, K.L. Lim,

R. Suhaili, M.S. Jalil, & A. Latiff (eds), *Sustainable Management of Matang Mangroves: 100 Years and Beyond* (pp. 1–20). Forest Biodiversity Series 4. Malaysia: Forestry Department Peninsular Malaysia.

Chong, V.C. (2007). Mangroves-fisheries linkages—The Malaysian perspective. *Bulletin of Marine Science*, 80(3), 755–772.

Gill, S.K., Ross, W.H., & Panya, O. (2009). Moving beyond rhetoric: The need for participatory forest management with the Jakun of South-east Pahang, Malaysia. *Journal of Tropical Forest*, 21(2), 123–138.

Ibn Battuta (2012). *The Travels of Ibn Batûta: With Notes, Illlustrative of the History, Geography, Botany, Antiquities, etc. Occurring throughout the Work* (S. Lee, Ed. and Trans.). Cambridge, MA: Cambridge University Press Reissue Edition.

Inoue, M. (2009). Design guidelines for collaborative governance (kyouchi) of natural resources. In T. Murota (ed.), *Local Commons in Globalized Era* (pp. 3–25). Kyoto: Minerva-shobou (in Japanese).

Inoue, M. (2011). *Prototype Design Guidelines for 'Collaborative Governance' of Natural Resource.* Presented at 13th Biennial Conference of the International Association for the Study of the Commons, Hyderabad, India. Retrieved on 12 December 2012 from http://dlc.dlib.indiana.edu/dlc/bitstream/handle/10535/7321/146.pdf?sequence=1.

International Tropical Timber Organization (ITTO) (1992). *Criteria for the Measurement of Sustainable Tropical Forest Management.* ITTO Policy Development series N3. Yokohama, Japan.

Gill, S.K., Ross, W.H., & Panya, O. (2009). Moving beyond rhetoric: The need for participatory forest management with the jakun of South-east Pahang, Malaysia. *Journal of Tropical Forest Science*, 21(2), 123–138.

Lim, H.F. & Mohd Parid (2002). *Sustainable Development of the Matang Mangroves and Its Impacts on Local Socio-economic Livelihood* (FRIM Report No. 74). Kuala Lumpur: FRIM.

MTC (2007). *Malaysia: Sustainable Forest Management.* Kuala Lumpur: Malaysian Timber Council.

——— (2009). *Malaysia: Forestry and Environment (Facts and Figures).* Kuala Lumpur: Malaysian Timber Council.

Mittermeier, R.A., Gil, P.R., & Mittermeier, C.G. (1997). *Mega-diversity: Earth's Biologically Wealthiest Nations. Mexico, DF:* Conservation International, Cemex.

MIDA (2010). *Wood-based Industry.* Retrieved on 2 February 2010 from http:www.mida.gov.my.

MTIB (2009). *NATIP: National Timber Industry Policy of Malaysia (2009–2020).* Ministry of Plantation Industries and Commodities Malaysia. ISBN 978-983-99606-3-1, 127 pp.

NBDP (1998). *National Biodiversity Policy.* Putrajaya: Ministry of Science, Technology and Environment.

Noramly, S.E. (2005). Bird diversity at Matang mangroves. In M.I. Dalam Shaharuddin, M. Azahar, U. Razani, A.B. Kamaruzaman, K.L. Lim, R. Suhaili, M.S. Jalil, & A. dan Latiff (penyunting) (eds), *Sustainable Management of Matang Mangroves: 100 Years and Beyond* (pp. 288–295). Kuala Lumpur: Jabatan Perhutanan Semenanjung Malaysia.

Norini, H. & Ahmad Fadli, S. (2007). The importance of Air Hitam Forest Reserve (AHFR), Puchong, Selangor to the Temuan Ethnic Subgroup. Pertanika Tropical Agricultural. *Science*, 30(2), 97–107.

Rosta, H., Azizah, S., Yip, H.W., Tengku Hanidza, I., Mohd Kamil, Y., Latifah, A.M., & Hafizan, J. (2010). Impacts of forest changes on indigenous people livelihood in Pekan District, Pahang. *Environment Asia*, 3(special issue), 156–159.

Rusli, M., Awang Noor, A.G., & Abdul Rahim, O. (1997). *Indigenous People Dependence on Non-wood Forest Produce: A Case Study*. Paper presented at International Workshop on Non-wood Forest Products, 14–17 October, Universiti Putra Malaysia.

SFD (2006). *Sabah Forest Department Annual Report 2006* (p. 202). Sandakan: Sabah Forest Department. Sandakan.

——— (2007). *Sabah Forest Department Annual Report 2007* (p. 265). Sandakan: Sabah Forest Department.

——— (2008). *Sabah Forest Department Annual Report 2008* (p. 317). Sandakan: Sabah Forest Department.

——— (2009). *Sabah Forest Department Annual Report 2009* (p. 362). Sandakan: Sabah Forest Department.

——— (2010). *Sabah Forest Department Annual Report 2010* (p. 409). Sandakan: Sabah Forest Department.

——— (2011). *Sabah Forest Department Annual Report 2011* (p. 423). Sandakan: Sabah Forest Department.

Shahwahid, O. & dan NikMustapha, R.A. (1991). *Economic Valuation of Wetland Plant, Animal and Fish Species of Tasek Bera and Residents' Perception on Development and Conservation*. A sub-project of AWB/WWF project 327.

Smith, J.S. (1959). Past, present and future—A review. *Malayan Forester*, 22(4), 272–280.

Thang, H.C. (1987). Forest management systems for tropical high forest, with special reference to Peninsular Malaysia. *Forest Ecology and Management*, 21, 3–20.

UNDP (2008). *Malaysia Sustainable Community Forest Management in Sabah*. Malaysia: United Nations Development Programme (UNDP).

Wyatt-Smith, J. (1963). *Manual of Malayan Silviculture for Inland Forests* (Malayan Forest Records No. 23, 2 Vols). Malaysia: Forestry Department Malaya.

# 11

# Philippines: Multi-tiered Forest Governance System on Uneven Playing Field

*Juan M. Pulhin, Rose Jane J. Peras and Maricel A. Tapia*

## Introduction

The claimed paradigm shift in the governance of natural resources has continued to disappoint the marginalised rural poor. The earlier optimism towards having more far-reaching positive impacts associated with forest devolution policies has yet to be realised and at times produced paradoxical outcomes (Pulhin, 1996; Pulhin & Inoue, 2008).

Historically, Philippine forest policies followed a highly regulatory, centrally controlled and industry-based approach to forest management. The turn of events in the past three decades forced the central government to follow the forest devolution path being supported by international donors in many developing countries. The continuous onslaught of forest resources and the resulting environmental degradation, glaring inequity in the access to and benefits from the utilisation of the forest resources and erosion of the state's political legitimacy to manage the nation's forest resources are among the key drivers in the adoption of the devolution approach (Pulhin & Inoue, 2008)

Other contributing factors include the increasing manifestation of common property regimes being better than state institutions, limited resources available to the government to implement its institutional mandates, and changing priorities for international funding institutions in favour of community-based approaches to forest management.

In the past decade, there was high confidence in the achievement of social justice and sustainable forest management (SFM) for people-oriented programmes and projects. The government invested significantly in this type of approach in the hopes of addressing environmental degradation and upland poverty. However, forest policies and the multi-tiered governance system continue to frustrate efforts at the local level. Success in the adoption of the community-based forest management (CBFM) strategy has not been realised despite close to three decades of continuous interventions at the policy, programme and project levels. The diversity and multiplicity of problems confronting the implementation of CBFM in the country continue to frustrate local communities.

This chapter will illustrate how the multi-tier governance system in the country constraints the effective implementation of CBFM programmes/projects at the grass-roots level. Two popular CBFM sites in the northern part of Luzon and in Mindanao were used as case study sites to shed light on the issue. The chapter is presented in five parts covering the general context of CBFM, the multi-tiered governance system, the case analyses, the implications of the design principles and the conclusion.

## Community-based Forest Management: Policy and Institutional Context

The country's forest and natural resources have been in continuous decline. Forest policies have evolved to address the problem of deforestation but have not moved an inch further to realise these objectives. The country's governance system is fashioned using the devolution approach promoted over the past 20 years (1990s). Major policy reforms have materialised in the past decade highlighting the most recent forms of state-initiated devolution in the forestry sector. At least four types of forest devolution in the Philippines relate to the present discussion. The first type is reflected in the different people-oriented forest management programmes and projects embodied under the CBFM policy, a mechanism by which certain management rights and responsibilities were transferred from the Department of Environment and Natural Resources (DENR) to local communities. A tenure instrument was granted through the CBFM Agreement (CBFMA), allowing local communities to access and benefit from forest resources for 25 years in exchange for forest rehabilitation, protection and conservation. The second type follows the devolution codified in Republic Act 7160, or 'The Local Government Code' of 1992, whereby certain environmental functions of the DENR were devolved to local government units (LGUs), especially the Integrated Social Forestry Projects. The third type provided space for local communities to participate in the management of protected areas through the enactment of the Republic Act 7586, or the NIPAS Law (National Integrated Protected Areas System). The last type is embodied in the 1997 Indigenous People's Rights Act (IPRA), which provides for the recognition, protection and promotion of the rights of indigenous cultural communities/indigenous people (ICCs/IP) to their ancestral lands through the issuance of a Certificate of Ancestral Land Title (CADT). IP are in turn entrusted with the responsibility of maintaining, developing, protecting and conserving these areas with support and assistance from government agencies.

Figure 11.1 shows the general trend in forest devolution policies in the Philippines as far back as pre-colonial period (Sajise, 1998). The pre-colonial time exhibited highly localised access and control of forest

**Figure 11.1:**
*General trend in forest devolution policies in the Philippines*

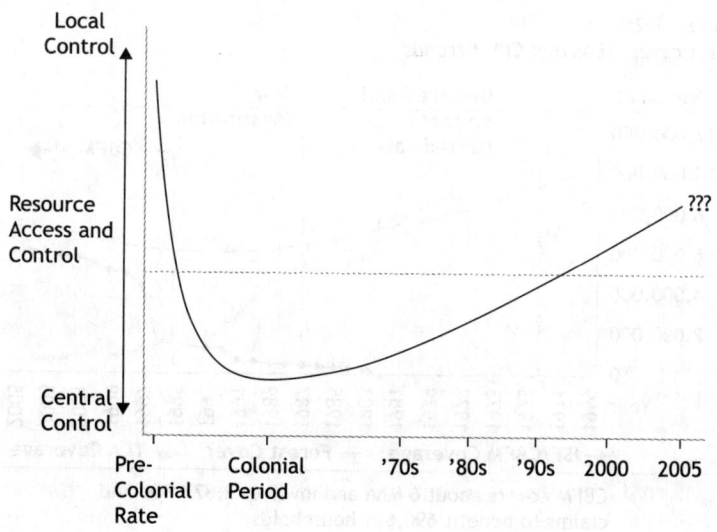

*Source:* Sajise (1998).

resources in the hands of 'Datu' and other local leaders. During the colonisation period, access and control of forest resources came under the jurisdiction of the state, characterised by the consolidated power of the state through the enactment of laws (Magno, 2003). The pattern of centrally controlled access to and management of the country's forest resources continued until recently. Hence, forest management in the country is strongly centrally determined, top-down and non-participatory.

Over the last five decades, the participation of local communities in forest management has gradually increased in terms of the forest area coverage under ISF/CBFM from 1982 to 1995. Participation increased especially abruptly after the adoption of the CBFM strategy in 1995. By contrast, the Timber License Agreement (TLA) coverage area has decreased from more than 10 million hectares (Mha) in the 1970s to less than 200,000 ha in 2005. In the same manner, the country's forest cover has gradually declined over the years (Figure 11.2). The challenge is now with the devolution approach to forest management, recognising the important contributions of local communities to forest development as forest managers.

To provide a better view of the political and institutional context of CBFM implementation in the Philippines and the evolution of forest policies, the historical development is presented below, showcasing the

**Figure 11.2:**
*Forest cover, TLAs and CBFM trends*

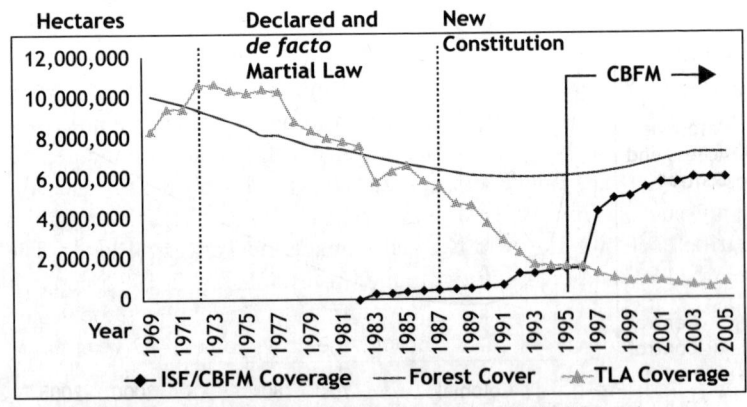

CBFM covers about 6 Mha and involves 2,877 POs and claims to benefit 690,691 households

*Source:* Pulhin et al. (2008).

different periods and highlighting their important features and characteristics in terms of policy reforms in community forestry (adopted from Pulhin, Inoue, & Enters, 2007).

## Pioneering Period (1971-1985)

This period was marked by the glaring inequitability of access to natural resources and the benefits derived from its utilisation, skewed towards the more favoured and influential holders of TLA. The marginalised upland poor was the centre of criticism and considered the main culprit for deforestation because they practised shifting cultivation. Forest policy formulation favoured short-term economic benefits for the government and forest industry. The logging policy led to the rapid exploitation of timber from virgin forests, which led to conversion into unsustainable upland agriculture. The alarming rate of forest destruction coupled with exacerbated upland poverty surpassed initiatives to increase the forest recovery rate by both natural and artificial means.

As a result, the government's promotion of the punitive legal approach as a strategy to halt deforestation by fining, imprisoning and evicting shifting cultivators from forest areas did not materialise due to the non-recognition of the socio-economic dimension of the problem. This realisation inspired the government to adopt people-oriented policies and programmes primarily addressing people's socio-economic needs and concerns.

Three 'people-oriented forestry' programmes were implemented between 1973 and 1979, namely the Family Approach to Reforestation (FAR) Programme, the Forest Occupancy Management (FOM) Programme, and the Communal Tree Farming (CTF) Programme. The Kalahan Educational Foundation, Inc., an organisation of indigenous people in Sta. Fe, Nueva Vizcaya, through the Ikalahan Tribe, led the way in securing from the government a 25-year agreement that provides the tribe exclusive rights to use and manage their ancestral lands. The Integrated Social Forestry Programme was established in 1982, consolidating the FAR, FOM and CTF programmes and recognising the vested interests of the forest occupants through the provision of a 25-year tenure security.

The policies and programmes developed during the pioneering period opened limited space to accommodate forest occupancy and the involvement of upland communities in forest rehabilitation activities. Local communities' involvement served as the source of cheap paid labour. Rebugio and Chiong-Javier (1995) considered this period's departure from a purely punitive nature of forest governance to a more accommodating form for forest occupants, emphasising their important role in forest management as 'pioneering'. The period also exhibited an experimentation stage in which the individual farmer, the family and the community were the focus of various alternative approaches. At this time, the integration of all socially oriented approaches to forest management was very vital.

## Experimentation and Infusion of Massive External Support (1986-1994)

President Corazon Aquino's administration introduced radical reforms in the forestry sector. DENR's re-organisation in 1992 removed corrupt officials, changed perspectives on forestry and significantly reduced the number of timber licenses. These changes were necessary to make DENR attractive to the donor community. Likewise, the presence of a civil society that calls for resource access and democratisation of resources, and people's participation in natural resource management offered great potential for policy and institutional reforms, which led to a pouring in of external funding for forestry projects in the country.

Five big forestry-related loans were implemented between 1988 and 1992, amounting to USD 73 million (Korten, 1994). There was also an enormous undetermined amount of external assistance in the form of grants and technical support from the Ford Foundation, the United States Agency for International Development (USAID), the German and Swedish Governments and other agencies.

External assistance was directed at 'people-oriented' forestry programme as the government advanced social justice and equity in the natural resources sector and the need to maintain DENR's political legitimacy in forest governance (Pulhin, 2004). Between 1988 and 1993, a total of nine forestry-oriented programmes were initiated through external support emphasising the core concerns of sustainable development, that is the advancement of social equity, poverty alleviation and environmental sustainability (Pulhin, 1996). The entry of new players such as non-government organisations (NGOs), people's organisations (POs), LGUs, academies and research agencies into the forestry sector was inspired by the application of different types of land tenure instruments, experimentation with different project components and strategies, and the various institutional and collaborative arrangements. The 1990 Philippine Master Plan for Forestry Development placed 1.5 Mha of residual forests and 5.9 Mha of 'open-access' area under community forest management for 10 years (DENR, 1990), whereas approximately 24% (682,000 ha) of the total forests were allocated towards commercial timber harvesting.

## Institutionalisation and Expansion (1995 to Present)

President Fidel Ramos' issuance of Executive Order 263[1] institutionalised the adoption of CBFM as the government's strategy to attain SFM in the country. It provided for long-term tenure rights of local communities to forest land with the condition of employing environmentally friendly, ecologically sustainable and labour-intensive harvesting methods. The ICCs/IPs were also encouraged to participate in the recognition of their ancestral domains' rights and land rights and claims. In addition, former DENR Secretary Victor Ramos issued DENR Administrative Order No. 96-29 in 1996, which provides the rules and regulations in the implementation of Executive Order 263 and Memorandum Circular No. 97-13 in 1997, which adopted the DENR Strategic Action Plan for CBFM. The plan envisioned placing 'open access' areas, vacated by TLAs, under proper management, with 9 Mha of forest lands under community management by 2008.

Republic Act 8371, or the IPRA Law, recognised the vested rights of the ICCs/IP over their ancestral lands and thus entitled them to the issuance of their CADT. However, ICCs/IP have the final decision if they choose to retain the CBFMA and Certificate of Ancestral Domain Claim (CADC) issued to them prior to the passage of IPRA Law over CADT.

There was a massive increase in CBFM areas after the promotion of DENR Strategic Action Plan in 1997. The CBFM coverage area increased from less than 1 Mha to around 5.97 Mha. Around 4.904 Mha are under various forms of land tenure arrangements [i.e., CADC (2.5 Mha), CBFMA (1.57 million ha), Certificate of Stewardship Contract (CSC) (0.631 Mha) and CFSA—Community Forest Management

---

[1] Known as the 'Adopting Community-Based Forest Management as the National Strategy to Ensure the Sustainable Development of the Country's Forestlands Resources and Providing Mechanisms for Its Implementation'.

Agreement (0.196 Mha) (FMB, 2006)]. Holders of these land tenure arrangements have the right to occupy, cultivate and develop their areas and utilise the existing forest resources subject to government rules and regulations.

Pulhin (1998) viewed CBFM as a radical and progressive structural policy reform in the forestry sector, replacing the century-old TLA approach of forest utilisation where benefits were skewed in favour of the elite minority. CBFM is an attempt to democratise access to and benefits from forest management by transferring certain management rights and responsibilities to forest communities. Timber concessions decreased from 10 Mha in 1973 under the control of 422 TLA holders to 253,000 ha with only four license holders remaining. As of 2009, the CBFM coverage area grew to around 1,633,891 ha of forest lands awarded to 1,790 CBFMAs, benefitting over 320,000 households (FMB, 2009). Overall, the CBFM programme covers some 6 Mha of forest lands, including tenured areas under CBFM and the different 'people-oriented' forestry projects, including those awarded to the IP. In addition, 16 protected areas have been awarded with Protected Area Community-Based Resource Management Agreements, covering 21,905.79 ha and involving 3,887 families and 10,897 individuals.

The presence of foreign-assisted funded projects and donor funds encouraged CBFM expansion. Its success was short-lived because it infused the belief in many people that a CBFM programme was something like a project that will end after a few years. This mentality had negative implications; project completion and the pull out of NGOs led to the discontinuation of many activities and initiatives in the CBFM areas.

President Gloria Arroyo's issuance of Executive Order No. 318, known as 'Promoting Sustainable Forest Management in the Philippines', in 2004 reiterated the government's confidence in CBFM towards SFM. Furthermore, DENR Secretary Elisea Gozun issued DENR Administrative Order No. 29 in the same year, which revised the 1996 implementing rules and regulations for the adoption of the CBFM strategy, providing more flexibility to participating communities by reducing some bureaucratic requirements.

The blooming development of CBFM strategy and programmes was tremendously affected by the decline in the number of foreign-assisted projects, particularly affecting the participation of NGOs. By contrast, only a limited number of LGUs have just started playing an active role in the implementation of CBFM since the enactment of the Local Government Code and the strengthening and institutionalisation of the

DENR–DILG–LGU partnership for devolved and other forest management functions. This observation is attributed to the LGU marginalisation during the early years of CBFM, which led to their inability to provide support to community organisations, especially when funds from international donors dried up. LGUs' limited capabilities and ineffectiveness in providing support for extensions, capacity building and social infrastructure hinder the promotion of successful CBFM implementation on the ground.

In 2009, the passage of the Climate Change Act gave way for the mainstreaming of climate change adaptations, specifically calling for the harmonisation of development programmes and projects towards climate change adaptation and mitigation. The LGUs act as the frontline agencies in the formulation, planning and implementation of climate change action plans in their respective areas, consistent with the provisions of the Local Government Code, the Climate Change Framework and the National Climate Change Action Plan. The local communities, through their respective *barangay*s, should be directly involved in prioritising climate change issues and in the identification and implementation of best practices and other solutions to be able to adapt to and mitigate the future impacts of climate change.

In 2011, President Benigno Aquino III signed Executive Order 23 'banning logging in natural and residual forests' and strictly prohibiting harvesting activities in these areas. This act also gave way for the creation of the Anti-Illegal Logging Task Force. President Aquino's administration hopes to plant 1.5 billion trees on 1.5 Mha of degraded lands within a span of six years. This aspiration was institutionalised through the issuance of Executive Order 26, or the 'National Greening Programme' (NGP), in the same year. The programme's unique feature is the call for all government agencies to unite and harmonise their environment-related initiatives into the NGP. The programme focuses its initiatives in all CBFM areas in the country towards sustainable development and re-greening all barren and open access lands.

This historical overview of the policy and institutional context of CBM implementation highlights the recognition and acceptance of local people as forest managers and that policy was shaped by the convergence of actors with diverse interests at various levels of governance system (Rebugio et al., 2010).

Challenges remain for the implementation of CBFM in the Philippines despite forest tenure reform efforts in the past decades. Success remains patchy and miniscule compared to the magnitude of problems in the forest lands. After three decades of initiatives, every step forward can easily be followed by one or more steps in the opposite direction, further threatening the achievements and more meaningful success of community forestry in the country.

At present, CBFM's successful implementation is hindered still by an uneven playing field on the ground, where various factors constrain the successful achievement of CBFM goals and objectives. Even with the organised, viable POs codifying the 'design principles' advocated by Ostrom (1990, 1998) and Inoue (2011), the success of CBFM on the ground remains unclear due to the multi-level or multi-tiered governance system.

## CBFM's Multi-tiered Governance System

The hierarchical structure of a bureaucratic organisation still dominates the country—from the national, to the regional, provincial and municipality/community levels. The CBFM areas follow the multi-level or multi-tiered governance system. The International Tropical Timber Organisation (ITTO)–CBFM Project site in Vista Hills, Bayombong, Nueva Vizcaya and the Ngan Panansalan Pagsabangan Forest Resources Development Cooperative (NPPFRDC) are used as case studies to shed light on the use of the design principles suggested by Ostrom (1990, 1998) and Inoue (2011).

Table 11.1 emphasises the multi-tiered governance system in CBFM implementation. The different key players in CBFM implementation

**Table 11.1:**
*Key players and their roles in the community-based forest management process*

| Key players | Major role in the CBFM process |
|---|---|
| **National** | |
| The Philippine Congress (Senate and House of Representatives) | Passage of three important pieces of legislations that provide the legal foundation for the adoption of community-based forest management in the country: (1) the 1992 National Integrated Protected Area System Act of 1992 (Republic Act No. 7585 and (2) Indigenous People's Right Act of 1997 (Republic Act No. 8371). Despite the presence of a strong policy framework, Congress still needs to enact a single comprehensive piece of legislation that embodies community rights, tenure and participation in forest governance |

*(Table 11.1 Continued)*

*(Table 11.1 Continued)*

| Key players | Major role in the CBFM process |
| --- | --- |
| The President of the Philippines | Of the four presidents who have governed the country since the EDSA I revolution, President Fidel V. Ramos' Administration appeared the most supportive of community-based initiatives. He issued the landmark policy, Executive Order No. 263, adopting CBFM as the national strategy to achieve sustainable forestry and social justice. EO 23 remains the basis for the current administration in formulating forestry rules, regulations and programmes concerning the devolution of forest management to local communities. By contrast, the issuance of Executive Order 318 by President Gloria Macapagal-Arroyo in 2004 promoted sustainable forest management in the Philippines and advanced community-based forest conservation and development as one of the six principles to attain sustainability of forest resources. On February 1, 2011, President Benigno S. Aquino III issued EO 23 declaring a moratorium on the cutting and harvesting of timber in the natural and residual forests and creating the anti-illegal logging task force. The order mandated the DA–DAR–DENR Convergence Initiative to develop a National Greening Programme. At the end of the same month, President Aquino issued EO 26, implementing the National Greening Programme of the government. It calls for the planting of some 1.5 billion trees covering approximately 1.5 Mha over a period of six years (2011–2016). CBFM areas are the target of its implementation, where planting will be conducted with the arrangement of CBFM people's organisations maintaining and nurturing the trees planted under the programme |
| Funding institutions | Includes multilateral and bilateral funding institutions that act as global drivers of forest policy in the Philippines (Malayang, 2001). Their instrument of influence includes the provision of funds and budgetary and technical support. The Ford Foundation Inc., USAID, Asian Development Bank and the World Bank appear to have greater influence in shaping the country's policy direction towards local forest management and control. The 15 years of experience from the Upland Development Programme funded by the Ford Foundation Inc. contributed significantly to the refinements of the earlier policy on social forestry, a major forerunner of the present CBFM programme. By contrast, the Natural Resources Management Programme implemented through a financial grant from USAID was instrumental for the crafting and approval of EO 263 as well as its implementing rules and regulations. Similarly, experiences from forestry projects funded by the WB and ADB contributed to the development of policies that provide upland communities land tenure security and access to forest resources as well as the participation of civil society in forest management. More recently, USAID, through the Philippine Environmental Governance (EcoGov) Project, was very instrumental in strengthening and institutionalising the DENR–DILG–LGU partnership through policy support and in the ground implementation of forest devolution initiatives |

*(Table 11.1 Continued)*

*(Table 11.1 Continued)*

| Key players | Major role in the CBFM process |
| --- | --- |
| Department of Environment and Natural Resources | Promulgates appropriate rules and regulations that translate the generalities of law into concrete terms. The DENR Secretary is responsible for the issuance of the various Administrative Orders and Memorandum Circulars that guide the implementation of forest laws or decrees issued by the Philippine President. At the implementation level, the outcomes of community-based forest management are largely influenced by the dedication and competence of local support, their ability to mobilise and resources from the DENR field offices and staff at the regional, provincial and municipal levels |
| Academic and other research institutions | Their main contributions lie in the promotion of science-based forest policy formulation; the provision of technical assistance and support to CBFM and related projects; project monitoring and evaluation; serving as critique of government forestry policies, programmes and projects; and the production of a new breed of 'people-oriented foresters' responsive to the needs of local communities |
| NGOs | NGOs are involved in advocacy work and providing technical support |
| **Regional** | |
| a. *DENR— Regional Office* | The DENR-Regional Office facilitates the processing of papers related to CBFM implementation. The Regional Executive Director (RED), assisted by the Regional Technical Director (RTD) for Forestry and the RTD for Environmental Management and Protected Areas System, is responsible for the effective implementation of CBFM programme in the region. The CBFM Division was created under the Forest Management Services and is directly supervised by the RTD for Forestry. This unit acts as the regional repository of all data and information on CBFM programme. The RED shall submit periodic reports to the Secretary, through the Undersecretary for Field Operations, on the implementation of programmes, including monitoring and evaluation |
| b. *DENR— Environmental Management Bureau— Regional Office* | The Regional Director (RD) efficiently and effectively implements policies, programmes, and projects for environmental management and pollution control. The RD approves the Environmental Compliance Certificate required for the issuance of the Resource Use Permit (RUP) of the CBFM POs. The same is coursed through the RTD for Forestry of the DENR Regional Offices for the issuance of the RUP before any harvesting activity is to be done |
| **Provincial** | |
| a. *Provincial Environment and Natural Resources Office—DENR* | The PENRO is responsible for the effective implementation of CBFM programme in the province, including the submission of periodic reports and the maintenance of a database for all CBFM projects in the province |

*(Table 11.1 Continued)*

*(Table 11.1 Continued)*

| Key players | Major role in the CBFM process |
| --- | --- |
| b. *Provincial Environmental Management Office* | The PEMO facilitates the processing of papers required in the approval of Environmental Compliance Certificate and directly recommends approval to the RD of the Environment and Management Bureau |

**Local**

| | |
| --- | --- |
| a. *Community Environment and Natural Resources Office (CENRO)DENR* | The CENRO is directly responsible for implementing the CBFM pro- gramme within its jurisdiction, in coordination with concerned LGUs, other government agencies and non-government organisations/private entities. The CENRO submits periodic reports of CBFM programme implementation to the PENRO for evaluation. Under the CENRO are five technical divisions: Land Management Services, Forest Manage- ment Services, Environmental Management Services, the Protected Areas Services and the Administrative and Support Services. Once the POs completed the necessary paper requirements for the Community Resource Management Framework, Annual Work Plan and RUP, processing of papers starts from this office. The Project Management Officer stationed in CENRO works closely with the CBFM POs, especially in the preparation of the requirements for CBFM |
| b. *Local government units* | The Local Government Code empowers LGUs to enforce forestry laws and engage in the implementation of CBFM and related projects in partnership with the DENR and the local communities. The Department of Interior and Local Government issued three circulars in the period of 1995–1996, enjoining all LGUs to help strengthen the CBFM programme of the government. Some LGUs in Luzon and Mindanao have passed provincial/municipal resolutions appropriat- ing funds to finance CBFM projects in their localities. Some of the successful initiatives on CBFM that have been backed up by LGU legislation include those established by the provincial governments of Nueva Vizcaya in Northern Luzon and Bukidnon in Mindanao |
| c. *The civil society* | Civil society constitutes the local communities themselves, the NGOs and POs that operate at the national and local levels, international NGOs, and the media. Their influence in CBFM ranges from the provision of funds, time and human resources; policy advocacy; the provision of legal assistance, especially to indigenous people; implementation, monitoring and evaluation of CBFM and related DENR projects; and community-level action and demands. The Local Government Code allows for the representation of civil society in the governmental and multi-sectoral policy-making bodies, such as the municipal, provincial and regional development councils and the Protected Area Management Board in the case of the NIPAS areas. The advocacy work of civil society in forestry was instrumental in the issuance of EO 263 and implementing its rules and regulations, the NIPAS Act of 1992, and the IPRA of 1997. More recently, national |

*(Table 11.1 Continued)*

*(Table 11.1 Continued)*

| Key players | Major role in the CBFM process |
|---|---|
| | NGOs, POs and academia entered into partnership with DENR to craft the new CBFM Strategic Action Plan to guide the nationwide implementation of the CBFM programme in the next 10 years or so |
| d. *Private sector* | During the initial conception of community forestry in the early 1980s, there was considerable resistance from the wood industry to allow local communities to utilise timber on a commercial scale. However, having been affirmed by the government of its support to CBFM, including that of other sectors, members of the private sector have increasingly accommodated the CBFM approach as the country's strategy for sustainable forest management. For instance, the private and other sectors continue to lobby for the passage of a proposed bill on sustainable forestry that singles out CBFM as the 'principal strategy' to achieve sustainable forest management |

**Source:** Adapted from Pulhin (2002).

are identified and categorised from the national, regional, provincial and local levels with varying degrees of influence as they perform their major roles and responsibilities as mandated by law.

## The Case Studies

### Case 1: Federation of Vista Hills, Kalongkong, Kakilingan Upland Farmers Inc. (FVHKKUFI)

The CBFM project was launched in 1995 in Barangay Buenavista and Bayombong, Nueva Vizcaya and is managed by the Federation of Vista Hills, Kalongkong, Kakilingan Upland Farmers Inc. (Figure 11.3). It covers an area of 3,000 ha with a total population of 2,764 and 595 households in 2008. The project site earned notable distinction by being acclaimed as a 'Model Sustainable Development Project'—Upland Category in Region 2 given by the Regional Development Council in 2003 and 2004. It was also promoted as a 'Model Reforestation Site' under the Clean Development Mechanism. ITTO sponsored the CBFM project in the area together with the DENR. The project's Phase I (1995–1997) concentrated on the establishment of 177 ha of tree plantation. The success the first phase of implementation earned led ITTO to sustain their effort

**Figure 11.3:**
*Location site of ITTO-CBFM*

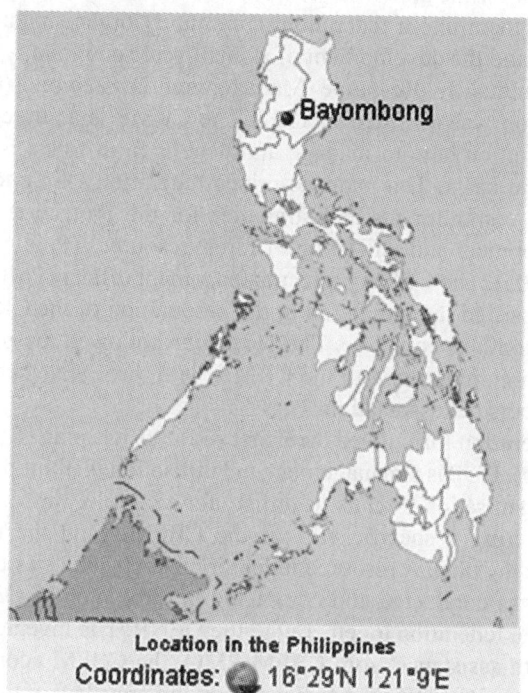

Location in the Philippines
Coordinates: 16°29'N 121°9'E

(Disclaimer: This image has been redrawn by the authors and is not to scale. It does not represent any authentic national or international boundaries and is used for illustrative purposes only.)

through Phase II (1998–2001) while expanding the tree plantation. The federation, composed of the three associations from Barangay Vista Hills, Kalongkong and Kikilingan, is responsible for the overall management of the CBFM.

Project interventions implemented in the area include community organising; capacity building and training; information, education and communication; land tenure and resources access security; enterprise development; tree plantation establishment; agro-forestry; and protection and conservation of dipterocarp forests. Among the outputs were the establishment of 100 ha of new plantations; the protection and management of 100 ha of regenerating natural forests; the protection and conservation of 1,500 ha of mature and secondary forests; the management of 1,300 ha of grasslands and brushlands; the security of the tenure and access rights for the 3,000-ha project area; the strengthening of the POs and uniting them into a federation; community organising and training conducted; and the development of a small-scale community enterprise.

The Community Resource Management Framework (CRMF) of the federation was affirmed in 1998. The CRMF is a strategic plan of the federation on how to manage and benefit from forest resources on a sustainable basis. This plan describes federation's long-term vision, aspirations, commitments and strategies for the protection, rehabilitation, development and utilisation of forest resources (Pulhin, Amaro, & Bacalla, 2005). The CBFM project management officer (PMO), based in CENRO, assisted the federation in the preparation of the CRMF, which took one month to complete. The federation followed the participatory approach in coming up with the CRMF, with three teams composed of three members per association.

The federation formulated their first resource use plan (RUP) after 12 years (2010). RUP is a management and utilisation plan for the resources that the organisation intends to utilise, for example, timber, rattan and resins, covering a specific area of the CBFMA and the time period covered by the plan. A resource inventory, which serves as the basis for the RUP, was conducted and completed in three months (May–August 2010) by the federation together with the DENR. The inventory was performed with assistance from CBFM PMO, the CBFM coordinator, the scaler at the CENR Office, a forest ranger, the president of the federation and the presidents of the three associations.

All planted *Gmelina* trees with diameter-at-breast height of 30 cm and above in both the CSC and communal area in the three *sitios* (small villages) where the three associations came from were included in the plan to harvest. The federation approved a board resolution soliciting the DENR for immediate approval of their RUP. Specifically stated in the resolution was the harvesting of 7,007 matured *Gmelina* trees with an approximate volume of 3,597.66 m³ based on the activities set forth in the Five-Year Work Plan and Annual Work Plan (AWP) for 2010–2011.

The total volume was planned to be harvested in seven years with a target annual allowable cut (AAC) of 500 m³ per year. The estimated cost and income based on 500 m³ of RUP issued was P1.696 million for 212,000 bd ft, with an anticipated net income of P318,000 (Table 11.2).

The standard mode of sharing the benefits derived from the forest activity is 3:1; that is, 75% goes to the organisation, and 25% goes directly to the government through the DENR. The government's share

**Table 11.2:**
*Estimated cost and income based on the 500 m³ of harvest*

| Activity | Cost/board foot (PhP) | Total cost (PhP) | No. of people |
|---|---|---|---|
| Harvesting | 00.50 | 106,000 | 9 people |
| Transporting using draft animals | 3.00 | 636,000 | 15 people |
| Sizing | 3.00 | 636,000 | 3 people |
| Grand total cost | | 1, 378,000 | |

Total volume = 500 m³ × 424 board feet/1 m³ = 212,000 bd ft.
Income PhP: 212,000 bd ft × P8.00 = 1,696, 000–1,378,000.
Anticipated net income: PhP 318,000.
**Source:** Authors.

**Table 11.3:**
*Mode of sharing the income derived from harvesting*

| Sharers | % share (remaining 75%) | Net income (PhP) | Remarks |
|---|---|---|---|
| Federation share | 25% | 59,625 | *The net income* |
| Association share | 25% | 59,625 | *amounting PhP* |
| Owner | 45% | 107,325 | *107,325 divided by* |
| Barangay share | 5% | 11,925 | *12 members = PhP* |
| Total | 100% | 238,500 | *8,943.75* |

Mode of sharing: Government—25% of PhP 318,000 = PhP 79,500. The remaining 75%
will be distributed to following PhP 238,500.
**Source:** Authors.

from the anticipated net income is around P79,500; the remaining 75%
was divided among different sharers as presented in Table 11.3.

The RUP served as the permit of the federation to utilise the timber
resources. However, the federation's application for RUP took a total of
eight months before its approval. The processes undertaken by the federation for RUP application is illustrated in Figure 11.4.

As per DENR Administrative Order 29 Series of 2000, the duration
for the issuance of RUP should be within 60 calendar days from the
issuance of the AWP. However, it took eight months for the federations
of Vista Hills, Kalongkong and Kakilingan to have their RUP approved.
The field offices of the DENR, that is, the regional, provincial and community offices, have already done their share to fast track the process of
RUP application and approval. The problem lies with the Environment
and Management Bureau (EMB), a line bureau of the DENR responsible
for the issuance of the Environmental Compliance Certificate (ECC), a

**Figure 11.4:**
*The process undertaken by the Federation in the approval of their resource use permit*

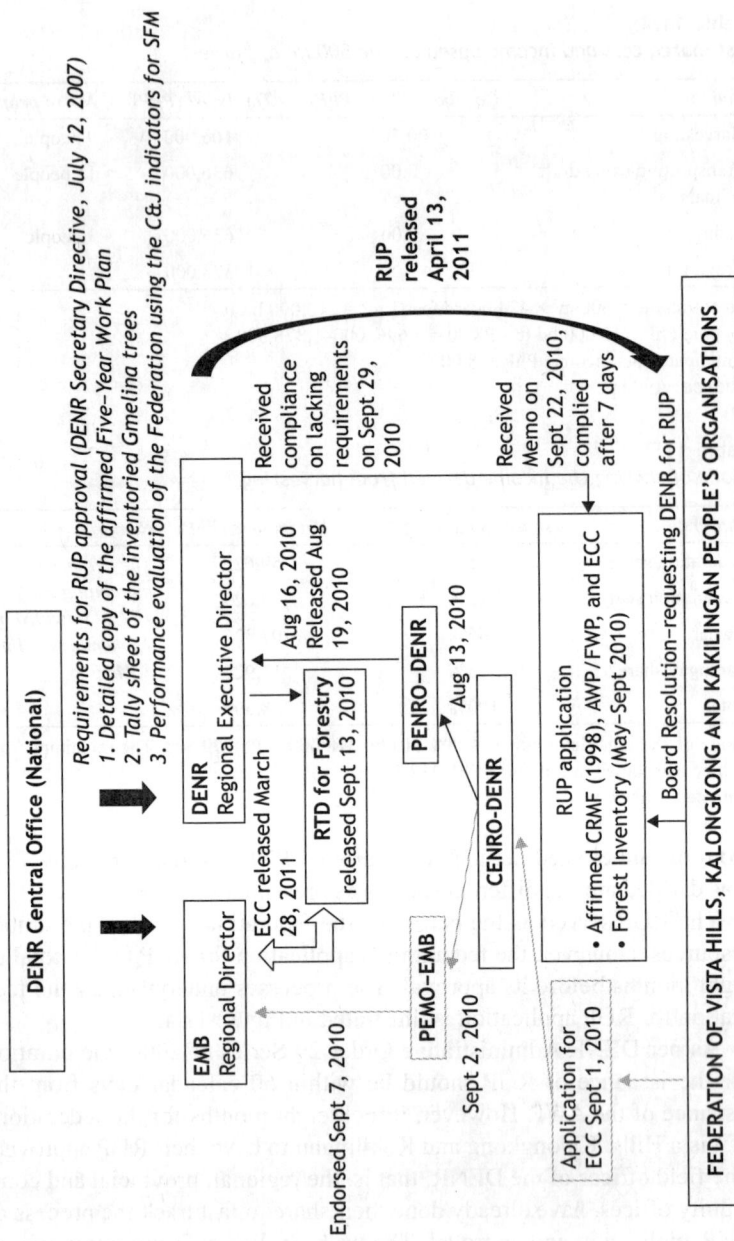

**Source:** Authors.

requirement stipulated by law under the Environmental Impact System. Based on the sequence of processing the papers, the application for approval seemingly slept for six months on the table of the Regional Director of the EMB for reasons unknown. The long-awaited approval of the RUP was made possible using 'political patronage'. Apparently, the federation had connections with a high-ranking official from the DENR, and, with the help of this influential person, the RUP was finally approved in March 2011 and received by the Federation in April 2011. Hence, only thereafter did the timber harvesting commence.

The total volume of 3,597.66 m³ allowed to be cut by the federation is planned to be harvested in seven years, with a target AAC of 500 m³. After four months of harvesting operation, the federation has only processed around 20 m³ of round logs, which is considered very low vis-à-vis their target volume of 500 m³ in a year. The officers of the federation attributed this problem to slow sizing.[2] The circular saw does not suffice for the needed volume of sized logs in a day. Accordingly, the saw also incurs a lot of wastage. As an alternative option, the federation wishes to install a band saw in the area, but they were constrained legally because installation of such equipment is prohibited by law in the CBFM areas. The federation is far behind their target volume because a week's felled cut took them at least five months to resize.

One of the sources of sustainable livelihood for the local people in a CBFM area is timber harvesting operations. The living conditions of the people are envisioned to be uplifted by employing labour-intensive harvesting methods. This aspiration is true according to the assumptions made by Dugan and Pulhin (n.d.), where huge estimates of assets are realised with two-person teams. However, the experience by the federation with their harvesting did not coincide with the assumptions. They lagged behind their supposed target of 500 m³.

The estimates of huge assets[3] made by Dugan and Pulhin (n.d.) in the CBFM areas would have been true if the assumptions made earlier were still true in the CBFM areas today. Considering the present labour force of the federation in their harvesting operation, even with three teams (two persons/team), the final output would only be 270[4] m³ in

---

[2] Refers to cutting the logs into desired sizes.

[3] The 'two-person teams using manual flitching saws can produce an average 0.25 m³ day (Bagong Pagasa Foundation 2006), or a potential daily income of US$ 7.50 per person day (0.25 m² × US$ 60 ÷ 2 persons = US$ 7.50 per person day)'.

[4] Three teams × 0.25 m³/team × 30 days/month = 22.5 m³ × 12 months/yearr = 270–500 m³ = 230 m³ (another 10 months are needed to finish the 500 cum AAC/year, or 22 months total).

a year. To meet the desired AAC, there should be a total of 11 persons[5] doing the sizing alone. The federation lacks a skilled labour force who can do log sizing. Most of the members of the federation are unable to help in the harvesting operation because they also have regular employment outside the CFBM areas/premises. The harvesting operation also requires skilled workers due to the intricacy of the work, which cannot be learned overnight.

## CASE 2: Ngan Panansalan Pagsabangan Forest Resources Development Cooperative

The cooperative named the NPPFRDC is a CBFM site in Mindanao spanning an area of 14,800 ha (Figure 11.1). It was a former logging concession of the Valderama Logging Company (VALMA). The 11th CBFMA was awarded to the cooperative in 1996 with support from USAID. The CBFMA provided tenure security to 324 members (including both migrants and indigenous people). The cooperative was also among the first community forestry programmes in Asia that was granted a SmartWood certification. The PO members have significant experience on how to operate a community-based timber enterprise, and the CBFM-PO is also regarded as advanced in timber harvesting operations in the Philippines (Pulhin et al., 2012).

A major benefit of the CBFM among the organisation members is the employment it generated through the logging operations. This employment resulted in significantly improved living conditions for the members. However, the positive impacts were not sustained because the logging operations were adversely affected and sometimes halted by a series of national RUP suspensions and delays in the issuance of RUPs. Nevertheless, despite the suspensions and irregularities in the RUP issuance, forest charges remitted to government coffers from timber harvesting amounted to PhP 8 million (1997–2007). In addition, the forest also benefited because profits from timber harvesting were used to support forest development and protection and livelihood activities and maintain forest guards and checkpoints, even during RUP suspension. Forest cover increased by 1,672 ha over 12 years (an average of approximately 139 ha/year).

---

[5] Desired: 5.5 teams × 0.25 m³ × 30 days = 41.666 m³ × 12 months = 500 m³ (this means that 11 people will be doing the sizing stage to reach the target AAC).

**Figure 11.5:**
*Location site of NPPFRDC*

Ngan, Panansalan,
Pagsabangan Forest
Resources Development
Cooperative (NPPFRDC),
Compostela Valley,
Southern Philippines

**Source:** Pulhin et al. (2012).
(Disclaimer: This image has been redrawn by the authors and is not to scale. It does not represent any authentic national or international boundaries and is used for illustrative purposes only.)

Although CBFM achieved forest protection, development and rehabilitation at the grass-roots level, local initiatives are, in most cases, hampered by the greater institutional and political systems that govern the programme's implementation. The RUP suspension and logging bans continuously beset timber harvesting in CBFM areas.

Figure 11.6 highlights the impact of policy issuances on the NPPFRDC timber harvesting operations. All policy issuances on timber harvesting

**Figure 11.6:**
*NPPFRDC net profit from logging and national resource use permit suspension*

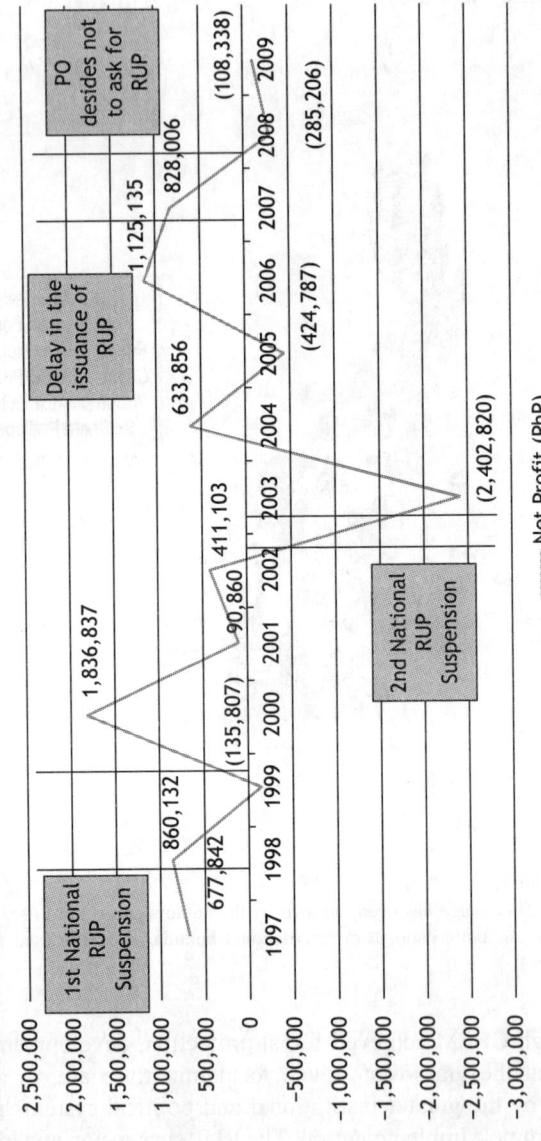

*Source:* Pulhin et al. (2012).

had a negative impact on the income of the PO. The most dramatic impact was during the second national RUP suspension in 2002, when the net profit of the PO plunged to—P2,042,820. This kind of policy environment does not favour successful CBFM implementation in the Philippines.

## Implications of the Design Principles

Governing natural resources is not an easy task. Processes have to be devised to follow the path of sustainable development. Local-level governance systems, such as those exemplified by the cases of ITTO–CBFM and NPPFRDC, clearly highlight that the design principles advocated by Ostrom (1990, 2005) and Inoue (2011) are working. However, the playing field in the Philippines does not suit such applications.

Ostrom's design principles include the following: (1) well-defined boundaries, which the federation has; (2) proportional equivalence between the benefits and costs that the community shares; (3) collective-choice arrangements; (4) monitoring; (5) graduated sanctions when there are violations, where sanctions have been defined and implemented; (6) conflict resolution mechanisms; (7) minimal recognition of rights to organise and (8) nested enterprises. Inoue (2011) added three principles to the list to ensure more robust management of common resources. These principles are as follows: 'graduated membership', in which some of the local people act as 'core members' with the strongest authority cooperating with other graduated members with relatively weaker authority; the 'commitment principle', which recognises an authority to make decisions in a capacity that corresponds to a commitment to forest use and management; and a 'fair benefit distribution', in which the benefit distribution is not necessarily equal but is fair in accordance with cost bearing. These are the most critical factors in the success of collaborative forest governance.

The above principles are designed for an organised local community to realise a governance system that will work best given the different parameters of success. These principles are exemplified by the above cases presented; both POs have well-defined boundaries and roles and responsibilities for their members. There is an equitable benefit–cost sharing agreement entitled to the members. The collective-choice arrangements normally happen during the conduct of the general assembly, which all members are required to attend, occurring at least once a year. At the general assembly, pressing matters/problems confronting the POs are deliberated upon. A monitoring system is also set up, which is overseen by the PMO as well as the officials of the POs, to determine the appropriateness of the actions of its members towards the goal of SFM. Once a violation is made, sanctions are accorded to individual(s) or group of individuals committing the violation. Conflicts, whenever they arise, are resolved within the confines of the POs; however, higher authorities are also sought for appropriate action resolving problems that go beyond the control of the POs. The POs also devise their own extensive rules to define their use of resources within the CBFM area through CRMF and AWP. They also follow a nested enterprise that falls within the concerns of the higher authorities such as the DENR and LGUs. The individual associations of the federation also conform to the rules that govern the management of the entire CBFM through its federation.

Table 11.4 assesses the current potential for the successfully application of the design guidelines proposed by Inoue et al. (2011). Both POs (ITTO–CBFM and NPPFRDC) have a high rating for 'graduated membership', where the dynamic PO leaders serve as 'core members' of the organisation, making everything to the best of their capabilities. The commitment principle is considered low for both POs because decision-making power still rests outside the community.

This includes the greater socio-political and institutional system that operates beyond the bounds of the community and where the community has no direct power, such as the multi-tiered governance system from the

**Table 11.4:**
*Current potential for successfully applying the design guidelines*

| Design guidelines | ITTO–CBFM | NPPFRDC | Remarks |
|---|---|---|---|
| Graduated membership | High | High | PO leaders, in principle, currently serve as 'core members' |
| Commitment principle | Low | Low | Decision-making power rests outside the community |
| Fair benefit distribution (within the community) | High | Low | Issue of fairness in the case of IPs in NPPFRDC may be vested on rights instead of cost* |

*But flow of benefits to community affected by government policies and higher governance structure.
*Source:* Authors.

national, regional, provincial and local levels. In terms of the fair benefit distribution (within the community), ITTO–CBFM has high rating because it is guided by an equitable sharing arrangement. NPPFRDC receives low rating because the issue of fairness in the case of indigenous people may be primarily vested in their rights over the CBFM area and its resources instead of simply the issue of cost. It should also be mentioned that for NPFRDC, the flow of benefits to the local community from timber harvesting is affected more by government policies and higher governance structures, like the cancellation of RUP and overly bureaucratic procedures, than by the issues of local arrangements among community members.

What is obvious, however, in the cases presented above is that serious problems are beyond the bounds of the federation or organisation. The uneven playing field of the multi-tiered governance system, especially the external socio-political and institutional factors, has a tremendous impact on the successful realisation of CBFM objectives.

Unlike any other CBFM PO, the case study federation of ITTO–CBFM exemplifies a model organisation by being the most sought CBFM site for most of the pioneering works and projects in CBFM areas. It has gained numerous linkages between capability and assistance. The enormous amount of technical assistance coming not only from the DENR but also from the various funding agencies and NGOs makes the federation capable of and empowered to handle the problems of their organisation. However, the multi-tiered governance system of CBFM, as exhibited in Table 11.1, makes it difficult for most CBFM POs to realise the fruits of their labour.

Under the revised rules and regulations governing the implementation of CBFM, the processing of papers on the activities of the CBFM POs, especially timber harvesting, has been fast tracked, providing the government agencies at least 60 calendar days to approve/confirm an application. The DENR, through its field offices, has already learned its lesson from the bureaucratic complexities of the process; clamour arouse from consultation with local implementers for second decade implementation of CBFM to make it more people-friendly and less bureaucratic. However, not all DENR offices have done their share. In the case of the ITTO-CBFM's application for RUP, where an ECC is required from the Regional Director of the EMB, the application did not see a decision for eight months. This has already hampered the federation's activities, as outlined in their AWP for the year 2010–2011. As mentioned in the previous section, the federation used its connection with a higher authority to facilitate ECC issuance. The presence of bureaucracy at different levels of the governance system in the country, in most cases, rendered the implementation of the CBFM programme inefficient.

In the case of the NPPFRDC, the cost associated with the timber harvesting application and operation was too much for the PO to bear; hence, they have no other recourse but to secure the support of a financier, who is usually a buyer of logs or a *padrino*, a politician who will exert pressure on DENR officials (Pulhin et al., 2012). This financier could make life easier for the NPPFRDC in terms of securing the necessary permits faster than normal, but there is a considerable cut from the income potentially derived out of the harvested logs.

In addition, the changing context of the socio-economic condition of the local populace has also affected CBFM implementation. Given the ITTO–CBFM allowable cut, their present capacities in harvesting operations do not match the expected output (volume) required of them. After four months of harvesting, the Vista Hills Federation was only able to resize approximately 22 m³ of logs. This is very slow considering their annual target of 500 m³. The circular band saw is being blamed for the slow process, but this is also coupled with the fact that only a small number of the members are skilled in harvesting operations. Most members also have to attend to their regular employment. Only three teams (two-person team) are involved in the sizing stage, with an average output of 0.06 m³ per team, which is far behind the estimated average of 0.25 m³ (Dugan & Pulhin, n.d).

The unstable policies of and related to CBFM add to the complexity of its implementation. The requirement of the submission of the ITTO–CBFM's performance evaluation using the criteria and indicators for SFM for RUP approval is considered new. With assistance from a committed PMO, the federation has been able to subscribe to the new requirement, given that they are the pilot site for such an initiative. However, most CBFM POs are not in the same boat as the ITTO–CBFM federation, which means that others may not be able to comply with such requirements, resulting in losses in their eagerness to participate in the programme and, worse, their trust in the government.

## Conclusion

The ITTO–CBFM and the NPPFRDC are two CBFM POs that continuously uphold the ideals and goals of CBFM through the activities they implement. Their dynamism in terms of implementation can be facilitated and constrained by the different socio-political and institutional factors working towards the attainment of SFM.

The evolution of the CBFM initiative in the country has recognised the vital contribution of forest communities in the attainment of SFM. Policies have been devised to suit the emerging trend in CBFM in the country. However, despite recent efforts to reform forest tenure, rural people remain marginalised in the policy formulation and implementation processes.

The decentralisation approach popularised by the CBFM initiative has continued to dismay its local implementers due to its bureaucratic impediments. The key design principles formulated by Ostrom (1990) and further developed by Inoue (2011) offer solutions to the problems and challenges faced by the governors of common pool resources. Currently, the Philippine CBFM areas have manifestations of adoptions of the prototype design guidelines proposed by Inoue (2011). The 'graduated membership' and 'fair benefit distribution' guidelines are already being practised to a certain degree, although there is still room for improvement. However, the realisation of the 'commitment principle' is still very much wanting because decision-making power lies in the hands of powerful people and institutions outside the community. A multi-tiered governance system employed on an uneven playing field limits the potential for successful employment of Inoue's prototype design guidelines. The confidence of the local community to make CBFM successful will subside if no radical institutional and structural reforms are made at the different forest governance levels. This strategy is vital for levelling the playing field and providing an enabling environment, thereby increasing the chance for more successful employment of the design guidelines on the ground.

# References

DENR (1990). *Philippine Master Plan for Forestry Development: Main Report*. Quezon City, Philippines: Department of Environment and Natural Resources.

Dugan, P. & Pulhin, J.M. (n.d.). Forest harvesting in community based forest management in the Philippines: Simple tools versus complex procedures. Retrieved from www. fao.org .

FMB (Forest Management Bureau) (2006). *Philippine Forestry Statistics*. Quezon City, Philippines: Department of Environment and Natural Resources.

——— (2009). *Philippine Forestry Statistics*. Quezon City, Philippines: Department of Environment and Natural Resources.

Inoue, M. (2011). *Prototype Design Guidelines for 'Collaborative Governance' of Natural Resource*. Presented at 13th Biennial Conference of the International Association for the Study of the Commons, Hyderabad, India, 12 January.

Korten, F. (1994). Questioning the call for environmental loans: Critical examination of forestry lending in the Philippines. *World Development*, 22(7), 971–981.

Magno, F. (2003). Forest devolution and social capital: State-civil society relations in the Philippines. In A. Contreras (ed.), *Creating Space for Local Forest Management in the Philippines* (pp. 17–35). Manila: De La Salle Institute of Governance.

Ostrom, E. (1990). *Governing the Commons: The Evolution of Institutions for Collective Action*. New York, USA: Cambridge University Press.

———— (1998). A behavioral approach to the rational choice theory of collective action. *American Political Science Review*, 92(1), 1–22.

———— (2005). *Understanding Institutional Diversity*. Princeton, NJ: Princeton University Press.

Pulhin, J.M. (1996). *Community Forestry: Paradoxes and Perspectives in Development Practice* (Ph.D. Dissertation). Canberra, Australia: The Australian National University.

———— (1998). Community forestry in the Philippines: Trends, issues, and challenges. In Proceedings of international seminar on community forestry at a crossroads: Reflections and future directions in the development of community forestry, 17–19 July 1997, Maruay Garden Hotel, Bangkok, Thailand, 201–215.

———— (2002). *Trends in Forest Policy of the Philippines*. Policy Trend Report, 29–41.

———— (2004). *Devolution of Forest Management in the Philippines*. Paper presented during the Workshop on evolution of devolution: contribution to participatory forestry in Asia Pacific, 25–26 March 2004, Bangkok, Thailand.

Pulhin, J.M., Dizon, J.T., Cruz, R.V.O., Gevaña, D.T., & Dahal, G. (2008). Tenure reform and its impacts in Philippine forest lands. Policy Brief. November 2008. University of the Philippines Los Baños, Center for International Forestry Research (CIFOR) and Rights and Resources Intiatives (RRI), 10 pp.

Pulhin, J.M. & Inoue, M. (2008). Dynamics of devolution process in the management of the Philippine Forests. *International Journal of Social Forestry (IJSF)*, 1(1), 1–26. Retrieved from www.ijsf.org.

Pulhin, J.M., Inoue, M., & Enters, T. (2007). Three decades of community-based forest management in the Philippines: emerging lessons for sustainable and equitable forest management. *International Forestry Review*, 9(4), 865–883.

Pulhin, J.M., Amaro, M.C., Jr., & Bacalla, D. (2005). *Philippines Community-based Forest Management*. A country report presented during the Community Forestry Forum organized by the Regional Community Forestry Training Center (RECOFTC) held on 24–26 August 2005 in Bangkok, Thailand.

Pulhin, J.M., Ramirez, M.A.M., Tapia, M.A., & Peras, R.J.J. (2012). *Enabling Forest Users to Exercise Their Rights: Rethinking Regulatory Barriers to Communities and Smallholders Earning Their Living from Timber*. Policy Brief May 2012 (unpublished), UPLB-CFNR, College, Laguna.

Rebugio, L.L. & Chiong-Javier, M.E. (1995). Community participation in sustainable forestry. *Review of Policies and Programs Affecting Sustainable Forest Management and Development (SFMD)*. A policy report submitted to the Department of Environment and Natural Resources, pp. 2–1 to 2–27.

Rebugio, L.L., Carandang, A.P., Dizon, J.T., & Pulhin, J.M. (2010). Promoting sustainable forest management. In G. Mery, P. Katila, G. Galloway, R. Alfaro, M. Kanninen, M. Lobovikov, & J. Varjo (eds), *Forests and Society: Responding to Global Drivers of Change*. Vienna: IUFRO World Series Volume 25, 509pp.

Sajise, P. (1998). Forest policy in the Philippines: A winding trail towards participatory sustainable development. In *IGES: A Step Toward Forest Conservation Strategy (1): Current Status on Forests in the Asia-Pacific Region* (Interim Report 1998). The Institute for Global Environmental Strategies (IGES): Tokyo.

# 12

# Thailand: Context and Outcomes of Community Forestry in Two Western Provinces

*Edward L. Webb and Demis Galli*

## Introduction

Local communities in Thailand often claim use and protection rights over forests to maintain repositories of goods and services for local livelihoods (e.g., Kabir & Webb, 2006; Kijtewachakul, Shivakoti, & Webb, 2008; Sudtongkong & Webb, 2008). The government of Thailand has experienced continuous challenges in legislating support for community-based forest management (CBFM)—as demonstrated by the inability to pass a Community Forestry Bill (Sato, 2003; Webb, 2008), so the formal recognition of a wide suite of rights has remained elusive. Nevertheless, decentralisation of power through policy implementation that delegates people to manage their own resources, recognises people's participation in forest management. Articles 46 and 56 of the Thai Constitution, as well as the Decentralisation Act of 1988, provide substantial room for local communities to engage in natural resource management and for local governments to facilitate such engagement. The Tambon Administration Organisation Act of 1992 strengthened the role of village government in forest planning and utilisation (Poffenberger, 1999). While this legal framework is clearly supportive of community management of degraded lands, there remains no legal support for communities to engage in community forestry activities within protected areas.

Despite the lack of legislation related to community forestry, it was estimated that by 2014 more than 8,000 community forests (CFs) exist in Thailand (RECOFTC, 2014). The Royal Forest Department (RFD) is the principal CF administrative body, both at the national and local (district) levels. The RFD may provide communities with technical support and/or economic aid for CF activities. Many of the CFs in Thailand are self-initiated; that is, communities have long-standing, recognised self-proclaimed rights over forest management and utilisation. The RFD often seeks to register these forests as CFs. Alternatively the RFD may actively engage with villages out new locations to establish CFs with community participation.

The outcomes of community forestry, both globally and within Thailand, are still poorly evaluated (Charnley & Poe, 2007; but see Pagdee, Kim, & Daugherty, 2006) and are influenced by a complex interface of physical, biological, economic, social and political dimensions. We took the first step in this process by testing the general hypothesis that CBFM improves forest condition, using a set of case studies from the provinces of Ratchaburi and Kanchanaburi, western Thailand. Replicated, long-term quantitative forest monitoring data in both protected and unprotected forests are absent in western Thailand; therefore, studies must rely on qualitative information provided by people who access, protect and utilise the forest on a frequent basis. Moreover, most of the research on community forestry in Thailand has been quite limited in geographical scope (e.g., Kabir & Webb, 2006; Kijtewachakul et al., 2008; Sudtongkong & Webb, 2008).

We utilised qualitative perception data from 10 community-managed forests to compare forest conditions 15 years before the initiation of CBFM with conditions at the present day. This provided us with a time-series analysis to test whether forest condition trends changed as a function of protection by local users. We follow the analysis with a discussion of the factors that are likely to explain variation in forest outcomes, through the lens of 'prototype design guidelines' of *graduated membership, commitment principle* and *fair benefit distribution*.

## Study Site Description

The research took place in the provinces of Kanchanaburi (19,483.2 km$^2$) and Ratchaburi (5,196.5 km$^2$) in western Thailand. Kanchanaburi province is home to part of the Western Forest Complex, a massive amalgamation of protected areas covering over 18,000 km², one of the largest protected area complexes in Southeast Asia. Much of the land area in both provinces is deforested and under agriculture, especially rice, sugarcane and cassava.

The climate is seasonally monsoonal and the undisturbed forest types in lowlands (below approximately 700 m above sea level [asl]) include deciduous with bamboo, mixed evergreen + deciduous, and deciduous dipterocarp + oak (Maxwell, 2004). In Kanchanaburi and Ratchaburi provinces, essentially all forests outside the national parks have been heavily disturbed by logging and fire. Thus, most of the CFs of this study are degraded, fire-influenced mixed deciduous and bamboo forest.

Amongst this agricultural setting, small hillocks of limestone or sandstone remain forested because they are unsuitable for agriculture (Figure 12.1). These hillocks often provide forest resources to the surrounding villages and may come under CBFM systems. These situations comprise much of the sites used in the study.

Most of the CFs in the study area began conservation voluntarily through the actions of groups or local individuals because of concerns

**Figure 12.1:**
*Most of the community forests in this study, and throughout much of Thailand, are small hillocks—often consisting of limestone—covered in degraded, deciduous forest*

This photo was taken at Baan Kao Lem in Ratchaburi Province (Photo credit: Edward L. Webb).

over forest degradation, decrease of fauna, decrease of non-timber forest products (NTFPs) and loss of environmental services. Every CF is a registered body with the RFD, but the RFD does not dictate management because Thai law allows village to produce their own rules and management regimes (aside from cutting trees). The executive leadership forming the management committee is made of a number of local people who may be proposed and selected directly by villagers through voting. The management committee has power to change all matters related to management, but always with the final approval of villagers. Mobilisation of users varies amongst the villages as does commitment and interest in forest conservation.

All of the villages in this study had low dependence on the forest for livelihood. Most of the inhabitants were engaged in agriculture, and young people had often moved to urban areas in search of different job opportunities. On occasion, members would collect NTFPs.

At the time of the study, there were 38 and 57 legally registered community managed forests in Ratchaburi and Kanchanaburi provinces, respectively. A stratified random sampling procedure was employed to select 10 legally registered community-managed forests for data collection. To prevent the influence of large forest area on governance, and the influence of elevation on forest condition (e.g., forest type), site selection was limited to forests with an area of < 96 ha (600 *rai*) and with an elevation < 300 m asl.

## Village (*Baan*) Descriptions

### Ratchaburi

*Baan Kao Hua Con (BHKC)*—Kao Hua Con CF is a self-initiated project that was legally registered with the RFD in 2000, although formal protection began 11 years prior (total of 18 years of protection). The area under community management is 24 ha, found on a limestone hill with the village temple at one side. The need for safeguarding the forest arose because people realised that degradation was influencing the quantity of resources they were collecting. The forest retreated from lowland to mountains and hills because of the need for farming land and wood for construction of the 1970s. Moreover, several individuals had been interested in exploiting limestone from hills, further degrading the forest condition. The village population is 340 people; however, only 200 participated in CF. Farming is the main activity that people engage with and the main planted crops are rice, sugarcane and cassava; however, some also grow alternative crops such as flowers (roses).

*Baan Kao Lem (BKL)*—People migrated in this area and started to clear land in order to gain farmland and wood for construction material, charcoal making and sale. Kao Lem is a self-initiated CF initiated by local people in 1988, and registered with the RFD in 2001. The CF is 72 ha and located on three hills, which are surrounded by a ring of planted and natural bamboo. Protection of this remnant forest began because of degradation perpetrated by loggers. The village consists of 507 inhabitants, of whom 395 participate in CF, mostly residing nearby the forest, and going occasionally to the forest to extract NTFPs. Most inhabitants are farmers growing mainly rice, sugarcane and cassava.

*Baan Nong Prue (Ratchaburi) (BNPR)*—This CF is a self-initiated project that began in 1999 and registered with RFD in 2002. The area is 47.36 ha and its history of degradation is similar to Kao Hua Con and Kao Lem. The village is home to 1,008 people of which 700 participate with CF. Similar to other forests, nobody in this village is dependent on forest resources; however, resources, especially herbs, are mainly used by elders and poorer members of the village. Most of the villagers are farmers and the main crops planted are rice, sugarcane and cassava. Many also rear cattle, sheep or chickens.

*Baan Kon Pra (BKP)*—This CF was initiated in 1994 and supported by the RFD since 2003. The area designated to CF is 82.56, located on a stony hill. The village is formed of 500 people of whom only 400 participate in CF. Similar to other sites, degradation of forest commenced because of local needs for land and wood. No one in the village is dependent totally on forest resources and NTFPs are collected occasionally. The village is about 500 m from the forest, which is surrounded by farms. The main crops are rice, cassava and sugarcane.

*Baan Kao Noi (BKN)*—Ban Kao Noi CF has an area of 24 ha. It is a self-initiated project that began in 1967, and subsequently registered with RFD in 2003. The village, which surrounds the forest, is formed of 1,010 people, of which only 150 participate in CF activities. The initiation of CF occurred because of degradation associated with local needs for land and wood.

## Kanchanaburi

*Baan Nong Prue [Kanchanaburi] (BNPK)*—This CF has an area of 18.4 ha, located on a limestone hill. Local protection of the forest commenced in 2000 and it was registered with the RFD by 2002. Degradation of the forest started after 14 families migrated to this area approximately in 1970 and started agriculture. The forest was cut in order to acquire farming land and wood for construction, charcoal production and sale. At the time of the study, 840 people lived in the village and only 400 participated into protection of CF. Most of the villagers are farmers and main crops planted are rice, cassava, pineapple and vegetables such as asparagus.

*Baan Ang Hin (BAH)*—This is a self-initiated CF that began in 2001 and registered with RFD in 2002. The CF covers 20.48 ha on a very steep hill, which is connected to a bigger, relatively well-preserved diptero-carp forest of about 1,440 ha, in which residents claim tigers and bears are still present. The CF is about 4 km from the village. Forest degradation commenced when people migrated to this area and started to cut the forest. The village has 425 inhabitants, of which 64 participate in CF. The villagers are not reliant on forest resources and are mainly farmers growing rice, cassava and sugarcane.

*Baan Lan Kao (BLK)*—*Ban Lan Kao* is a CF initiated through advice and influence of an RFD officer in the year 2000. The forest covers 8.96 ha on a limestone hill surrounded by farming land and settlements. In 1961, many families migrated from Kanchanaburi and Suphanburi to this area initiating land development and causing degradation of forest resources in order to acquire land, wood for construction and charcoal production. Currently there are 160 household and 860 inhabitants, of which 500 participate in CF. Most of the villagers are farmers and grow sugarcane, corn, cassava, asparagus and chillies.

*Baan Chon Kaep (BCK)*—*Ban Chon Kaep* is a CF that was self-initiated in 1979, and registered with the RFD in 2002. The area covers 56 ha mostly on hilly land. The first family inhabiting this area settled here in 1955, consequently more families migrated from Kanchanaburi, Ratchaburi, Nakon Phatom and Suphanburi. They cultivated corn, millet, rice, soybean and fruit trees such as mango, jackfruit and tamarind, and also raised cattle. Forest resources were utilised to give space to agricultural land, and wood utilised for construction, charcoal production and sale. Currently the village is composed of about 2,000 people of whom 900 participate in CF. All the inhabitants receive their main income from activities not related to the CF; therefore, they are not dependent on forest resources for their livelihood.

*Baan Tung Prong (BTP)*—This CF covers an area of 72 ha, distributed on a rocky mountain and connected to a larger forest complex. The forest is 2 km from the village. Protection was self initiated in 1997 and was registered with the RFD in 2003. The village is formed of about 700 people of whom 350 participate in CF. Degradation began in 1977 when many people migrated to this area and started to exploit forest resources. Most of the inhabitants are farmers growing mainly sugarcane, corn and cassava.

## Research Methods

Group interviews were carried out in each village in the presence of the village head (*poo-yai baan*) and other available stakeholders. Informal group discussion was conducted to collect data about history, management and use of forest resources (Table 12.1). No formal structure was selected for group interviews, and key informant interviews were conducted to provide supplemental data. The village heads, elders and government officials were interviewed separately.

Data were also collected through a perception questionnaire where several forest resources were assessed in three points in time. People were asked to recollect information about forest resources: 15 years before CFM started, at the beginning of management activities and at the time of the study. In order to remember past information, historical events were used as benchmarks. They were asked to give their own perception on forest resources conditions by scoring condition from zero (the worst possible condition) to 10 (the best possible conditions). These data allowed us to test the hypothesis that CBFM improved forest condition, by comparing the trends of forest condition and biodiversity before and after CBFM. To determine whether forest parameters increased or declined between two points in time (15 years prior vs. inception and inception *vs.* present), a sign test was applied between parameter scores. Statistical significance indicated whether a trend was common across the 10 sites. To test for differences in the direction and

**Table 12.1:**
Characteristics of the forest and communities in 10 community forests of western Thailand*

| | Ratchaburi province | | | | | Kanchanaburi province | | | | |
|---|---|---|---|---|---|---|---|---|---|---|
| | BKHC | BKL | BNPR | BKP | BKN | BNPK | BAH | BLK | BCK | BTP |
| **Forest size and boundaries** | | | | | | | | | | |
| CF size (ha) | 24 | 72 | 47.4 | 82.6 | 24 | 18.4 | 20.5 | 9 | 56 | 72 |
| Total people involved in CFM | 200 | 395 | 700 | 400 | 150 | 400 | 64 | 500 | 900 | 350 |
| Forest area:user ratio | 0.12 | 0.18 | 0.07 | 0.21 | 0.16 | 0.05 | 0.32 | 0.02 | 0.06 | 0.21 |
| Well-defined boundaries | Yes | Yes | No | No | Yes | No | Yes | No | Yes | Yes |
| Presence of a signboard delineating boundary | No | Yes | No | No | No | No | No | No | Yes | Yes |
| Road around CF | Partially | Yes | Partially | No | No | No | No | No | Partially | No |
| **Group characteristics** | | | | | | | | | | |
| Clearly defined membership | No | No | No | No | Yes | No | No | No | Yes | No |
| Number of people typically joining meetings | 120 | 130 | 100 | 20 | 10 | 70 | 40 | 100 | 100 | 60 |
| Number of people typically joining provision activities | 120 | 30 | 60 | 20 | 20 | 30 | 20 | 40 | 150 | 60 |
| **Shared norms** | | | | | | | | | | |
| Past successful experiences | Yes | No | No | No | No | No | No | No | Yes | No |
| Appropriate leadership | Yes | No | No | No | No | No | Yes | No | Yes | Yes |
| Interdependence between group members | Yes | No | No | No | No | No | Yes | No | Yes | Yes |

(Table 12.1 Continued)

(Table 12.1 Continued)

| | Ratchaburi province | | | | | Kanchanaburi province | | | | |
|---|---|---|---|---|---|---|---|---|---|---|
| | BKHC | BKL | BNPR | BKP | BKN | BNPK | BAH | BLK | BCK | BTP |
| Homogeneity of endowments | No | No | No | No | No | No | No | No | No | No |
| Homogeneity of identities | Yes | Yes | Yes | Yes | Yes | Yes | Yes | Yes | Yes | Yes |
| Homogeneity of interests | No | No | No | No | No | No | No | No | No | No |
| **Resource-group relationships** | | | | | | | | | | |
| Distance from village to forest (m) | 100 | 300 | 2,000 | 500 | 0 | 500 | 4,000 | 1,000 | 0 | 2,000 |
| Overlap between user group residential location and resource location | No | No | No | No | Yes | No | No | No | Yes | No |
| High level of dependence by group members on resource system | No | No | No | No | No | No | No | No | No | No |
| Perceived fairness in allocation of benefits from common resources | Yes | No | No | No | No | No | No | No | Yes | No |
| **Institutional arrangements** | | | | | | | | | | |
| Rules are easy and simple to understand | Yes | Yes | Yes | Yes | Yes | Yes | Yes | Yes | Yes | Yes |
| Locally devised access and management rules | Yes | Yes | Yes | Yes | No | Yes | Yes | Yes | Yes | Yes |
| Ease in enforcement of rules | Yes | Yes | Yes | Yes | Yes | Yes | Yes | Yes | Yes | Yes |
| Graduated sanctions | Yes | No | Yes | No | Yes | Yes | Yes | Yes | Yes | Yes |

| | | | | | | | | | | | |
|---|---|---|---|---|---|---|---|---|---|---|---|
| Availability of low cost adjudication | Yes | Yes | Yes | Yes | Yes | Yes | Yes | Yes | Yes | Yes | Yes |
| Accountability of monitors and other officials to users | Yes | Yes | Yes | Yes | Yes | Yes | Yes | Yes | Yes | Yes | Yes |
| *External technical assistance* | | | | | | | | | | | |
| Presence of technical assistance | Yes | No | No | No | Yes | No | Yes | No | No | Yes | Yes |
| Frequency of visits/year by RFD | 12 | 1 | 1 | 2 | 5 | 1 | 2 | 2 | 12 | 25 | 12 |
| People request more RFD visits | No | Yes | Yes | Yes | Yes | Yes | Yes | No | No | Yes | Yes |
| RFD participation in meetings | Yes | No | No | Yes | No | Yes | Yes | Yes | No | Yes | Yes |
| RFD participates actively to meetings giving new ideas | Yes | No | No | Yes | No | Yes | Yes | Yes | No | Yes | Yes |
| Participation of RFD to provision activities | Yes | No | No | No | No | No | No | No | No | Yes | Yes |
| *Provision activities* | | | | | | | | | | | |
| Total days spent for provision activities each year | 6 | 10 | 6 | 6 | 3 | 3 | 3 | 6 | 3 | 5 | 10 |

*Community forest names are as follows: *Baan Kao Hua* Con (BKHC), *Baan Kao Lem* (BKL), *Baan Nong Prue* [Ratchaburi] (BNPR), *Baan Kon Pra* (BKP), *Baan Nong Prue* (Kanchanaburi) (BNPK), *Baan Ang Hin* (BAH), *Baan Lan Kao* (BLK), *Baan Chon Kaep* (BCK), *Baan Tung Prong* (BTP).

**Source:** Authors.

magnitude of score changes before vs. after CBFM, a paired-sample *t*-test was applied to the mean score changes between the two time periods.

Forest transects were utilised to collect information on forest condition at the time of the study. Parallel transects were walked across the entire CF. A random number generator was used to choose a length of time to walk (essentially a distance measurement) between transects, ranging from 120 to 360 seconds. This operation was repeated until the entire CF patch had been traversed. On each transect, points were located at random intervals on the transect line. At each point, several forest structural parameters were evaluated and recorded within a 10-m radius from the point: canopy cover, the number of trees with a diameter above 40 cm (canopy trees) and the number of trees that had been cut. Fire severity and soil erosion were also assessed by categorising evidence of fire into one of four categories: none (no burned materials), minor (lightly burned bark with burned branches with diameter <2 cm), moderate (lightly burned bark and burned branches with diameter 2–5 cm) and major (severely burned trees or burned branches with diameter > 5 cm). Soil erosion was categorised as none (no erosion), minor (sheet erosion), moderate (rill erosion) or major (gully erosion).

We also evaluated technical assistance and external economic support as exogenous parameters. These parameters are closely tied to the incentives of local people to participate in an institutionalised form of CBFM.

## Results and Discussion

### Village Characteristics

The CF groups we studied can be characterised as loosely structured, and loosely managed, organisations. Eight of the 10 groups had no clear rules for membership. Typically, recognition as a member of the CF group is *de facto* when one is a member of the village that claims usufruct rights over the forest. Historical experiences varied among the sites, but most villages did not have previous successful experiences in managing resources or working in collective activities. Although all the villages were ethnically homogeneous, their interests were never completely aligned and therefore there was a diversity of perceptions and interests in how the forest should be used and managed.

Villages in Thailand have a head, called the *poo-yai baan*, who serves as a village leader and who spearheads activities such as CF organisation, discussions and decisions over CF management and protection, and seeking community consensus over such actions. In these sites, there were variable strengths of leadership. Although the mechanisms behind such variability is impossible to explain without in-depth analysis about the complex social, economic and political dynamics of each village, we found that only 4 of 10 villages had leadership that was clearly adequate in galvanising community action and compliance over forest management, such as bridging the differences among members over the objectives of CF management.

Most forests had clearly defined boundaries, owing to the fact that they were often isolated patches of forest surrounded by agricultural land (e.g., Figure 12.2). Perhaps because of this, 7 out of 10 villages

**Figure 12.2:**
*Relationship between average fire severity index and average estimated canopy cover in 10 community forests of western Thailand*

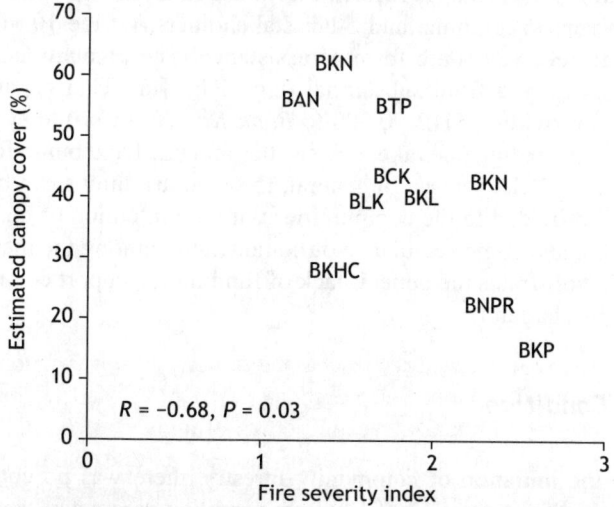

**Note:** Acronyms refer to village names. Results of a Pearson correlation are presented in the lower left of the graph.
**Source:** Authors.

did not feel the need to erect a signboard to substantiate rights over the forest. Furthermore, most CF groups did not claim exclusion rights over the forest. This typifies the general view of accessibility in many of the communities in the survey; people are granted access to enter the forest whether they are village members or not, but only for harvesting of products such as mushrooms, herbs or fruits.

All villages had institutional arrangements indicating that village members were aware of rules and regulations over the forest. Whether this translated into behaviour that conformed to those rules and regulations, however, is another matter.

The greatest variability among villages was in the level of technical and economic assistance to villages (Table 12.1). The number of RFD visits per village ranged from 1 to 25, indicating that some villages appeared to be favoured by the RFD, while others were either too inaccessible (e.g., too far from the office) or were considered to be of less importance. Of the six villages that received fewer than 10 visits per year, five villages had requested more frequent visits. The RFD also exhibited varying levels of participation in CF meetings; however, this may be a function of the RFD officers, the community members or both.

In terms of economic assistance, Table 12.2 details the specific contributions that villages received from the RFD, the United Nations Development Programme and other stakeholders. Of the 10 villages, seven had received some form of assistance. The amounts provided to villages ranged from substantial (e.g., PTT [the Thai oil and gas company] providing THB 90,000 to *Baan Kao Hua Con*) to symbolic (e.g., PTT providing one rake, one swatter and one hand pump for fire-fighting in a 72 ha forest). In general, these contributions can be seen as tokens provided to the communities with the intention of maintaining interest and some semblance of homogeneity among the members. It also demonstrates the general lack of funding to support community forestry in Thailand.

## Forest Condition

Prior to the initiation of community forestry, there was no consistent trend in the direction of 12 of 14 forest condition parameters across the 10 villages (Table 12.3). For example, tree density was seen to decrease in five villages, but increase in five others, suggesting no aggregate trend prior to CF across the villages. Similarly, bird density had declined in

**Table 12.2:**
*Direct economic assistance provided to villages from external sources, according to village surveys using a recall method\**

| | Royal Forest Department | United Nations Development Programme | Other |
|---|---|---|---|
| **Ratchaburi** | | | |
| *Baan Kao Hua Con* | | | THB 90,000 (1999) Local organisation gave THB 30,000 at CFM inception<br><br>Firefighting equipment:<br>1 rake<br>1 swatter<br>1 hand pump |
| *Baan Kao Lem* | THB 50,000 at beginning of CFM (180,000 was promised) | | |
| *Baan Nong Prue* (Ratchaburi) | | | |
| *Baan Kon Pra* | | | |
| *Baan Kao Noi* | | | |
| **Kanchanaburi** | | | |
| *Baan Nong Prue* (Kanchanaburi) | | (1) THB 4,500 for *buat paa* (symbolic ordaining/ protection of forest by monks)<br>(2) THB 1,500 to clean forest (every year)<br>(3) THB 300 every meeting<br>(4) THB 1,500 (2005–2006) for conservation activities<br>(5) Money to buy firefighting equipment:<br>3 rakes<br>3 baskets<br>3 water pumps<br>3 swatters | |

*(Table 12.2 Continued)*

(Table 12.2 Continued)

| | Royal Forest Department | United Nations Development Programme | Other |
|---|---|---|---|
| **Baan Ang Hin** | | (1) THB 300 every meeting<br>(2) THB 2,500 for fire prevention measure (on demand)<br>(3) THB 2,000 for signboard<br>(4) THB 1,700 for *buat paa* (symbolic ordaining/protection of forest by monks) | |
| **Baan Lan Kao** | | (1) THB 300 every meeting<br>(2) THB 3,000 for fire prevention measure (every year)<br>(3) THB 2,500 for growing new plants<br>(4) Give seedlings<br>(5) THB 800 for conservation activities (each year) | |
| **Baan Chon Kaep** | | (1) THB 300 every meeting<br>(2) THB 3,000 every year<br>(3) An amount every year depending on the needs<br>(4) Gave money to buy jackets, signboards, fire control tower | |
| **Baan Tung Prong** | (1) THB 12,000 to buy signboard, uniforms, *buat paa* (symbolic ordaining/protection of forest by monks), firefighting tools | | Miasawa (Japan) in 1999 gave THB 130/person to 100 people to carry out provision activities |

*Note that three villages in Ratchaburi received no assistance from any organisation since the CF inception.
**Source:** Authors.

**Table 12.3:**

*Changes in forest condition relative to the initiation of community forestry in 10 villages of western Thailand**

| Condition parameter | 15 years prior until inception of CF (period 1) | | | | Inception of CF until present (period 2) | | | | Average score change | | | | |
| --- | --- | --- | --- | --- | --- | --- | --- | --- | Period 1 | | Period 2 | | Outcome |
| | DEC | IC | NC | P | DEC | IC | NC | P | Mean | Std | Mean | Std | |
| Tree density | 5 | 5 | 0 | 1.000 | 0 | 9 | 1 | **0.004** | 0.10 | 3.96 | 2.90 | 1.91 | **Marginal improvement** |
| Large tree abundance (>40 cm dbh) | 5 | 2 | 3 | 0.453 | 0 | 5 | 5 | **0.063** | -1.30 | 3.23 | 0.70 | 0.82 | **Marginal improvement** |
| Logging occurrence | 9 | 0 | 1 | **0.004** | 6 | 0 | 4 | **0.031** | -4.30 | 3.27 | -1.40 | 1.43 | **Improvement** |
| Canopy cover | 3 | 6 | 1 | 0.508 | 0 | 9 | 1 | **0.004** | 1.20 | 4.02 | 2.50 | 1.72 | |
| Tree diversity | 2 | 6 | 2 | 0.289 | 0 | 7 | 3 | **0.016** | 0.80 | 3.68 | 1.30 | 0.95 | |
| Mushroom density | 5 | 5 | 0 | 1.000 | 2 | 8 | 0 | 0.109 | -0.50 | 3.27 | 1.30 | 2.21 | |
| Bamboo density | 4 | 6 | 0 | 0.754 | 3 | 6 | 1 | 0.508 | 0.20 | 3.08 | 1.10 | 2.51 | |
| Medicinal plant density | 1 | 5 | 0 | 0.219 | 0 | 6 | 0 | **0.031** | 2.17 | 2.32 | 2.83 | 1.47 | |
| Useful plant density | 6 | 4 | 0 | 0.754 | 1 | 8 | 1 | **0.039** | -0.80 | 4.18 | 2.60 | 2.37 | **Marginal improvement** |
| Bird abundance | 7 | 2 | 1 | 0.180 | 4 | 4 | 2 | 1.000 | -2.30 | 3.47 | 0.60 | 2.72 | **Improvement** |
| Mammal density | 7 | 3 | 0 | 0.344 | 2 | 4 | 4 | 0.688 | -1.30 | 3.47 | 0.00 | 2.16 | |
| Total wildlife density | 8 | 2 | 0 | 0.109 | 5 | 5 | 0 | 1.000 | -2.70 | 2.75 | 0.00 | 2.31 | **Large improvement** |
| Fire occurrence | 8 | 1 | 1 | **0.039** | 7 | 1 | 2 | **0.070** | -2.40 | 3.86 | -2.00 | 3.20 | |
| Soil erosion | 2 | 2 | 5 | 1.000 | 4 | 0 | 5 | 0.125 | 0.22 | 2.49 | -0.78 | 0.97 | |

*The values in the cells refer to the number of villages recording a decrease (DEC), increase (IC) or no change (NC) in the condition parameter. A non-parametric sign test was applied to each data set to compare whether there were trends in forest parameters within each time period; the p value is denoted as 'P'. A paired-sample t-test compared whether the average (n = 10) score change was different before vs. after the initiation of community forest. The interpretation of the t-test is given in the last column, where 'marginal improvement' is given for p < 0.10, 'improvement' is given for p < 0.05 and 'large improvement' is given for p < 0.01. Bold type for p values and outcomes indicates statistical significance.

*Source:* Authors.

seven villages, increased in two villages, and had not changed in one village; statistically, this distribution was not significantly different from random (Table 12.3). Logging occurrence and fire occurrence, however, showed a consistent directionality among the 10 villages prior to CF initiation. In terms of logging, a decrease had been seen in nine villages, whereas in one village there was no change in occurrence. This is likely the result of the 1989 logging ban imposed across the entire Kingdom. Secondly, the occurrence of fire had been decreasing in eight villages, increasing in one, and unchanged in one. This was also indicative of declining fire prior to CF.

Trends were observed for eight of the parameters after the inception of CF. All of the eight trends suggested an improvement in forest conditions: increases in tree density, large tree abundance, canopy cover, tree diversity, medicinal plant diversity and useful plant diversity; meanwhile the decrease in logging fire occurrence continued (Table 12.3).

To test whether changes in forest conditions were different between pre- and post-CF, the difference between the average score change was calculated across the 10 villages for each parameter. The paired-sample *t*-test revealed that the mean score change before vs. after was significantly different for tree density, logging occurrence, useful plant density, bird abundance, total wildlife density and large tree abundance.

Improvements were seen in tree density, large tree abundance and logging occurrence. These three conditions are probably all interrelated, and a function of the logging ban that led to a decline in the removal of large trees and opportunity for forest regeneration. This result could also be linked with reductions in fire occurrence: although there was no difference in the rate at which fire occurrence decreased between the two time periods, fire occurrence decreased in at least seven villages during both time periods. Reducing fire frequency has important positive effects on forest regeneration, which could over the long term, led to increases in adult tree densities.

Possibly, continued reduction in fire occurrence had positive impacts on non-tree plant abundance as well. For example, useful plant diversity improved in eight villages after CF, as opposed to improvement in only four villages prior to CF. Similarly, medicinal plant diversity increased in six villages during both the first and second periods.

Finally, wildlife densities appeared to be rebounding across these CFs. Although only half of the villages reported increases in wildlife abundances, eight villages reported declines in wildlife abundance occurring prior to the implementation of CF.

**Table 12.4:**
*Main factors influencing forest conditions, ranked in each village by level of importance, with '1' being most important*

| | Ratchaburi | | | | | Kanchanaburi | | | | |
|---|---|---|---|---|---|---|---|---|---|---|
| | BKHC | BKL | BNPR | BKP | BKN | BNPK | BAH | BLK | BCK | BTP |
| Fire | | 1 | 1 | 1 | 2 | 1 | 2 | 1 | 2 | 2 |
| Drought | | | | 2 | 1 | | | | | 3 |
| Illegal logging | 1 | | 2 | 3 | | 2 | 1 | 2 | 1 | 1 |

**Source:** Authors.

While interviewing the villages, respondents were asked to rank the main factors affecting the forest conditions (Table 12.4). Fire was ranked as the most important factor in 5 of the 10 villages, and ranked second in 4 others. Illegal logging was the second most frequent response, being ranked first in four villages, second in three villages and third in one village. Drought was seen as an important factor in three villages; however, in *Baan Kao Noi* it ranked first.

It is interesting that illegal logging was considered to be such an important factor still affecting forest conditions, even though it has been on the decrease since 15 years prior to the inception of CF in most villages (Table 12.3). This may suggest that although logging was on the decline in many villages, the rate of decline was slow and that there were still significant levels of tree cutting taking place in those villages.

Similarly, fire was on the decline in most villages, yet it was considered by many to be the most important factor in shaping forest conditions (Table 12.4). Results from the forest surveys show that at the forest plot level, there was a significant negative correlation between the fire severity index and canopy cover (Figure 12.2). Respondents obviously recognised this relationship even though their experience may have been limited to their own CF. Fire control is therefore seen as a principal factor associated with the conditions of CFs. In this context, the pittance of external support provided to communities for firefighting equipment becomes all the more relevant, and requires improved investment.

In general, these results suggest that overall forest conditions have improved since the implementation of community forestry across the 10 study forests, supporting a cautiously optimistic view of the impacts of local protection over forest resources. However, one caveat of this analysis is that forest dependency of villagers has concomitantly decreased through the last decades, providing the opportunity for a slow but constant amelioration of the forest conditions and biodiversity in general. Therefore, while we argue that CF was at least partially responsible for improved forest conditions, we must also recognise an unaccounted influence of on- and off-farm income opportunities.

## Forest Management Outcomes

We scored 11 variables across the 10 villages according in order to gain an understanding of how variable overall management performance was in our sites (Table 12.5). Overall performance scores were low, with a maximum of 8.0 (*Baan Chon Kaep*) and a minimum of 3.2 (*Baan Nong Prue [Ratchaburi]*). The average score was 5.2, meaning that typical overall performance was weak (5.2/11.0 = 0.47).

The best performing village was *Baan Chon Kaep* with a score of 8.0, followed by *Baan Tung Prong* (6.6) and *Baan Kao Hua* Con (6.4). What appears to set *Baan Chon Kaep* apart from other villages is a combination of very strong leadership, well-articulated rules, high proximity to the forest, high participation and very strong technical assistance. Conversely, the worst-performing village, *Baan Nong Prue* (Ratchaburi), had a very weak leadership, low proximity to the forest, and received neither technical assistance nor external economic assistance (Table 12.5).

## Through the Lens of 'Prototype Design Guidelines'

Here, we consider the three prototype guidelines: graduated membership, commitment principle and fair benefit distribution. Thailand serves as an interesting case study because of the loose, but recognisable, organisational structure surrounding community forestry.

*Graduated membership* was observed at all these sites. Indeed, a core group of leaders, who are typically elected, are instrumental in making decisions regarding rules and regulations, and are responsible for communicating those decisions to the rest of the community. In fact, the core members were the ones who were the first 'point of contact' for the researchers during this study. Often, group meetings would include only those core members, while not necessarily excluding non-core members.

**Table 12.5:**

*Major parameters influencing forest management and their scoring for each village\**

| Parameter | Definition/justification | Ratchaburi | | | | | Kanchanaburi | | | | |
|---|---|---|---|---|---|---|---|---|---|---|---|
| | | BKHC | BKL | BNPR | BKP | BKN | BNPK | BAH | BLK | BCK | BTP |
| Leadership | Researcher's assessment of strength of leadership, taking into account involvement in decisions, organisation, knowledge of the forest, respect given by the villagers, etc. | Strong | Weak | Very weak | Weak | Weak | Very weak | Moderate | Weak | Very strong | Moderate |
| Rules | Existence and specificity of rules, especially regarding quantity and specificity of harvesting, which are understood by all | Weak | None | Weak | Weak | Strong (RFD rules) | Moderate | Strong | Weak | Strong | Strong |
| Sanctions (appropriate to offence) | Whether a clear system was in place to apply sanctions to rule breakers (e.g., collection time, fire, hunting) | Weak | None | Weak | Weak | Strong | Strong | Strong | Weak | Strong | Strong |

*(Table 12.5 Continued)*

*(Table 12.5 Continued)*

| Parameter | Definition/justification | Ratchaburi | | | | | Kanchanaburi | | | | |
|---|---|---|---|---|---|---|---|---|---|---|---|
| | | BKHC | BKL | BNPR | BKP | BKN | BNPK | BAH | BLK | BCK | BTP |
| Monitoring activities | Frequency of monitoring activities by ≥ 1 person. Because the vast majority of rural people held day jobs, we considered weekend monitoring to be sufficient (eight days/month) | Weak | Weak | Weak | Weak | Weak | Weak | Weak | Weak | Weak | Weak |
| Clearly defined forest boundaries | Whether the community-protected forest was easily recognisable by users and non-users, e.g., by posting signage | Strong | Strong | Weak | Weak | Weak | Weak | Moderate | Weak | Strong | Strong |
| Distance of forest from village | Villages closer to their forest have an easier time of monitoring, excluding and enforcing institutions | Strong | Moderate | Very weak | Weak | Very strong | Weak | Very weak | Very weak | Very strong | Very weak |

| | | | | | | | | | | |
|---|---|---|---|---|---|---|---|---|---|---|
| Provision activities | Frequency with which the village members undertook forest improvement activities. Provision activities of at least once per month considered sufficient | Weak | Moderate | Weak | Weak | Very weak | Very weak | Weak | Very weak | Weak | Moderate |
| Participation | Proportion of people actively participating in provision activities and meetings. Participation was considered sufficient when at least 50% of the user group typically joined the provision activities or meetings | Strong | Weak | Weak | Weak | Weak | Very weak | Strong | Weak | Strong | Moderate |
| Technical assistance | Whether the Royal Forest Department visited the CF group to provide technical assistance on a frequent basis. A visitation rate of once per month was considered sufficient | Strong | None | None | None | None | Weak | Weak | Strong | Very strong | Strong |

*(Table 12.5 Continued)*

(Table 12.5 Continued)

| Parameter | Definition/justification | Ratchaburi | | | | | Kanchanaburi | | | | |
|---|---|---|---|---|---|---|---|---|---|---|---|
| | | BKHC | BKL | BNPR | BKP | BKN | BNPK | BAH | BLK | BCK | BTP |
| Economic help | Whether external support (government or otherwise) was sufficient to subsidise forest management and protection activities. Since no one in the villages was dependent on the forest as a primary source of income, financial incentives must be in place to maintain interest in CF | Weak (at beginning only) | Weak (at beginning only) | None | None | None | Weak (but every year from UNDP) | Weak (but every year from UNDP) | Weak (but every year from UNDP) | Weak (but every year from UNDP) | Weak (but every year from UNDP) |
| Forest condition | Heuristic assessment of the forest condition, based on tree density, diversity, canopy cover, etc. | Weak | Weak | Weak | Very weak | Strong | Moderate | Moderate | Weak | Moderate | Moderate |
| **Score** | | 6.4 | 4.0 | 3.2 | 3.4 | 5.2 | 4.8 | 6.0 | 4.4 | 8.0 | 6.6 |
| **Rank** | | 3 | 8 | 10 | 9 | 5 | 6 | 4 | 7 | 1 | 2 |

*Classification and scoring was as follows: non-existent (0.0), very weak (0.2), weak (0.4), moderate (0.6), strong (0.8), very strong (1.0). Summed scores are listed at the bottom of the table. Minimum total score (theoretically) was 0.0, and maximum possible score was 11.0.

**Source:** Authors.

However, the core is often given substantial recognition in representing the views of the community.

Given the fact that there was consistency in the *structure* of graduated membership, but high variability in overall CF performance scores, this particular feature appears to lack power in predicting management performance. Rather, it may be the *quality* of the core leadership group that is most predictive of the management score, and likely sustainability outcomes. For example, a study on the governance, management and conservation of mangroves by communities in Southern Thailand (Sudtongkong & Webb, 2008) showed that high-quality leadership was essential in not only internal institutional arrangement, but also in terms of liaising with external entities such as the forest department, non-governmental organisations and local political leaders. Such a result has also been seen in other terrestrial CF governance systems in Thailand (e.g., Kabir & Webb, 2006). The importance of quality of leadership for Thai CFs can fully be appreciated when one considers that assassinations of community forestry leaders occur on a surprisingly frequent basis. At the CF site investigated by Kabir and Webb (2006), the village head was allegedly assassinated by external business interests because he defended the conservation of the *ca.* 500 ha forest patch claimed by the community, in the face of strong interest to convert the forest to cash crops. In the area studied by Sudtongkong and Webb (2008), the village head in one of the four villages was allegedly murdered for issues directly related to mangrove conservation. Thus, quality of leadership in Thailand, perhaps more than simply the presence of a core structural group, may be the most important feature leading to long-term sustainable outcomes.

The *commitment principle* suggests that forest users who have a greater commitment to forest use and management have greater authority in making decisions over the resource. 'Commitment' to forest use and management is assumed to arise out of either greater direct dependency on the forest and its products (subsistence) or from a vested interest in the economic potential of certain forest products. In these 10 CFs, virtually none of the households was dependent on forests for its livelihood, indicating that subsistence-based commitment was absent from the villages. However, a total of 46 useful species were listed as being collected from the 10 CFs of this study (Table 12.6). One of the listed species was *Melientha suavis* Pierre (*pak waan*), a small tree with edible leaves used widely in soups and curries. This species has commercial value. Secondly, in most of these forests a highly valuable mushroom locally called *het kon* could be harvested, which claims a substantial market price.

**Table 12.6:**
*Species commonly collected and utilised by rural people in the 10 community forest villages of western Thailand*

| Family | Species name | Thai name | Habit | Villagers use | Other uses |
|---|---|---|---|---|---|
| Acanthaceae | *Andrographis paniculata* Nees | Fa thalai chan | a/h | To cure flu, cough, blood pressure and malaria, as tonic | 1,2 |
| Acanthaceae | *Nelsonia canescens* (Lam.) Spreng. | Salet phang phon | a/h | | |
| Anacardiaceae | *Spondias pinnata* Kurz | Makok | d/t | Food (leaves and fruit) | 1 |
| Anacardiaceae | *Spondias bipinnata* Airy Shaw & Forman | Makkak | d/t | Food, fuel wood | 1,4 |
| Anacardiaceae | *Swintonia Schwenckii* Teijsm. & Binn | Khan thong | e/t | | |
| Anacardiaceae | *Buchanania lanzan* Spreng. | Mamuang hua maeng wan | d/t | | |
| Aquifoliaceae | *Ilex umbellulata* (Wall.) Loes. | Baraphet | e/t | To treat diabetes, muscle pain | 1,2 |
| Araceae | *Colocasia esculenta* (L.) Schott | Phueak | a | | |
| Araceae | *Pseudodracontium siamense* Gagnep. | Buk I rok | d/h | Edible leaves | 1 |
| Bignoniaceae | *Millingtonia hortensis* L.f. | Pip | d/t | | |
| Bombaceae | *Bombax anceps* Pierre | Ngeu bah | d/l | Edible flowers | 1,2 |
| Combretaceae | *Terminalia mucronata* Craib & Hutch. | Ma kluea lueat | d/t | | 2,3 |
| Dioscoraceae | *Dioscorea hispida* Dennst. | Kloi | d/v | Edible tuber | 1 |

*(Table 12.6 Continued)*

*(Table 12.6 Continued)*

| Family | Species name | Thai name | Habit | Villagers use | Other uses |
|--------|--------------|-----------|-------|---------------|------------|
| Dioscoraceae | *Dioscorea birmanica* Prain & Burkill | Man nok | d/v | Edible tuber | 1 |
| Dioscoraceae | *Dioscorea alata* L. | Man sao | d/v | Edible tuber | 1 |
| Ebenaceae | *Diospyros rubra* L. | Phaya rak dam | d/t | | 2 |
| Euphorbiaceae | *Phyllanthus amarus* Schumach. & Thonn | Luk tai bai | a/h | To cure diabetes, for kidney stones | 2 |
| Euphorbiaceae | *Phyllanthus emblica* L. | Makham poam | d/t | To cure cough and diarrhoea, edible fruits | 1,2,3,4,5, 6,8 |
| Graminae | *Thyrsostachys siamensis* Gamble | Phai ruak | h | Edible shoots | 1,3,7 |
| Graminae | *Dendrocalamus membranaceus* Munro | Phai nuan | p/h | Edible shoots, construction | |
| Gramineae | *Dendrocalamus strictus* Nees | Phai ta dam | p/h | Edible shoots, construction | 1,3,7 |
| Gramineae | *Gigantochloa hasskarliana* Backer ex K. Heyne | Phai phaak | p/h | Edible shoots, construction | 1,3,7 |
| Lauraceae | *Cinnamomum subavenium* Miq. | Cha aem | a/t | Food, and medicine (extracted from bark) | 1,2 |
| Leguminosae, Caesalpinioideae | *Senna siamea* (Lam.) H.S. Irwin & Barneby | Khi lek | e/t | | 1,2,3 |
| Leguminosae, Caesalpinioideae | *Caesalpinia sappan* L. | Fang | d/w/c | | 2,3,4,6,7 |
| Leguminosae, Caesalpinioideae | *Cassia fistula* | Khun | d/t | Bark used as dye, pods are used for traditional medicine | 2,6 |
| Leguminosae, Papilionoideae | *Endosamara racemosa* (Roxb.) R. Geesink | Hang lai | d/w/c | | |

*(Table 12.6 Continued)*

*(Table 12.6 Continued)*

| Family | Species name | Thai name | Habit | Villagers use | Other uses |
|--------|--------------|-----------|-------|---------------|------------|
| Leguminosae, Papilionoideae | *Christia vespertilionis* (L.f.) Bakk.f. | Non tai yak | a/h | | |
| Leguminosae, Papilionoideae | *Pterocarpus macrocarpus* Kurz | Pradu | d/t | Construction | 3,7 |
| Leguminosae, Papilionoideae | *Dalbergia oliveri* Gamble ex Prain | Ching chan | dt | Construction, firewood, charcoal | 3 |
| Meliaceae | *Azadirachta indica* A. Juss | Sadao | dt | Extract from leaves used as insecticide | 1,2,3,5 |
| Menispermaceae | *Tiliacora triandra* Diels | Thao yanang | e/v | | 1 |
| Moraceae | *Ficus rumphii* Blume | Poo ki nok | d | Bark is utilised as medicine | 2 |
| Myrtaceae | *Eugenia cumini* (L.) Druce | Wa | e/t | Edible fruits | 1,6 |
| Olacaceae | *Schoepfia fragrans* Wall. | Khi non | e/l | Edible leaves | |
| Opiliaceae | *Melientha suavis* Pierre | Pak wan | d/t | Edible leaves | 1 |
| Passifloraceae | *Passiflora foetida* L. | Ka thok rok | a/v | Edible fruits | 1,2 |
| Rubiaceae | *Mitragyna hirsuta* Havil. | Kra tum | d/t | Construction, firewood, charcoal | 2,4,7,8 |
| Rubiaceae | *Morinda coreia* Buch.-Ham. | Yo pa | d/t | Medicine (fruits) | 2 |
| Smilacaceae | *Smilax corbularia* Kunth subsp. *synandra* (Gagnep.) koyama | Khao yen nuea | e/v | Food, boiled roots to treat for menstrual pain | 1,2 |
| Verbenaceae | *Vitex pinnata* L. | Samo pa | e/v | | 3 |
| Vitaceae | *Ampelocissus martini* Planch. | Sam kung | d/w/c | | |
| Zingiberaceae | *Curcuma comosa* Roxb. | Wann chak modlookk | d/h | Treatment of uterus after birth (rhizome) | 1,2 |

*(Table 12.6 Continued)*

*(Table 12.6 Continued)*

| Family | Species name | Thai name | Habit | Villagers use | Other uses |
|--------|-------------|-----------|-------|---------------|------------|
| Zingiberaceae | *Curcuma longa* L. | khamin | d/h | Food, rhizome as a spice | 1,2 |
| Zingiberaceae | *Kaempferia rotunda* L. | Phro | d/h | Food, rhizome as a spice | 1,2 |
| Zingiberaceae | *Zingiber zerumbet* (L.) Sm. | Kra Thue | d/h | Relief from indigestion, rhizome as a spice | 1,2 |

*Habits:* a: annual; p: perennial; t: tree; l: treelet; wc: woody climber; d: deciduous; e: evergreen; h: herb; v: vine.
*Use:* 1, food; 2, medicine; 3, construction; 4, fuel wood; 5, fodder; 6, dye; 7, crafting; 8, charcoal.
**Source:** Authors.

Strong institutional arrangements existed in all villages with regards to *het kon*, in order to regulate the harvesting of this valuable product. However, few if any institutional arrangements existed to exclude non-village members from collecting in the forest, as long as that collector received permission (which typically was not denied). Therefore, those people with the greatest stake in the resource, and therefore 'commitment' to maintaining sustainability in the system, included core members, fringe members as well as non-members of the community or the forest management group. Hence, the authority to make decisions over resources—which lies with the elected core leadership group—did not appear related to one's personal stake in, or commitment to, the forest resources.

Our argument that these villages lack a 'commitment principle' should not be taken as a negative comment on the groups' management style. On the contrary, this may be viewed as the outcome of not only lack of dependency on forests for critical livelihood services as described above, but also a reflection on a combination of Buddhist philosophy and/or the somewhat egalitarian *mai pen rai* (colloquially, 'do not worry about it') attitude espoused by Thai culture. Such a culture provides a large space for sharing of limited resources, allowance of people to enter into and access forests, as well as offering (and requesting) assistance in times of need. Such a philosophy extends beyond village boundaries and encompasses humanity in general.

In other systems, such as mangrove CF management, the situation is far different. Fishing-based villages may rely heavily on mangrove resources, especially mud crabs, for economic gain and therefore have a strong incentive to exclude non-village members (Sudtongkong & Webb, 2008). In that case, the concept of the commitment principle is more likely to be supported. However, in terrestrial forests managed by suburban villages with external sources of income, for example, the villages of this case study, such a concept is probably less applicable.

The concept of *fair benefit distribution* is that distribution of benefits is commensurate with the cost borne by the harvesters. In these CFs, benefit access is equal amongst all community members. Therefore, the concept of fair benefit distribution applies; each person harvesting the resources keeps the benefit, according to her/his input. However, again the high variability across village performance scores, while benefit distribution mechanisms remained constant, implies that the benefit distribution mechanism did not explain the variance in performances across villages.

## Conclusion

One of the main shortcomings of this study is that the study was not designed specifically to test the hypotheses of prototype design guidelines set forth in this book. However, the 10 study sites of this paper, along with evidence from other community forestry settings the authors have worked in, suggest that Thailand may be a unique case among Southeast Asian community forestry systems. A core group of decision makers is present in all villages, making the community forestry governance bodies somewhat homogeneous in structure; thus, the concept of graduated membership exists in Western Thailand but does not vary sufficiently to explain variance in CF management outcomes. We suggest that the quality of the core group is essential in providing the appropriate leadership, under which CF management may succeed or fail. Secondly, while institutional arrangements exist over resource harvesting and forest management, the very loose membership requirements embodied by the fact that permission to enter the forest and extract a limited set of resources is easily obtained and generally applies equally to village members as well as non-village members, suggest that the commitment principle is also limited in its application to understanding outcomes of forest management. Core decision makers may or may not have a greater stake in the sustainability of the resource, for example, *pak waan*, or *het kon*, than other village members, or even collectors from other villages. Finally, the concept of fair benefit distribution does exist across the 10 sites in this study; however, it exhibits homogeneity in structure similar to that of the governance institutions: collectors gain benefit depending upon their personal input into the system. Yet, it is the consistent recognition of those rights across all villages, which prevents us from concluding whether variation in fair benefit distribution has had any influence in the overall outcomes of CF management in Western Thailand.

These case studies are instructive in that overall they provide evidence that the concepts of graduated membership, commitment and fair benefit distribution may have limitations, and be applicable under a set of conditions not present in Western Thailand. In particular, the forest resource was not a cornerstone of villages' livelihoods, and therefore the institutional arrangements of exclusion were far more relaxed than in other systems where forest resources make up a large proportion of users' livelihoods. Under conditions of greater dependency and/or economic value, it has been shown in other Thai systems that less egalitarian institutions are enforced. Thus, there may ultimately be framework conditions that support the emergence or evolution of the prototype design guidelines.

# References

Charnley, S. & Poe, M.R. (2007). Community forestry in theory and practice: Where are we now? *Annual Review of Anthropology*, 36, 301–336.

Kabir, Md. E. & Webb, E.L. (2006). Saving a forest: The composition and structure of a deciduous forest under community management in northeast Thailand. *Natural History Bulletin of the Siam Society*, 54, 63–84.

Kijtewachakul, N., Shivakoti, G.P., & Webb, E.L. (2008). Evolution of community-based management and forest health in Northern Thailand: Case study of Nahai and Huai-muang villages in Sopsai Watershed, Thawangpa District, Nan Province. In E.L. Webb & G.P. Shivakoti (eds), *Decentralization, Forests and Rural Communities: Policy Outcomes in South and Southeast Asia* (pp. 232–268). New Delhi: SAGE Publications.

Maxwell, J.F. (2004). A synopsis of the vegetation of Thailand. *Natural History Journal of Chulalongkorn University*, 4, 19–29.

Pagdee, A., Kim, Y., & Daugherty, P.J. (2006). What makes community forest management successful: A meta-study from community forests throughout the world. *Society & Natural Resources*, 19, 33–52.

Poffenberger, M. (1999). *Communities and Forest Management in Southeast Asia*. Gland: IUCN.

RECOFTC (2014). *Community Forestry Adaptation Roadmap to 2020 for Thailand.* Bangkok, Thailand: The Center for People and Forests.

Sato, J. (2003). Public land for the people: The institutional basis of community forestry in Thailand. *Journal of Southeast Asian Studies*, 34, 329–346.

Sudtongkong, C. & Webb, E.L. (2008). Outcomes of state- vs. community-based mangrove management in southern Thailand. *Ecology and Society*, 13(2), 27. Retrieved from http://www.ecologyandsociety.org/vol13/iss2/art27/.

Webb, E.L. (2008). Forest policy as a changing context in Asia. In E.L. Webb & G.P. Shivakoti (eds), *Decentralization, Forests and Rural Communities: Policy Outcomes in South and Southeast Asia* (pp. 21–43). New Delhi: SAGE Publications.

# 13

# Vietnam: Implications of Community-based Forest Management for Sustainable Forest Governance

*Tran Nam Thang and Ganesh P. Shivakoti*

## Introduction

There is a large body of research demonstrating that the conservation of forests in situations where local people dependent on forest resources requires a degree of participation from those communities. Crafting the overall framework of participatory management and forest conservation for large-scale policy setting requires an understanding of the actual use and management of forest resources and even of individuals' dependence on forest resources. Therefore, understanding the site-specific property rights structure, forest use pattern and forest dependence is fundamental for efforts to establish co-management programmes that may be sustainable over the long term. In-depth attention on property rights is crucial in situations where property rights offer incentives for management; provide authorisation and control over the resource; reinforce collective action; and assign rights to users, demonstrating the commitment of the government to devolution (Meinzen-Dick & Knox, 2001). Therefore, property rights are a central issue in policy development that alters the governance structure and the rights of individuals to use forest resources.

Current policy in Vietnam provides for the handover of the property rights to natural forests to communities, similar to the community forestry programmes of other Asian countries (The Vietnam Forestry Development Strategy for 2006 to 2020, issued by Decision 18/2007/ QD-TTg). This policy, initiated in 1995 through pilot programmes and experiments and, later, the Law on Forest Protection and Development (2004), has dramatically changed the people–forest relationship and is the result of an evolution of policy that began in 1975 with the reunification of Vietnam. This programme allocates forest and forest land to local people and communities and makes them responsible for its management.

Understanding property rights to natural resources helps to identify incentives, disincentives and, ultimately, the prospects for sustainable management and conservation of forests by communities. With this concern in mind, in this chapter, we evaluate the forest-related property rights of rural, forest-accessing communities of the Katu, an important ethnic group in the remote regions of Thua Thien Hue Province, central Vietnam. Our focus is to examine how local people manage, use and protect forest resources allocated to them through changes in their property rights towards forest resources.

Based on these findings, we examine the '*prototype design guidelines*' (Inoue, 2011) that derived from and evolved out of the design principles of common pool resources (CPRs) (McKean, 1999; Ostrom, 1990, 2005; Stern, Dietz, Dolsak, Ostrom, & Stonich, 2002), and we consider their applicability to this case study.

## Study Area

This study was conducted in the Nam Dong district of Thua Thien Hue province, northern central Vietnam. The total area of Nam Dong is 650.5 km². Sixty-four percent of the area (416 km²) is covered by forest, of which the majority (399.6 km²) is considered to be natural forest. Approximately 3,219 ha (5.31%) of Nam Dong is under permanent, registered agricultural use. Swidden agriculture was historically an important land management system that was practised along the margins of natural forest and permanent agricultural land, but has declined in Nam Dong since the government banned swidden agriculture in 1997. The government of Vietnam, through programme 327, initiated extensive reforestation, and in Nam Dong this programme resulted in the extensive establishment of exotic species (*Acacia* spp., *Eucalyptus*, *Cinnamomum* spp. and *Hevea brasiliensis*) on former swidden and degraded forest lands. Moreover, many former swidden fields have been converted into permanent agricultural areas. In addition to the effect of programme 327, the distribution of forestry land to households, new forest management practices and food crop intensification combined to create 'push and pull' effects, decreasing the agricultural footprint on hillsides (Meyfroidt & Lambin, 2008). Forests expanded as a result, mainly due to the liberalisation of agricultural output markets and the availability of new technologies (Sikor, 2001).

The programme 327 was implemented in Nam Dong with most of the investment funds coming from the government, so that the state owned the entire plantation forest. Local people only participated in forest management through labour and were paid in the form of rice or money. However, in the early 2000s, these rubber and forest plantations reached the age of harvest, and the benefits of plantation forests looked attractive to local people. Local people started to convert their swidden fields and hill gardens to plantations, and received land use certificates (LUCs) for that land by registering with the local authority. The government's allocation of forest land is one of the important incentives inducing people to participate in reforestation.

All forests and forest land (land designated for forestry purposes) are claimed by the state. Oversight and management of forests is the responsibility of the Ministry of Agriculture and Rural Development and the Ministry of Resources and Environment. These two ministries have provincial offices, the Department of Agriculture and Rural Development (DARD) and the Department of Resources and Environment (DoRE). These two departments are in charge of the protection, management and allocation of forest land and forests to households and communities (DoRE is in charge of land planning and administration). Thus, two agencies are responsible for the management of forests in Hue Province. Under DARD, there are Forest Protection Divisions and Forest Development Divisions at both the provincial and district levels. These divisions have specific tasks related to the protection and development of forest resources. The main owners of the forest and forest land are state forest enterprises (SFEs), making them important actors in forest allocation (prior to 2005). However, due to the SFEs' weak and inefficient resource management, they were converted into Watershed Management Boards and Forestry companies (2005–2007), and a large portion of the forest land under their management was taken back by the state for allocation to local people.

**Figure 13.1:**
*Study communes in Nam Dong district, Thua Thien Hue Province, Vietnam*

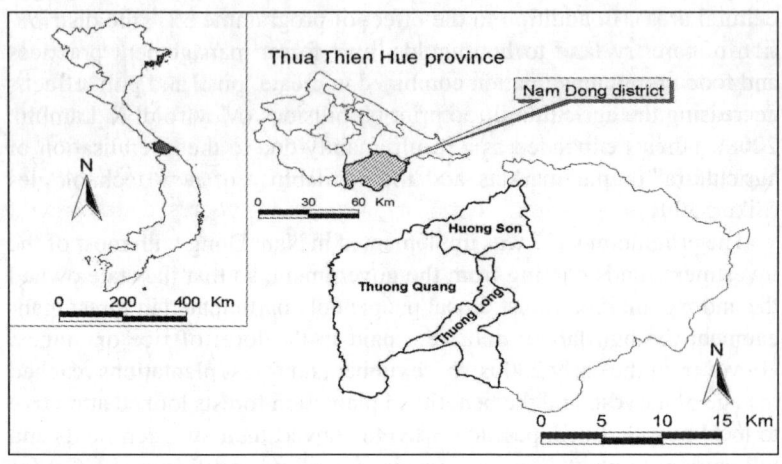

*Source:* Authors.
(Disclaimer: This image has been redrawn by the authors and is not to scale. It does not represent any authentic national or international boundaries and is used for illustrative purposes only.)

Three villages in three different communes of Nam Dong district, Hue province, were chosen as the study sites (Figure 13.1). Huong Son Commune is in the southeast region of the district, approximately 8 km from the district centre of Khe Tre. Thuong Quang lies in the southwest region of the district, approximately 17 km west of Khe Tre, and Thuong Long is approximately 14 km southwest of the district centre. All communes are accessible by asphalt roads, although Thuong Quang and Thuong Long were historically fairly remote villages until 1996. Huong Son consists of seven Katu villages, Thuong Long has eight Katu villages, and Thuong Quang has four Katu and three Kinh villages. The forest allocation process began in these three villages in 2004, although the red book certificates were only released in late 2006.

The Katu people are a large minority group in Thua Thien Hue province and they are the major ethnic group in Nam Dong district, accounting for 10,292 out of the total population of 23,875 people in 2008 (Nam Dong statistical book 2008). As with many other ethnic groups in Vietnam, they have a history of rotational swidden cultivation. The Katu people have usually lived in the upstream area, and their lives have been heavily dependent on forest resources. Since the sedentarisation programme (1968) and the ban on swidden cultivation (1997), the Katu people have settled down and begun to live permanently in arranged areas. They have learned to cultivate paddy fields and gardening, changing from hunting and gathering to cultivating and creating conditions for major crops (sustainable and intensive farming). Most Katu villages lie in and around the forests, and these communities still harvest forest products for their daily life, despite the fact that forest resources belonged to the state before the forest land allocation (FLA) process. On average, the Katu people are poor, and Nam Dong is one of the poorest districts in the country (Wetterwald, Zingerli, & Sorg, 2004).

Economic indicators suggest that Nam Dong has an average income far below the national average. The rural populations of Nam Dong, particularly the Katu households, depend to a significant degree on the forest for their subsistence and livelihood (Tran, 2004; Wetterwald et al., 2004) because there is shortage of flat areas for paddy cultivation and a low diversity of livelihoods in the area.

## Methods

Study communes were selected (1) containing a high percentage of Katu households, all of which have a high dependence on the forest for their livelihoods; (2) possessing a high percentage of natural forest cover and (3) with good access to both the forest and the markets.

Several preliminary reconnaissance visits were made to the three study communes to conduct informal interviews with key informants and establish a sampling design. Interviews were held with commune leaders, officers, village headmen and key Katu village elders. These preliminary discussions about property rights, institutions, historical and current forest use, and forest dependence revealed a high degree of similarity across Katu villages within a commune and across the three study communes. Based on the preliminary survey, we concluded that surveying one village in each commune would be sufficiently representative of the Katu villages in Nam Dong. Within each commune, one Katu village with allocated forest was randomly chosen for the study. This chapter aggregates the results from all the three villages and considers the results to be representative of the Katu people.

We randomly chose the households to interview from the list of households in each village. A total of 96 households were surveyed out of the total 148 households in the three villages. Semi-structured interviews were conducted to investigate historical and current patterns of Katu forest use with a total of 20 key informants. Key informants were village headmen, village elders, commune leaders, and representatives of the DARD, DoRE and FPD. Finally, a group discussion was held in each village to discuss the information that had been collected and to build consensus on its reliability. Final revisions were made to ensure that it represented the overall trends in Katu villages over the past several years.

Qualitative analysis was used to analyse the legal rights associated with the process and the *de facto* rights generated and practised by local people. The data on forest use and forest dependency were also quantitatively analysed using SPSS software.

# Results

Before describing the major changes that have occurred in the rights systems of the Katu, one important ethnic minority society of central Vietnam, and how these people use and manage forest resources at the grass-roots level, we briefly review the existing literature on property rights, forest devolution and community forest management (CFM) in Vietnam. Based on that review, we describe the forest use and forest dependence of the Katu people before forest allocation, and then examine the current status of those CFM parameters. Finally, we discuss the implications of those changes on property rights for the future of community-based forest management in Hue in particular, and on the overall governance of community forestry in general.

## *Property Rights, Forest Devolution and CFM in Vietnam*

Property commonly refers to things or assets (Bromley, 1989; Hann, 1998; MacPherson, 1978), while its formal usage refers to rights to things (Bromley, 1989; Bruce, 1998; MacPherson, 1978). Bruce (1998) stressed that the term 'property rights' is used to clarify the meaning of the word property. Schlager and Ostrom (1992) classified property rights to resources into different types. Access rights to a forest allow the rights holder to enter the forest and enjoy non-subtractive benefits. Withdrawal rights allow rights holders to obtain resource units of products from the forest. Management rights allow the right to regulate internal use patterns and transform the forest by making improvements (e.g., thinning and planting). Exclusion rights grant the right to determine who will have forest access and withdrawal rights, and how those rights are determined. Alienation rights allow rights holders to sell or lease the rights listed above. These five fundamental rights are the main rule structures under which individuals formulate 'ownership', and therefore shape relationships between people and the forest and amongst themselves. Property rights are *de jure* when resource users are officially granted the rights by the government and are lawfully recognised by formal, legal instruments. Property rights can also be *de facto* when resource users cooperate to define and enforce rights among themselves. *De jure* and *de facto* property rights can overlap, complement or even conflict with one another for a single resource (Schlager & Ostrom, 1992).

Devolution is the transfer of authority over forest management decision making from central government bureaucracies to local civil society actors, generally forest users and user organisations not created or controlled by the government (Fisher, 1999). Devolution is under consideration in many countries throughout the world that are trying to transfer property rights and responsibilities from the central government to local people (Edmunds & Wollenberg, 2003; White & Martin, 2002) with the goals of forest conservation (Balooni & Inoue, 2007) and livelihood improvement (Castella, Boissau, Thanh, & Novosad, 2006). The actual characteristics of the implementation of devolution differ from place to place for many reasons, and the benefits derived from forest devolution vary widely (Edmunds & Wollenberg, 2001, 2003; Shackleton & Campbell, 2001). However, forest devolution commonly leads to limited gains (Balooni & Inoue, 2007) that are often distributed unequally. The poor are usually neglected, leading to the loss of their livelihood (Edmunds & Wollenberg, 2003), while wealthier groups receive the benefits of devolution at the expense of the poor (Kumar, 2002). Nurse and Malla (2005) also questioned the contribution of forest devolution to rural development because these processes in South and Southeast Asia are restricted to secondary and degraded forests (Balooni & Inoue, 2007).

Community forestry has been considered as a new approach to natural resource conservation and livelihood improvement in Vietnam because forest resources were not sustainably managed by the state forest agencies (De Koninck, 1999) and because the best remaining forest resources remain under the control of SFEs (Sunderlin, 2006). In addition to the successful pilot programmes undertaken by the non-governmental organisations (NGOs) and government agencies at different levels, community forestry has been connected to the FLA through the Land Law of 2003 and the Forest Protection Law of 2004. Both these laws recognised the efforts of those pilot programmes and established the legal basis for community forestry in Vietnam.

Several studies have been performed related to forestry and FLA in Vietnam. One study conducted an in-depth examination of the incentives of stakeholders in the FLA process (Ngo & Webb, 2007). Several studies have reported that in Vietnam, legal rights do not translate into analogous changes in actual rights and practices (Sikor & Nguyen, 2007; Tran & Sikor, 2006). Changes in land use, for example, not only transfer responsibility to an individual but also give farmers incentives to rationally use the land to protect the resource because they receive the benefits of ownership and pay the costs associated with the degradation of the resource (Castella et al., 2006; Tran & Sikor, 2006). In contrast, the allocation of forest land played a major role in land use changes not because it provided the right incentives for farmers to reforest, but rather because the forest land allocation disrupted local institutions and collective land use systems (Clement & Amezaga, 2009). Moreover, forest re-growth in Vietnam was not due to a single process or policy but rather to a combination of economic and political responses to forest and land scarcity, economic growth, and market integration at the country scale, while destruction of old-growth forest continues (Meyfroidt & Lambin, 2008). The total forest area remained stable but major transitions from forest and non-forest categories occurred (Thiha, Webb, & Honda, 2007).

Nguyen and Noriko (2008) found a high dependency on forest resources in one community in Nghe An province after forest allocation, with the poor households receiving more than 65% of their income from the forest, richer households receiving 40% and the poorest households reporting 75% of their income from the forest (Nguyen & Noriko, 2008). In some successful cases, the forest allocation creates benefits for forest recipients and increases livelihood diversification in the community (Bao, 2003; Do, Hoang, & Le, 2007; Le et al., 1996; Sikor, 2001). However, there are some cases in which local livelihoods were not significantly improved and the benefit from forestry has decreased because the forest allocation policy does not allow people to practice slash-and-burn cultivation and raise livestock in allocated forest areas (Castella, Boissau, Thanh, & Novosad, 2002). The reduction of swidden cultivation outweighed the slow increase in paddy production, thus reducing labour productivity (Jakobsen, Rasmussen, Leisz, Folving, & Nguyen, 2007).

## Local Context and Forest Governance at the Grass-roots Level

The Katu people have a long tradition of dependence on forest resources. They live in upstream areas, and almost all of their daily food and food-stuffs come from the forest. In the past, they lived as self-sufficient communities, living separately in remote areas with very little contact with outsiders. Products harvested from the forest by the Katu people before the initiation of FLA include subsistence products for house-hold usage such as fuel wood, fruits, vegetables, bamboo, mushroom, honey, bush meat, medicinal plants, material for household construction and tools for production. The Katu people have also obtained several market-oriented products for additional family income, including honey, rattan, broom-making plant, hat-making plant, fruits, mushrooms, veg-etables and bamboo. The products harvested by the local Katu people are usually sold to the Kinh traders living in the villages, or communes or mobile traders who come to Katu households. Men are usually in charge of hunting and harvesting products that require intensive labour, while women and children work in the hill gardens and are involved in harvesting available and abundant products such as the broom-making plant, the hat-making plant, mushrooms, vegetables and bamboo. Until 2004, the Katu people in Nam Dong had a very high forest dependency for subsistence and additional income (Wetterwald et al., 2004), with the majority of Katu households obtaining 11–50% of their total income from the forest (Tran, 2004).

Note that villages are chosen for forest allocation based on the geo-graphic distribution of forest resources and the traditional possession of forest resources. This means that within a single commune, only a few villages receive forest allocations (these allocations are actually still in the form of experimental forest allocations). This has created two groups at the commune scale: forest recipients and non-recipients. This has changed the relationship between the villages.

Since the commencement of FLA, the rights and property rights of local people to forest resources have changed. Before forest allocation, all forest resources belonged to the SFEs. Thus, villagers did not have any legal rights towards these forest resources except for access rights. The most common way in which SFEs hired labourers was for pro-tection contracts, with local people patrolling and protecting forest resources, making them protectors of the SFE's forest. However, local people still harvested non-timber forest products (NTFPs) in protection

**Table 13.1:**
*Changes in* de jure *rights and property rights in the Katu villages under the FLA*

| Rights | Before the FLA | | After the FLA | |
|---|---|---|---|---|
| | *Allocated village* | *Neighbouring villages* | *Allocated village* | *Neighbouring villages* |
| Access | Yes | Yes | Yes | Yes |
| Withdrawal | No | No | Yes | No |
| Management | No | No | Yes | No |
| Exclusion | No | No | Yes | No |
| Alienation | No | No | Yes, but restricted | No |

*Source:* Tran (2011).

and production forests, and the state agencies implicitly allowed local people to exercise withdrawal rights.

In the forest allocation process, forest rights are transferred from the state to forest recipients, changing the relationship between villages that have received forest land and those that have not. Forest recipients received forest land and LUCs or *red books* for the period of 50 years, which are renewable if they wish and if they manage their forest well. As stated by the Land Law of 2003 and the Forest Protection and Development Law of 2004, forest recipients have the basic rights and responsibilities of forest management and a benefit-sharing mechanism regulated by the Decision 178/2001/QD-TTg. However, communities are not allowed to transfer their rights over forest resources to others. The only possibility for alternative uses of their rights is to mortgage, provide a guarantee or contribute capital in the amount of the added value of forest use rights. For this reason, Table 13.1 lists 'Yes, but restricted' for the alienation rights of the allocated village. Meanwhile, the rights of non-recipient villages (neighbouring villages) remain unchanged.

The forest management activities of recipient villages have undergone significant change. These villages have received support from local authorities or national and international NGOs such as The Green Corridor project, Tropenbos International (an NGO based in the Netherlands), the SNV (Netherland Development Organisation), ETSP/Helvetas (Swiss Association for International Cooperation; ETSP: Extension and Training Support Project for Forestry and Agriculture in the Uplands) and CORENARM (a local NGO: Consultative and Research Center on Natural Resource Management). Recipient villages have performed forest inventories to learn about their community forests, and from

**Table 13.2:**
*Changes in* de jure *and* de facto *rights in the Katu villages under the FLA*

| Rights | De jure *rights* | | De facto *rights* | |
| --- | --- | --- | --- | --- |
| | *Allocated village* | *Neighbouring villages* | *Allocated village* | *Neighbouring villages* |
| Access | Yes | Yes | Yes | Yes |
| Withdrawal | Yes | No | Yes | Yes, but limited |
| Management | Yes | No | Yes | No |
| Exclusion | Yes | No | Yes but loose | No |
| Alienation | Yes, but restricted | No | Yes, but restricted | No |

*Source:* Tran (2011).

that point developed detailed plans for forest management, building up nursery gardens for the replanting and enrichment of the forest, undertook rehabilitation projects, and formed patrol groups in each village that regularly patrol the forest (two times/month). During their patrols, these groups protect the forest by preventing violations by encroachers such as timber logging or animal hunting, performing management activities such as clearing climbers, performing fire prevention, and observing changes in the forest to enable appropriate management.

Under the FLA process, both the *de facto* and *de jure* rights of local people have changed, as described in Table 13.2.

Following the imposition of forest allocation, forest recipients have all rights over their forest resources, both *de jure* and *de facto*. For withdrawal rights, recipient villages have all rights to NTFP products except for the wildlife list regulated by the governmental Decree 32/2006/ND-CP. For timber, local people can remove the amount stipulated in Decision 178/2001/QD-TTg by the Prime Minister. However, this withdrawal requires the participation of the forest protection department and district functional agencies. Since the first implementation of CFM, only one village in Daklak province and one village in Thua Thien Hue province have experimented with harvesting timber resource (Oberndorf, Durst, Mahanty, Burslem, & Suzuki, 2006). For exclusion rights, recipient villages have full exclusion rights to prevent outsiders from encroaching on their forest. However, they only exercise this right to a certain extent. They still allow outsiders to come and harvest NTFPs from their forest. When asked why recipient villages are 'loose' about their exclusion rights, local people provided the following answers. (1) Different forest areas are adjacent to each other and there is no clear boundary or fence. (2) Local people used to have 'open access' to the state forest and people are used to harvesting NTFP products anywhere they want. (3) People in the village and throughout the commune all know each other, and they feel that it would be difficult to stop their neighbours from harvesting the NTFP products. In addition, these forest areas used to be the harvest areas of other villages, so those villages still carry out their traditional practices. (4) Small amounts of products are harvested due to resource depletion, and thus the villagers do not care if those small amounts of products are harvested by outsiders. In 2009, the three villages studied here found four violations related to timber logging in their forests. They managed two of those cases themselves and informed the commune forest protection unit to obtain support in dealing with the other two cases of armed encroachers.

Meanwhile, the non-recipient villages do not have *de jure* rights over the forest allocated to recipient villages except for access rights. However, non-recipient villages still enjoy *de facto* rights, including the benefits of NTFPs obtained from the allocated forest thanks to the loose exclusionary practices of recipient villages, although this right does not extend to timber resources. Because this right is limited to certain types of products and depends on the will of the recipient village, we call the *de facto* withdrawal rights of non-recipient villages as *limited* withdrawal rights.

Changes in patterns of forest use have led to changes in forest-derived income. The contribution of forest-derived income to the total income of local people has dramatically decreased (Thang, Shivakoti, & Inoue, 2010) over the years.

The downward trend of forest dependency of local people in the study area calls into question the often assumed panacea that allocating forest land to the community will increase forest destruction due to over harvest. Although recipients of forest resources are granted greater individual and community rights towards the forest, forest land and other products derived from the forest land, local people have become less dependent on forest resources for their livelihoods. The reason for this reduction in forest income is that forest products are becoming increasingly scarce due to the over extraction and poor status of allocated forest. Group discussion also revealed that forestry activities yield a low income for intensive labour, with risks associated with complex topographical conditions and unexpected threats. For these reasons, many of the local people switch to other livelihood options due to the alternative choices available for people as a result of livelihood diversification and the changes in local people's income structures due to opportunities for improved methods of agricultural production and animal husbandry activities. At present, the allocated forests are perceived by local people as a long-term low-cost investment to ensure future family security (Thang et al., 2010).

## Discussion

Local people have actively participated in the FLA process and have played the main role in most related activities, including discussion, decision making, implementation, monitoring and evaluation. This process not only requires the participation of local people but also demands the support and participation of local authorities and functional agencies to help in the implementation of all related activities in the field.

Before FLA, local people held *non-written* access and withdrawal rights over forest resources with *de facto* rights to harvest NTFP products in both protected and production forests. The forest allocation programme led to a significant change in the property rights of recipient villages over forest resources. This transition from access rights alone to almost full rights over forest resources was a great achievement of the FLA programme. From the point of view of resource management theory, this contributed to the protection and management incentives for local people and altered the governance structure and the rights of users over forest resources (Meinzen-dick & Knox, 2001). With these changes, the *de jure* rights of forest recipients and non-recipients are concretely different. Forest recipients hold almost full rights, while the rights of non-recipients remain the same, including access rights to the allocated forest resources.

The *de facto* rights of non-recipient villages are different from the *de jure* rights. People in adjacent villages or even different communes still have the rights of withdrawal of *legal* forest products from allocated forest (for legal NTFPs, with the agreement of the allocated village) and *non-written* access and withdrawal rights over state forest resources. This agreement allowing for legal harvesting by outsiders is prevalent in all three study villages, demonstrating the close relationships among local villages as well as the *loose* management and protection of their natural capital. Moreover, as most of the allocated forests are poor and degraded, the quantity and quality of forest products are low, reducing the incentives for local people to restrict access to those resources. For legal NTFP products in the study area, recipients and non-recipients currently enjoy the *same* withdrawal rights in practice for both allocated forest land and state forest land. Recipient villages have full rights over their allocated forest, and similar to non-recipients, have *non-written* access and withdrawal rights over the state forest land. In principle, this allows forest recipients to receive a greater legal share of benefit from the forest compared to non-recipients in the same commune. However, these changes have not brought about much short-term benefit or contributed to the livelihood of the local Katu people.

Until now, forest recipients have only practised access, withdrawal, management and exclusion rights over their forest resources. As specified in the Law on Forest Protection and Development (2004), allocated communities are not allowed to '*divide forests among their members; not to convert, transfer, donate, lease, mortgage, provide guarantee or contribute business capital with, the value of the use rights over the assigned forests*'. They are only allowed to '*mortgage, provide guarantee or contribute capital with, only the added value of forest use rights, which is brought about by forest owners' investments compared to the forest use right value determined at the time of being leased forests according to law provisions*'. It is difficult for local people to determine the added values of their forest resources and how to specify those as capital contributions. This makes it complicated for local communities to call for investment in or cooperation over their allocated resources.

In addition to exclusion and alienation rights, the other *de facto* difference between recipients and non-recipients is timber incentives. However, it will take a long time for forest recipients to benefit from these incentives due to the low quality of the allocated forest. There are several reasons for the degraded conditions of the allocated forest (Sunderlin, 2006), including poor management of valuable natural forests by SFEs (Balooni & Inoue, 2007), or worse, the delay of forest allocation by the SFE to maximise timber extraction before allocation (Ngo & Webb, 2007). Thus, in the short and medium term, there is not much difference between forest recipients and non-recipients in terms of the actual benefits derived from forest resources. This affects the incentives of local people and their participation in forest protection and management activities to some extent.

While recipient villages have received increased property rights over forest resources, the forest use and forest dependency of forest-allocated communities have decreased compared to the past. This is in contradiction to another study (Jakobsen et al., 2007) that found a very high level of forest dependency in one community in Nghe An province after forest allocation.

## Prototype Design Guidelines

We now consider the Prototype Design Guidelines for 'Collaborative Governance' of Natural Resources (Inoue, 2011), including graduated membership, commitment principles and fair benefit distribution. This case study in Vietnam can serve as an example for these guidelines due to their presence in the organisation, operation and management of the community forest.

*Graduated membership:* Graduated membership is observed in all study villages, with the community transferring the decision-making power to the elected Board of Management. The chairman (usually the village head) is also the representative of the community forest, and the regular community forest meeting mainly involves these core members of the Management Board. This current structure of a chairman and a management board is common among all community forests in Vietnam, and these bodies are mainly responsible for both internal and external interactions related to the management and protection of the community forest.

*Commitment principle:* Due to the current low amount and economic value of NTFP products, in addition to the low recent level of forest dependency, local people are not highly concerned about preventing outsiders from harvesting NTFP products. Thus, outsiders are not excluded from harvesting NTFPs. Outsiders still enjoy the benefits given to villagers and to the members of the core groups. We believe that the traditional cohesion of the ethnic groups and the fact that they live in close proximity is one of the main reasons for this lack of the commitment principle in the protection and management of the community forest. Thus, in the context of this case study, the commitment principle concept is infrequently applied, except to protect the long-term accumulation of timber value.

*Fair benefit distribution:* In the current system of management by local people, the benefits derived from the community forest are at the household level (for all NTFP products), and no major changes will occur in the way that local people harvest their NTFPs because local people currently have low interest in these low benefit resources. They even allow outsiders to come and harvest these products freely. The only possible benefit-sharing mechanism is timber, although this benefit will not be realised for a long time due to the current poor status of the forest. Therefore, the guideline of 'fair benefit distribution' seems to apply, as local people have achieved a fair benefit distribution in accordance with their own efforts to harvest forest products. For the timber products, the village management boards have established that these benefits will be fairly distributed among all members of the community forest over the long run.

However, creating recipients and non-recipients within the same communes will lead to a difference in the level of commitment to forest protection between the two groups, which could inhibit the protection and management of natural resources over the long run. There are no differences in the current benefit distribution between these two groups, reducing the incentive for forest recipients to participate in forest management and also creating low incentives for non-recipients to try to become forest recipients.

## Conclusion

In general, FLA has led to positive changes in the participation of local people in forest conservation and management, including changes in property rights and changes in the forest use and dependency of local people in the three study villages. The changes include the *de jure* and *de facto* rights of both forest recipient villages and non-recipient villages. Forest recipient villages now hold the full bundles of both *de jure* and *de facto* rights, confirming their absolute ownership over the allocated forest resources. In theory, this would provide incentives for local people to participate in and contribute to the protection and management of the allocated forest. Recipient villages almost have full *de jure* rights towards allocated resources except for the *restricted* right of alienation. The community rights cannot be transferred to others. The Katu people allow others to enjoy strong *de facto* rights to harvest NTFPs from their community forests. At present, the non-recipient villages enjoy the *limited* benefit of the defined and practice *de facto* rights of recipient villages. This implies *loose* management and protection of allocated forest resources by forest recipients.

Based on the differences in the *de jure* and *de facto* rights over forest resource of the Katu people, we can conclude that: (1) to some extent, the influence of the long tradition of forest management of the Katu people and the importance of community in their daily activities remains in effect. Community forestry is a tradition in Katu communities because they have lived on forest resources for centuries. (2) These resources are so poor that they do not create sufficiently strong incentives for recipient villages to care about protecting them.

Our findings confirm those of Tran and Sikor (2006) that legal rights and actual rights are not necessarily translated from laws and regulations to property rights and land use practices, and forest allocation and property rights transfers are only the first step in sustainable natural resource management.

Moreover, we argue that the creation of forest recipients and non-recipients with different property rights over forest resources during the FLA process will create long-run conflicts and constraints on achieving sustainable forest resource management. This should be considered by the government and local authorities in the formulation of forest development strategies and in the implementation of CFM to improve the conservation of forest resources and to improve the livelihoods of local people.

The results of this study suggest that the government of Vietnam should consider the differences in *de jure* and *de facto* rights in the local context. These differences in rights can sometimes contribute significantly to the improvement or reduction of the conservation of forest resources and the livelihoods of local people. Moreover, supporting policies and post-allocation programmes are needed to create incentives for local people to protect and invest in the allocated forest and to help forest recipients enjoy more benefits from the protection and management of the allocated forest. For example, these programmes could provide support in the form of silviculture techniques, investments and loans for long-term production models, payment for environmental services, REDD or upland watershed protection rewards. These supporting policies would create long-term incentives for local people to better protect and manage their forest resources. This is also in agreement with the conclusion of Nguyen (2006) that for people to benefit from forest devolution, the state policy should not only focus on how people obtain rights to the devolved forest but also on how people derive true economic benefits from that forest.

This case study provides proof that the 'prototype design guidelines' of graduated membership, the commitment principle and a fair benefit distribution exist in community forests in Vietnam and can be applied with some conditions. Forest protection and management and the satisfaction of local people could be better served by the allocation of a higher quality and larger quantity of forestland, the encouragement and mobilisation of greater investment in forest resources, the development of favourable policies for forest development by both the government and local authorities, a higher level of decentralised rights towards forest resources, improved local institutions and post-allocation programmes that embrace the local context.

# References

Balooni, K. & Inoue, M. (2007). Decentralized forest management in south and Southeast Asia. *Journal of Forestry*, 105(8), 414–420.

Bao, H. (2003). *Participatory Technology Development on Natural Forests Allocated to the M'nong Ethnic Community*. Tay Nguyen: Social Forestry Support Program, Tay Nguyen University.

Bromley, D.W. (1989). *Property Rights and Institutional Change. Economic Interest and Institutions: The Conceptual Foundations of Public Policy*. New York: Basil Blackwell Inc.

Bruce, J.W. (1998). Review of tenure terminology. *Tenure Brief*, 1, 1–8.

Castella, J.C. Boissau, S. Thanh, N.H., & Novosad, P. (2002). Impact of forestland allocation on agriculture and natural resources management in Bac Kan province, Vietnam. In J.C. Castella, & D.D. Quang (eds), *Doi Moi in the Mountains: Land Use Changes and Farmers' Livelihood Strategies in Bac Kan Province, Vietnam* (pp. 197–220). Hanoi: Agricultural Publishing House.

———— (2006). Impact of forestland allocation on land use in a mountainous province of Vietnam. *Land Use Policy*, 23(2), 147–160.

Clement, F. & Amezaga, J.M. (2009). Afforestation and forestry land allocation in northern Vietnam: Analyzing the gap between policy intentions and outcomes. *Land Use Policy*, 26(2), 458–470.

De Koninck, R. (1999). *Deforestation in Viet Nam* (101pp.). Ottawa: IDRC.

Do, D.S., Hoang, L.S., & Le, Q.T. (2007). Forest governance in Vietnam. In Henry Scheyvens, Kimihiko Hyakumura, & Yoshiki Seki (eds), *Decentralization and State-sponsored Community Forestry in Asia* (pp. 139–159). Japan: Institute for Global Environmental Strategies (IGES).

Edmunds, D. & Wollenberg, E. (2001). Historical perspectives on forest policy change in Asia: An introduction. *Environmental History*, 6, 190–212.

———— (2003). *Local Forest Management: The Impacts of Devolution Policies*. London: Earthscan Publication.

Fisher, R. (1999). Devolution and decentralization of forest management in Asia and the Pacific. *Unasylva*, 50(4), 3–5.

Inoue, M. (2011). *Prototype Design Guidelines for 'Collaborative Governance' of Natural Resource.* Presented at 13th Biennial Conference of the International Association for the Study of the Commons.

Jakobsen, J., Rasmussen, K., Leisz, S., Folving, R., & Nguyen, V.Q. (2007). The effects of land tenure policy on rural livelihoods and food sufficiency in the upland village of Que, North Central Vietnam. *Agricultural Systems*, 94(2), 309–319.

Hann, C.M. (1998). Introduction: The embeddedness of property. In C.M. Hann (ed.), *Property Relations: Renewing the Anthropological Tradition* (pp. 1–47). Cambridge, UK: Cambridge University Press.

Kumar, S. (2002). Does 'participation' in common pool resource management help the poor? A social cost-benefit analysis of joint forest management in Jharkhand, India. *World Development*, 30(5), 763–782.

Le, T.C., Rambo, A.T., Fahrney, K., Tran, D.V., Romm, J., & Sy, D.T. (1996). *Red Books, Green Hills: The Impact of Economic Reform on Restoration Ecology in the Midlands of Northern Vietnam.* CRES, Hanoi University. Southeast Asian Universities/ Agroecosystem Network. Berkeley: East-West Center/Program on Environment, University of California.

MacPherson, C.B. (1978). The meaning of property. In C.B. MacPherson (ed.), *Property: Mainstream and Critical Positions* (pp. 1–13). Toronto: University of Toronto Press.

McKean, M.A. (1999). Common property: What is it? What is it good for, and what makes it work? In Clark Gibson, Margaret McKean, & Elinor Ostrom (eds), *Keeping the Forest: Communities, Institutions, and the Governance of Forests* (Chapter 2). Cambridge: Massachusetts Institute of Technology Press.

Meinzen-dick, R. & Knox, A. (2001). Collective action, property rights, and devolution of natural resource management: A conceptual framework. In R. Meinzen-dick, A. Knox, & M. Di Gregorio (eds), *Collective Action, Property Rights and Devolution of Natural Resource Management.* Feldafing, Germany: German Foundation for International Development, International Food Policy Research Institute, and International Centre for Living Aquatic Resource Management.

Meyfroidt, P. & Lambin, E.F. (2008). The causes of the reforestation in Vietnam. *Land Use Policy*, 25(2), 182–197.

Ngo, T.D. & Webb, E.L. (2007). Incentives of the forest land allocation process: Implications for forest management in Nam Dong district, Central Vietnam. In E. Webb, & G.P. Shivakoti (eds.), *Decentralization, Forests and Rural Communities: Policy Outcomes in South and Southeast Asia.* New Delhi/Thousand Oaks/London: SAGE Publications.

Nguyen, Q.T. (2006). Forest devolution in Vietnam: Differentiation in benefit from forest among local households. *Forest Policy and Economics*, 8, 409–420.

Nguyen, V.Q. & Noriko, S. (2008). Forest allocation policy and level of forest dependency of economic household groups: A case study in Northern Central Vietnam. *Small-Scale Forestry*, 7(1), 49–66.

Nurse, M. & Malla, Y. (2005). *Advances in Community Forestry in Asia.* Bangkok, Thailand: Regional Community Forestry Training Center for Asia and the Pacific.

Oberndorf, R., Durst, P., Mahanty, S., Burslem, K., & Suzuki, R. (2006). *A Cut for the Poor.* Proceedings of the International Conference on Managing Forests for Poverty Reduction: Capturing Opportunities in Forest Harvesting and Wood Processing for the Benefit of the Poor. Ho Chi Minh City, Vietnam, 3–6 October 2006.

Ostrom, E. (1990). *Governing the Commons: The Evolution of Institutions for Collective Action*. New York: Cambridge University Press.

———— (2005). *Understanding Institutional Diversity*. Princeton, NJ: Princeton University Press.

Schlager, E. & Ostrom, E. (1992). Property-rights regimes and natural resources: A conceptual analysis. *Land Economics*, 68(3), 249–262.

Shackleton, S. & Campbell, B. (2001). *Devolution in Natural Resource Management: Institutional Arrangements and Power Shifts. A Synthesis of Case Studies from Southern Africa*. Harrare: Center for International Forestry Research.

Sikor, T. (2001). The allocation of forestry land in Vietnam: Did it cause the expansion of forests in the northwest? *Forest Policy and Economics*, 2, 1–11.

Sikor, T. & Nguyen, Q.T. (2007). Why may forest devolution not benefit the rural poor? Forest entitlements in Vietnam's Central Highland. *World Development*, 35(11), 2010–2025.

Stern, P.C., Dietz, T., Dolsak, N., Ostrom, E., & Stonich, S. (2002). Knowledge and questions after 15 years of research. In E. Ostrome, T. Dietz, N. Dolsak, P.C. Stern, S. Sonich, & E.U. Webber (eds), *The Drama of the Commons* (pp. 445–490). Washington, DC: National Academy of Sciences.

Sunderlin, W.D. (2006). Poverty alleviation through community forestry in Cambodia, Laos, and Vietnam: An assessment of the potential. *Forest Policy and Economics*, 8, 386–396.

Thang, N.T., Shivakoti, G.P., & Inoue, M. (2010). Change in property rights, forest use and forest dependency of Katu communities in Nam Dong district, Thua Thien Hue province, Vietnam. *International Forestry Review*, 12(4), 307–319.

Tran, N.T. (2004). *Forest Use Pattern and Forest Dependency in Nam Dong District, Thua Thien Hue Province, Vietnam*. MSc.thesis. Asian Institute of Technology (AIT), School of Environment, Resources and Development, Thailand.

———— (2011). *Payment for Environmental Services as Incentive for Sustainable Forest Management in Thua Thien Hue Province, Vietnam*. PhD thesis. Asian Institute of Technology (AIT), School of Environment, Resources and Development, Thailand.

Tran, N.T. & Sikor, T. (2006). From legal acts to actual powers: Devolution and property rights in the Central Highlands of Vietnam. *Forest Policy and Economics*, 8(4), 397–408.

Thiha, Webb, E.L. & Honda, K. (2007). Biophysical and policy drivers of landscape change in a central Vietnamese district. *Environmental Conservation*, 34(2), 164–172.

Wetterwald, O., Zingerli, C., & Sorg, J. (2004). Non-timber forest products in Nam Dong District, Central Vietnam: Ecological and economic prospects. *Ecological and Economic Prospect*, 155(2), 45–52.

White, A. & Martin, A. (2002). *Who Owns the World's Forests? Forest Tenure and Public Forests in Transition*. Washington, DC: Forest Trends and Center for International Environmental Law.

# 14

# Laos: Local Communities and Involvement of External Stakeholders

*Kimihiko Hyakumura*

## Introduction

In recent years, global-scale environmental destruction due to forest loss and deterioration has become a major environmental concern. Forests of the world are disappearing at the rate of approximately 13 million hectares (Mha) each year (FAO, 2010). The impacts of that loss are causing a reduction in biodiversity and vertebrate populations, which reportedly declined by 30% between 1970 and 2007 (Pollard, 2010). This loss is also believed to have negative impact on local people who rely on forest products (WRI, 2005). In recent years, forest loss appears to have had a major impact on climate change (IPCC, 2007); countermeasures are being discussed at international negotiations (Hyakumura & Scheyvens, 2012). Tropical countries are a major factor in the loss of forests (FAO, 2010). In Southeast Asia, the loss of forests continues today, although this overall trend differs from that of China; however, China is also located in Asia and is significantly increasing the amount of forest cover through mobilised reforestation policies.[1]

---

[1] An increase in the forest area does not necessarily mean that biodiversity is increasing or that the villagers' livelihoods are being improved (Seki, Ko Ko, & Furukawa, 2009).

Located in Southeast Asia, the Lao People's Democratic Republic, or Laos, is one of the countries that is experiencing a steady loss of forests. Forest cover has declined in Laos from 49% in 1982 to 47% in 1992 and further to 42% in 2002 (DOF, 2005b); the Lao government acknowledges the importance of measures to address forest loss (DOF, 2005a). The government has identified the direct causes of this forest decline as illegal logging, land-use conversion for commercial crops and unsustainable swidden agriculture, and the underlying causes that have been identified include poverty, rapid population growth and weak law enforcement (DOF, 2005a).

Since the 1990s, with assistance from aid organisations, Laos has extensively promoted the establishment of protected areas for the conservation of biodiversity, including rich ecosystems and important flora and fauna. Being promoted from the perspective of biodiversity protection, the protection of the target areas comes with the nuance of protecting not only the useful trees but also the precious flora and fauna and their forest habitats.

At the same time, because the protected areas encompass rich forests, they are also home to local people who have lived there for generations and who depend on those forests. For these local people, the forests are a place of livelihood and productive activities. When many of the protected areas were established, the protections brought restrictions on the access to or use of the forest, and in some cases, the local people were removed from or refused access to the land. The government tolerated some negative impacts on local people as a result of the demarcation of protected areas, and in some cases, this stance led to conflicts. It has thus been suggested that 'collaborative governance' can be used to build relationships among stakeholders, to mitigate the negative impact on local people that arises from establishing the protected areas and to enable sustainable forest management.

This study sheds some light on the status of land use changes that involve the local people of a village in one protected area, and it also illuminates the impact on forest management and the livelihoods of local people, which is caused by external stakeholders' activities and protected area management policies. This study also examines the potential for collaborative governance in forest resource management associated with protected areas.

Field research was conducted in Village K, which is located in the forest of a protected area in the Atsaphangthong District of Savannakhet

Province. The study included interviews with all of the households regarding their productive activities, and interviews with influential villagers such as the village chief, elder groups and forest volunteers, regarding the status of land and forest use, related rules and regulations and historical changes. To analyse historical changes in village land and forest use, the study also used aerial photos and topographic maps as well as interviews with village elder groups. In addition, to determine the land and forest forms in the target village, supplementary observations were also conducted for the village's swidden land, rain-fed paddy fields, dry dipterocarp forests and deciduous mixed forests.

The author also conducted interviews and literature research in the national capital city of Vientiane regarding agricultural and forest policies and project implementation, at the offices of the Department of Forestry (DOF), the Ministry of Agriculture and Forestry and the Swedish International Development Cooperation Agency (SIDA); in addition, the study included the Provincial Agricultural and Forestry Office (PAFO), offices for Savannakhet Province and the District Agriculture and Forestry Office (DAFO) for Atsaphangthong District. The field studies were conducted in November 2001 and November 2002, and additional studies were performed in January 2004 and March 2006.

## Protected Area Management System in Laos

The former Lao government issued a report in 1965 that proclaimed the need for national parks in 10 areas (Sigaty, 2003, p. 20). After Laos became a socialist country through the 1975 revolution, the idea of national parks gained momentum. In 1986, the government adopted what it called New Economic Mechanisms, and while strictly adhering to single-party control on the political dimension, they embarked on a path of economic reform, promoting a shift from socialism to a market economy. While momentum for reforms grew, in May 1989, the government held the First National Forestry Conference, which was held amid the backdrop of rapid forest loss and deteriorating environmental conditions. Some topics discussed at the meeting such as countermeasures for those problems included the production of commercial products, policies to reallocate land and forests to local people to manage the forest resources and promote local forest management, and the promotion of tree planting. Additionally, based on their projects, the Lao-SIDA Forestry Programme and the International Union for Conservation of Nature (IUCN) articulated the desire for the rapid establishment of protected areas (Salter & Phanhtavong, 1989; Sigaty, 2003, p. 20).

As a result, Laos made gradual progress in the creation of a system for protected area management policies that are aimed at the protection of biodiversity. The first protected area management policy that is worthy of special mention is Prime Minister Decree No. 164, which was enacted in 1993 and called the 'Prime Minister Decree on Establishment of National Biodiversity Conservation Areas (Protected Areas)'. This Prime Minister Decree set forth 18 protected areas (*Paa Sagwan Hensaat*) for the first time at the national level. This decree, however, did not translate into detailed management of the protected areas. Two more protected areas were added later (Prime Minister Decree No. 7 in 1995 and Prime Minister Decree No. 210 in 1996), and the area expanded (Prime Minister Decree No. 579 in 1998). Furthermore, by Prime Minister Decree No. 193 (promulgated in November 2000), two 'corridor zones' were established between the present protected areas. With these additions, the protected areas increased from the original 2,824,300 ha to a total of 3,313,596 ha.

From the time of its original establishment until 2000, forest management in the protected areas was conducted with the support from many aid organisations (Table 14.1). The overall picture shows clearly that the forest management in the many protected areas was conducted with support from aid organisations.

At the same time, the government of Laos became a party to the Convention on Biological Diversity in 1996, but the only sites registered under the World Heritage Convention (which Laos joined in 1987) are the Vat Phou and Associated Ancient Settlements within the Champasak Cultural Landscape and the Town of Luang Prabang. Laos became a party to the Washington Convention (CITES) in 2004.

The legal framework related to protected areas was formulated after 1996, when specific laws relating to forests were enacted, including the Forest Law and the Land Law. The Forest Law was the country's first comprehensive law relating to forests. The Forest Law delineated forest districts, which are defined as 'protected areas', and articulated the concept of zoning. This law also recognised the rights of local people to use the forests in protected areas, *albeit* with restrictions.

**Table 14.1:**
*Protected areas of Laos*

| | Protected area | Province | Area (ha) | Established | No. of affected villages | Population | Supporting organisation (period) |
|---|---|---|---|---|---|---|---|
| 1 | Dong Ampham | Attapeu | 200,000 | 1993 | 188 | 87,182 | IUCN (1995–2000) |
| 2 | Dong Houa Sao | Champasack | 110,000 | 1993 | 90 | 36,987 | FOMACOP (1995–2000)[6] Netherlands (2000–2002) |
| 3 | Dong Phouvieng | Savannakhet | 197,000[1] | 1995 | 75 | 17,843 | FOMACOP[6] (1996–2000) |
| 4 | Hin Nam No | Khammuane | 82,000 | 1993 | 48 | N/A | WWF (1999–2001) |
| 5 | Na Kai Nam Theun | Khammuane Borikhamxay | 353,200 | 1993 | 75 | 15,700 | IUCN/WCS[7]/World Bank (1998–2000) World Bank (2000–2002) |
| 6 | Nam Et | Huaphanh | 170,000 | 1993 | Minimal[2] | N/A | IUCN (1999–2000) |
| 7 | Nam Ha | Luang Nam Tha | 224,000[3] | 1993 | 104 | N/A | WCS[7]/DED[8] (1996–1999) UNESCO (2000–2002) |
| 8 | Nam Kading | Borikhamxay | 169,000 | 1993 | N/A[4] | N/A | None |
| 9 | Nam Poui | Xayabury | 191,200 | 1993 | 45 | 24,556 | 1993–2000 (SIDA/IUCN) |
| 10 | Nam Xam | Huaphanh | 70,000 | 1993 | Minimal[2] | N/A | None |
| 11 | Phou Den Din | Phongsaly | 222,000 | 1993 | Minimal[2] | N/A | None |
| 12 | Phou Hin Poun (Khammouan Limestone) | Khammuane | 150,000 | 1993 | 109 | 29,603 | FOMACOP[6] (1996–2000) |
| 13 | Phou Khao Khouay | Borikhamxay/Vientiane Vientiane Municipality Xaysomboum | 200,000 | 1993 | 49 | N/A | SIDA[10]/ADB (1993–2000) |

*(Table 14.1 Continued)*

(Table 14.1 Continued)

| | Protected area | Province | Area (ha) | Established | No. of affected villages | Population | Supporting organisation (period) |
|---|---|---|---|---|---|---|---|
| 14 | Phou Leui | Huaphanh, Louang Phabang Xieng Khouang | 150,000 | 1993 | Minimal[2] | N/A | IUCN (1999–2000) |
| 15 | Phou Phanang | Vientiane Municipality Vienntiane Province | 70,000 | 1993 | N/A (many) | N/A | Assistance from Canada (1997–1998) |
| 16 | Phou Xang He | Savannakhet | 109,900 | 1993 | 117 | N/A | SIDA[10]/IUCN (1994–2000) |
| 17 | Phou Xieng Thong | Saravane | 120,000 | 1993 | 78 | 29,743 | SIDA[10]/PDI[11] (1996–2000) |
| 18 | Xe Bang Nouan | Saravane/Savannakhet | 150,000 | 1993 | 65 | N/A | SIDA[10]/IUCN (1996–2000) |
| 19 | Xe Pian | Champasack/Attaphu | 240,000 | 1993 | 58 | 26,834 | FOMACOP[6] (1996–2000) DANIDA[9] (2000–2001) |
| 20 | Xe Sap[5] | Saravane/Xekong | 133,500 | 1996 | 63 | 14,670 | FOMACOP[6] (1996–2000) |
| | | | 3,313,596 | | | | |

[1]Established with 53,000 ha, by Prime Minister Decree No. 7 in 1995. Expanded to 197,000 ha by Prime Minister Decree No. 579 in 1998.

[2]Specific numbers have not been provided.

[3]First established with 69,000 ha. Expansion was approved by Prime Minister's Office in 1999, bringing the area to 224,000 ha.

[4]No villages are located in the most important sections of the protected area.

[5]Established by Prime Minister Decree No. 210 in 1996.

[6]Forest Management and Conservation Programme. This is the name of a joint assistance project involving Finland and the World Bank.

[7]Wildlife Conservation Society, an international NGO.

[8]German Development Service, a German aid organisation.

[9]Danish International Development Agency, Denmark's international aid organisation.

[10]Lao-SIDA Forestry Programme, a forestry programme of the Swedish International Development Cooperation Agency (SIDA, Sweden's aid organisation).

[11]Population and Development International, and aid organisation based in Thailand.

**Source:** Prepared by the author based on LSFP (2001).

## Description of Village K and the Surrounding Protected Area

Savannakhet Province, where Village K is located, is in southern Laos. The forest cover of 56.6% in this province is higher than the national level of 42% (DOF, 2005b). Much land suitable for paddy fields can be seen in the Savannakhet Plain, in the western part of the province, but the central to eastern parts are characterised by mountains and hills, and it is in these forested areas that protected areas and production forests have been designated at the national level. In the hilly areas, besides paddy fields, swidden agriculture is also evident, but in recent years, commercial plantations have been active for rubber, eucalyptus and other products (DOF, 2007; Fujita, 2012; NAFRI, 2009). The main ethnic groups, the *Lao* and the *Phu Thai* tribes, occupy the plains of the province, but a variety of minority groups live in the hilly areas and surroundings, including the *Bru* tribe, the *Katang* tribe and the *Xuey* tribe.

As mentioned earlier, Village K is in the Atsaphangthong District, in the central part of the province. This village is located in hills that have an elevation of 200–300 m, on the north side of a ridge of the Phou Xang He mountains, which are aligned East to West. The entire territory of the village is contained within the protected area (Figure 14.1). To get from the village to the Ban Nakoutchan of the district's central area requires travelling approximately 30 km to the south, crossing over the Phou Xang He mountains on a walking path. At the same time, approximately 5 km north of the village is provincial route No. 10, and access from here to the major provincial town is good. During the dry season, a vehicle can travel from Village K to the provincial road, and at Village D along the provincial road, people can procure necessities at a weekly market. The village is not served by infrastructure such as electricity, gas or water. The village population is 270 persons, in 49 households (as of 2002), and most villagers are of the *Bru* tribe, speaking the language of the *Mon-Khmer* tribes, and only one in five households is of the ethnic *Phu Thai* tribes.

The main livelihoods in the village are paddy cultivation and swidden agriculture. In the southern part of the town is a natural deciduous mixed forest connected to the Phou Xang He mountains, while in the hilly area of the northern part is a dry dypterocarp forest known as *ghok*. The traditional *Bru* tribe livelihood was swidden agriculture in the hills, but in recent years, they have also actively engaged in rain-fed

**Figure 14.1:**
*Protected area and Village K in Savannakhet Province*

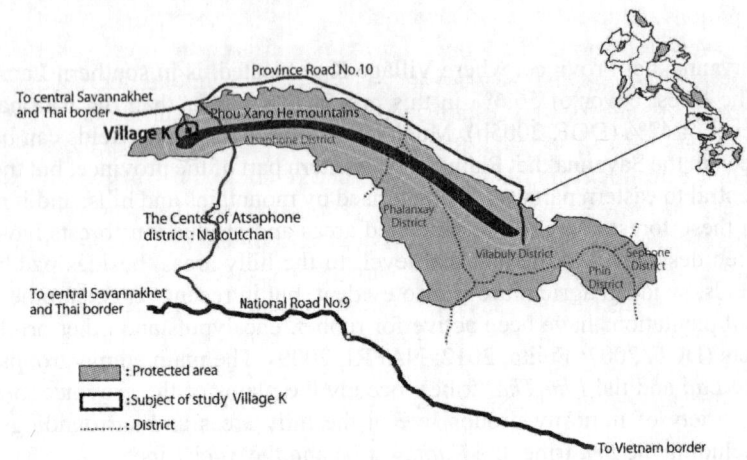

**Source:** Prepared by the author from topographic map.
(Disclaimer: This image has been redrawn by the author (Kimihiko Hyakumura) and is not to scale. It does not represent any authentic national or international boundaries and is used for illustrative purposes only.)

paddy cultivation. The village's agricultural land can be seen in the flat terrain of the water catchment along the north side of the Phou Xang He mountains. Paddies are located in the well-drained basin-like flat terrain, while swidden agriculture is conducted in the surrounding hills. Almost no agricultural land is located in the deciduous mixed forest, although villagers actively gather non-timber forest products (NTFPs) such as bamboo shoots and dammar resin. At the same time, villagers are engaged in swidden agriculture in some places in the dry dypterocarp forest south of the village, but large rocks in the soil make many spots unsuitable for cultivation.

Regarding the village decision-making process, in many cases, a discussion starts on a specific topic or issue with the leadership of the village organisation, such as the chief or vice-chief, and a decision is made by presenting opinions to the elder group meeting (*neohoom*). Important items, such as issues that significantly affect villagers' livelihoods or aid projects, are ultimately decided at village meetings that are held later, with all of the villagers participating. The elder group meetings are composed of influential members such as former village chiefs and former village board members. In Village K, the chief has the greatest authority, but the chief typically considers the village affairs in consideration of the opinions of the elder group meeting.

Work relating to the village forests is entrusted to forest volunteers (*asasamak pamaiban*), who are selected from among the villagers. Forest volunteers were established based on aid organisations' recommendations for protected area management projects during the 1990s. Their roles include tasks on projects that relate to the village protected area and forest management, acting as a conduit for focal points of aid organisations and provincial and district agricultural and forestry offices and communicating with villagers about the projects.

Of the three strategies in terms of the local people, which are indicated in Chapter 1, this case applies to the adjustment strategy. The reason is that Village K's territory, including the human settlement and agricultural land, is entirely contained in a protected area. Thus, a Protected Area Management Policy is being implemented by the government, and local people must comply with it.

The following stakeholders could be identified as relating to Village K forests (Table 14.2). First is the local people, who are the main users of the forest resources. Next are the central and provincial governments,

**Table 14.2:**
*Stakeholders involved in land and forest in Village K*

| Social level | Stakeholders | Role |
|---|---|---|
| National | Central government (Department of Forestry) | Formulation and implementation of protected area management policies |
| | Provincial Agricultural and Forestry Office (PAFO) District Agriculture and Forestry Office (DAFO) | Implementation of protected area management policies at the local level |
| Province/district | Aid organisations (SIDA) | Support and facilitation of implementation of protected area management policies |
| | Private sector companies, investors | Purchasing of NTFPs |
| | Villagers of Village K | Use of forest resources in Village K |
| Local | Local people of surrounding areas | Use of forest resources in vicinity of Village K |

*Source:* Author.

which implement protected area management policies in this area. Then, there are aid organisations that promote the protected area management policies; in addition, there are private companies and investors who attempt to benefit or profit from forest and land resources.

## Status and Changes of Land Use and Forest Use

### Livelihoods of Village Households

To begin with, the 49 households in the village were classified into three economic levels using a wealth ranking method (Grandin, 1988). Specifically, members from the village's influential class were asked to determine a standard of wealth in the village and to classify the households accordingly. Their standard of wealth was 'having enough rice to last throughout the year'. This system resulted in level A having 9 households (18.4% of all households), for which the amount of rice was fully sufficient, level B having 19 households (38.8%), which sometimes lacked rice, and level C having 21 households (42.9%), which regularly lacked sufficient rice.

Next, the author conducted interviews with each household about their livelihoods (Table 14.3). All of the households in levels A and B and 81% of the households in level C owned rice paddies. This observation is a clear sign that rice paddies constitute primary source of livelihood for the villagers. At the same time, whereas 77.8% of the households in level A owned swidden land, slightly more than 30% in both levels B and C owned any. Level A, in addition to rice paddies, could obtain rice from swidden agriculture. In addition, 44.4% of level A households conducted vegetable cultivation at the river bank during the dry season, compared to 15.8% for level B and 4.7% for level C; the higher the economic level, the more likely the household had this option available. One can see that the upper levels had more agricultural land, including rice paddies and swidden land.

A comparison of labour capacity reveals a large gap in the workforce[2] between the economic levels, with 4.3 persons per household in level A and 2.4 persons per household in level C, which is a disparity

---

[2] The workforce is counted as the number of villagers from 15 to 60 years of age.

**Table 14.3:**

*Livelihood of each household in Village K*

| Economic level | | Level A | Level B | Level C | Total |
|---|---|---|---|---|---|
| Number of households | | 9 (18.4%) | 19 (38.8%) | 21 (42.9%) | 49 |
| Gender of household head | Male | 9 | 18 | 14 | 41 |
| | Female | 0 | 1 | 7 | 8 |
| No. of persons | | 60 | 113 | 97 | 270 |
| No. of workers | | 39 | 66 | 50 | 155 |
| No. of workers per household | | 4.3 | 3.5 | 2.4 | 0 |
| No. of households depending on other households for livelihood | | 0 | 1 | 9 | 10 |
| Rain-fed rice paddies | Yes | 9 (100%) | 19 (100%) | 17 (81.0%) | 45 (91.8%) |
| | Many (1 ha or more) | 4 (44.4%) | 5 (26.3%) | 2 (9.5%) | 11 (22.4%) |
| | Few (less than 1 ha) | 5 (55.6%) | 14 (73.7%) | 15 (71.4%) | 34 (69.4%) |
| | No | 0 (0.0%) | 0 (0.0%) | 4 (19.0%) | 4 (8.2%) |
| Swidden agriculture land (ratio of household who has land) | | 7 (77.8%) | 6 (31.6%) | 7 (33.3%) | 20 (40.8%) |
| Vegetable garden along riverside | | 4 (44.4%) | 3 (15.8%) | 1 (4.8%) | 8 |
| Cattle (animals per household) | | 21 (2.3) | 17 (0.9) | 4 (0.2) | 42 (0.9) |
| Water buffalo (animals per household) | | 19 (2.1) | 30 (1.6) | 10 (0.5) | 59 (1.2) |
| Chickens (birds per household) | | 55 (6.1) | 76 (4.0) | 55 (2.6) | 186 (3.8) |
| Pigs (animals per household) | | 15 (1.7) | 21 (1.1) | 18 (0.9) | 54 (1.1) |
| Ducks (birds per household) | | 4 (0.4) | 0 (0.0) | 0 (0.0) | 4 (0.1) |

**Source:** Author interviews with each household.

of almost two persons per household. Disparities also appear for draft animals, such as the water buffalo, with 2.1 animals per household for level A vs. 0.5 animals for level C, which is a disparity of 1.5 animals. For cows, the number was 2.3 animals for level A vs. 0.2 for level C, which is a disparity of more than two animals. Considering both the workforce and the draft animals, the work capacity of the lower economic level was significantly lower. Additionally, for smaller animals such as chickens, pigs and ducks, the lower the level, the fewer animals each household owned.

Among the 21 households in the lower level, nine were surviving by receiving food from relatives, including parents and children who belonged to levels A and B. These households included five that owned no paddy fields or swidden land; thus, they probably depended largely on relatives for their livelihood. At the same time, all 12 households in the lower level that did not depend on other households owned paddy fields, but most of those sites were dry dypterocarp forests that were relatively recently cleared, which provide small harvests. Thus, approximately half of these households also owned swidden land.

## The Use of Non-timber Forest Products

Households at each economic level make active use of NTFPs (Table 14.4) (Hyakumura, 2003). Most households gather yam potatoes (*Dipsola hispida*), which are known as *goy* in the local language. The yam potato is a rice alternative; it is steamed and consumed together with rice during the rainy season and at other times when rice is in short supply (Hyakumura, 2002). The yam potato is eaten only when there is lack of rice. Thus, one could surmise that virtually none of the households had sufficient rice during the study years. Villagers stated that during normal years, more than half of the households would not eat the yam potatoes. In addition, all of the households would gather bamboo shoots (*Bambusa* spp.), which are known locally as *nomai*. Bamboo shoots are part of the routine diet, but rice was in short supply during the survey years; thus, the importance of bamboo shoots increased. Rattan shoots were also being harvested by all of the households. Most often they were used as food, but they were also used for other purposes, including barter and for sale. During the study years, rattan shoots were reportedly harvested in large quantities, for income

**Table 14.4:**
*Use of non-timber forest products gathered in Village K*

|  | *Use* | *Source location* | *Level A* | *Level B* | *Level C* | *Total* |
|---|---|---|---|---|---|---|
| Yam potato | Food | *Ghok* | 9 | 18 | 21 | 48 |
|  |  | *Turong, Turung* | 100.0% | 94.7% | 100.0% | 98.0% |
| Bamboo shoots | Food and for sale | *Ghok* | 9 | 19 | 21 | 49 |
|  |  | *Turong Turung* | 100.0% | 100.0% | 100.0% | 100.0% |
| Rattan shoots | Food and for sale | *Turung* | 9 | 19 | 21 | 49 |
|  |  | *Turong* | 100.0% | 100.0% | 100.0% | 100.0% |
| *Nyang* resin | For sale | *Turung* | 6 | 12 | 9 | 27 |
|  |  | *Taling turung* | 66.7% | 63.2% | 42.9% | 55.1% |
| Damar resin (*kisi*) | For sale | *Ghok* | 5 | 8 | 5 | 18 |
|  |  |  | 55.6% | 42.1% | 23.8% | 36.7% |
| Burmese mahogany resin bark (*shishiat*) | For sale |  | 1 | 1 | 1 | 3 |
|  |  |  | 11.1% | 5.3% | 4.8% | 6.1% |

**Source:** Author interviews with each household.

to purchase rice and for bartering. Thus, NTFPs provide an important food safety net for the villagers.

Yam potatoes are believed to be an essential famine food. It is important to take into consideration the harvesting process of potatoes. Potatoes grow deep in the ground, and it is important to not take the growing portion and to leave as much as possible growing in the ground. If this task is performed properly, then it will be easier to grow from the same rhizome a few years later. This knowledge was a sustainable method of resource use that villagers had acquired.

Villagers used to sell *nyang* resin (from *Dipterocarpus alatus*) to middlemen who visited the village, but demand for this product dropped rapidly after the government decided to ban the export of unprocessed *nyang* products. Thus, efforts to gather *nyang* dropped dramatically, leaving it to be used at home for torches or for sale to nearby villages (Figure 14.2). Damar resin (from *Shorea* spp.), known locally as *kisi*, is another source of revenue, and approximately 37% of all village households gather it. Damar resin is obtained from dry dypterocarp trees, but finding the trees that produce the resin requires considerable effort because one must search on foot; thus, only the households that have adequate labour and those that need the material would make the effort.

**Figure 14.2:**
*Torch Made Using* Nyang *Resin*

*Source:* Author.

## Right to Use the Land and Forest Products

This section describes the forms of land and forest use in Village K based on the interviews conducted (Table 14.5).[3] It is easy to focus on swidden agriculture as the major form of land and forest use in Village K. Agricultural land that can produce rice includes swidden land (*sharai*), abandoned swidden (*arui nyom*) and secondary forests that are swidden land in fallow (*arui* and *patensao*). If the stage of

---

[3] Village K is within the boundaries of the national protected area, and the land and forest use rights ultimately belong to the national government. However, because this area is remote, the government's protected area management policies are not being strictly applied (Hyakumura, 2010). The discussion here relates to the villagers' customary land and forest use rights.

Table 14.5:
Land types, forest types and rights to use forest products in Village K

| | Bru language | | Lao language | Forest type | Possessor of usage right | Right to gather NTFP | Timber rights |
|---|---|---|---|---|---|---|---|
| Paddy field | Taling turung | | Naa dong | MDF | Land user | Gatherer | Land user |
| | Taling ghok | | Naa khok | DDF | Land user | Gatherer | Land user |
| New paddy field | Taling ghok | | Naa sao mai | DDF | Person who clears land | Gatherer | Land user |
| Swidden | | Sharai pong | | | | | |
| | | Year 1 | Hai | MDF/DDF | Land user | Gatherer | Land user |
| | | Sharai kulay | | | | | |
| | | Year 2 | | | | | |
| | | Sharai kuluy | | | | | |
| | | Year 3 | | | | | |
| Abandoned swidden | Arui nyom | | Paa lao oon | MDF/DDF | Land user | Gatherer | Land user |
| Fallow swidden (secondary forest) | Sparse | Arui | Paa lao | MDF/DDF | Land user | Gatherer | Land user |
| | Dense | Patensao | Paa Lao Kae | MDF/DDF | Land user | Gatherer | Land user |
| Deciduous mixed forest (dense) | Sparse | Turong | Paa dong | MDF | Village | Gatherer | Village |
| | Dense | Turung | Paa dong dip | MDF | Village | Gatherer | Village |
| Dry dipterocarp forest | Rocky | Ghok | Paa kkok | DDF | Village | Gatherer | Village |
| | Not rocky | | Paa kkok hin | DDF | Village | Gatherer | Village |
| Spirit forest (protected) | Turung gian (turung put) | | Paa mahesak/ Paa sagwan | MDF | Village | Gathering prohibited | Use prohibited |
| Burial forest | Pingkhamui | | Paa saa | DDF | Village | Gathering prohibited | Only for cremation |
| Settlement | Kute wil | | | | | | |

MDF, mixed deciduous forest; DDF, dry dypterocarp forest.
**Source:** Author interviews in Village K.

transition proceeds further to the point that tall trees dominate and approach the height of a natural forest, the forest is known as *turong*, while a mixed deciduous natural forest at the climax stage is known as *turung*. On swidden land and swidden fallow, it is the cultivators who have the right to use the land and the portion of the trees above the ground. However, if the transition of *patensao* proceeds to the stage of *turong* and the site is abandoned, then the cultivators are seen to have abandoned their land use rights there. *Taling* (paddy field) that was cleared from what was originally swidden can be seen as a different form of use compared to swidden land. Thus, the usage rights of trees growing on a paddy field belong to the household using the site. Paddy fields can be classified as *taling turong* or *taling ghok*, depending on the original vegetation. Most of the old paddy fields were originally *turong*, while most of the paddy fields cleared relatively recently were *ghok*. Most *ghok* consists of land that was not yet claimed, but because it is on rocky soil, *ghok* land has few cultivable sites that are suitable for water paddies and swidden agriculture.

*Turong*, *turung* and *ghok* are considered to be communal forest of the village. The trees in a communal forest can be provided for public projects of the village, such as road repair, school construction and drainage. If a household wishes to use a tree in this forest, it can use the trees subject to having obtained the consent of the village chief, but only for household uses such as to build a dwelling. Meanwhile, any villager is permitted to gather NTFPs, not only from *turung*, *turong* and *ghok* but also from fallow forest and agricultural land. However, exclusive rights are attached to trees such as *nyang* (*D. alatus*), from which resin is extracted, *sii siat* (Burmese mahogany, or *Pentacme burmanica*), from which bark is taken, and only the household that owns the rights can use them.

Meanwhile, *turung gian* (spirit forest; Figure 14.3) and *pingkhamui* (cemetery forest) are communal assets of the entire village, and their mixed deciduous forest and dry dypterocarp forest (Figure 14.4), respectively, remain as climax forests. Each has its prohibitions. In the *turung gian*, no one is permitted to use trees or NTFPs. In *pingkhamui*, only deadwood can be used for cremation.

As mentioned earlier, with the exception of trees with exclusive rights and forests that have religious uses, any villager can gather NTFPs regardless of who has the land-use rights. As explained above, different rights can exist even on the same site, with regard to the right to use land and the right to gather NTFPs, and uses of the land can overlap.

**Figure 14.3:**
*View towards Turung dense forest of Phou Xang He Mountains from* Taling Dong
*paddy fields*

*Source:* Author.

**Figure 14.4:**
*Rocky dry dypterocarp forest, known locally as* Ghok Hin

*Source:* Author.

## Changes in Land and Forest Use

This section describes changes in land and forest use in Village K, based on interviews with elder groups and aerial photos. Village K is a village that was established about 80 years ago on flat terrain along the Dong River (location ☐1 in Figure 14.5). The village later moved slightly upstream (location ☐2 in Figure 14.5). The main livelihood at the time was reportedly swidden agriculture, plus the gathering of NTFPs. After many of the villagers died due to a disease *ca.* 1940, to avoid further losses, the survivors descended from the Phou Xang He mountains to level the terrain in the valley, close to Village S (☐3 in Figure 14.5). In Village S, the main group is *Phu Thai* tribes, and since paddy cultivation was their principle form of agriculture, the villagers from Village K survived by engaging in swidden agriculture near their settlement. However, conflicts with Village S arose after 10 years roughly, and as a

**Figure 14.5:**
*Study village and its past movements*

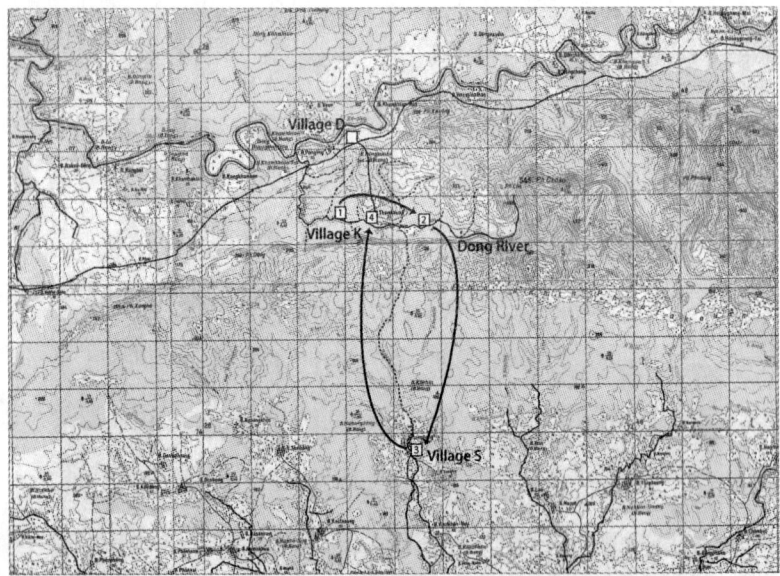

**Source:** Prepared by the author from topographic map.
(Disclaimer: This image has been redrawn by author (Kimihiko Hyakumura) and is not to scale. It does not represent any authentic national or international boundaries and is used for illustrative purposes only.)

result, the villagers returned to their original territory in the Phou Xang He mountains. Cultural differences were reportedly the cause of the conflicts between the tribes. To avoid returning to the original settlement where the disease had occurred, however, the villagers moved to the current location of Village K ([4] in Figure 14.5). It was apparently starting at this time that some Village K households began paddy field cultivation. The idea reportedly came from villagers who had observed paddy cultivation in Village S.

After the country's revolution in 1975, Vietnamese soldiers provided assistance to Village K, a major component of which was to help with paddy cultivation. After receiving that assistance, Village K increased the area that was under paddy cultivation, to the point that, by the 1980s, many households owned paddy fields. Later, from the 1990s onward, the government promoted policies against swidden agriculture, which resulted in a trend toward further expansion of paddy cultivation. Meanwhile, in inverse proportion to the growth of paddies, swidden land decreased in the area.

In 1993, external access improved with the completion of the provincial road No. 10 north of the village, which made the village accessible by motor vehicle during the dry season. This increased access reportedly made it easier for traders to arrive, after which the logging and sale of NTFPs began to flourish. In 1993, the government designated a protected area in the Phou Xang He mountains, including the territory covered by the village. At the time that the area was designated, however, the decision was only at the level of the central government; no outreach activities or concrete efforts that related to the management of the protected area were conducted around Village K. Thus, for some time thereafter, the protected area status of Village K and its surroundings existed only on paper. However, things changed in 1998, which was the year the PAFO and DAFO issued a notification that Village K territory had been designated as a protected area; villagers were prohibited from logging for timber, and they were requested to halt swidden cultivation.[4] A sign from the DAFO was installed in the village explaining the roles and rules of the protected area, and at this time, protected area management activities became more noticeable (Figure 14.6). Because the land suitable for paddy agriculture on *taling dong* along the Dong River had been mostly cleared already, increased paddy field cultivation by the village came mostly by gradually increasing the use of *taling ghok*,

---

[4] The area that contains Village K was designated as a protected area in 1993. Five years later, an effort was made to inform the village (Hyakumura, 2001).

**Figure 14.6:**
*Sign board in Village K settlement indicating rules of protected area*

**Source:** Author.

which exists only sparsely in dry dypterocarp forest. As a result of these shifts of agricultural land use towards paddy fields, most households came to own paddy fields.

Next is a review of the past 40 years of Village K's changes in land use, which is based on aerial photos and the author's interviews with village elder groups. Land use in 1958 was characterised by swidden and fallow land in the area along the Dong River, which runs east to west past the village (Figure 14.7). Because the northern portion of this agricultural land includes rocky dry dypterocarp forest, however, almost no farmland can be seen. Additionally, in the flat terrain that is adjacent to the Dong River, only a small area of paddy fields can be seen. Thus, in 1958, swidden agriculture was the main form of cultivation for Village K, and it was in the same year paddy fields were first introduced.

Moving forward in time, land use in 1998 is characterised by extensive paddy fields along the Dong River, running east to west past the village. In this area, one can also see swidden and fallow land. The area of swidden and swidden fallow land has dramatically decreased compared to 40 years earlier. The villagers have converted the swidden land to paddy fields

**Figure 14.7:**
*Land and forest use in Village K (1958)*

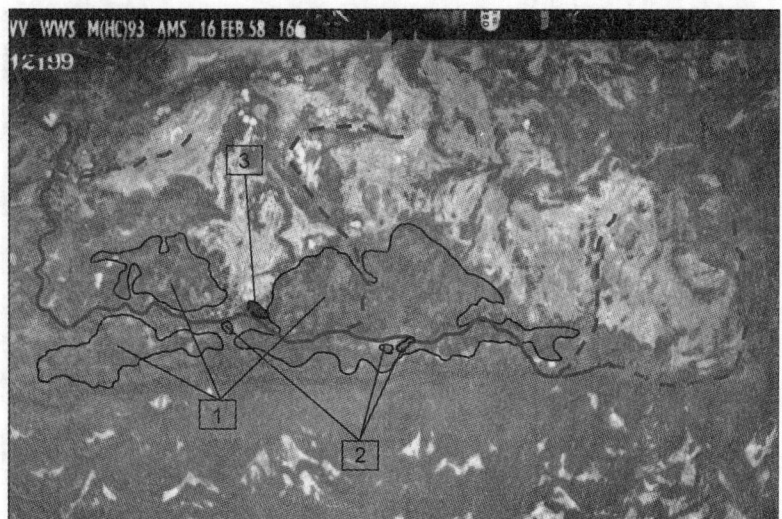

*Note:* The area marked with 1 shows swidden and fallow land, 2 shows paddy fields and 3 shows settlement.
*Source:* Prepared by the author based on 1958 aerial photos.

in this water catchment, which has good water and soil, and thus, they can count on large harvests. This arrangement means that the swidden and fallow land are broadly evident north of the village, and after laying fallow, they have almost entirely been abandoned (Figure 14.8).

Examining a cross-sectional view of the land and forest use in 1998, one can see dense *turung* forest south of the village in the Phou Xang He mountains (Figure 14.9). Additionally, *taling dong* paddy fields are located along the Dong River and are surrounded by *sharay* swidden land and *Arui* and *Patensao* swidden fallow land. Moving north from the village, one sees swidden land that is fallow, but in recent years, the fallow land has simply been abandoned in some places. In some locations that have good water conditions, villagers are attempting to convert to *taling ghok* paddy fields, but on a very small scale.[5] Further north, the land is dominated by rocky dypterocarp forest (*ghok*), which is not suitable for agriculture. However, this land is being used as a source for gathering NTFPs.

[5] Some sites are abandoned even after a paddy field has been established due to poor production results. Primary problems include water conditions, including the inability to secure sufficient water during the rainy season.

**Figure 14.8:**
*Land and forest use in Village K (1998)*

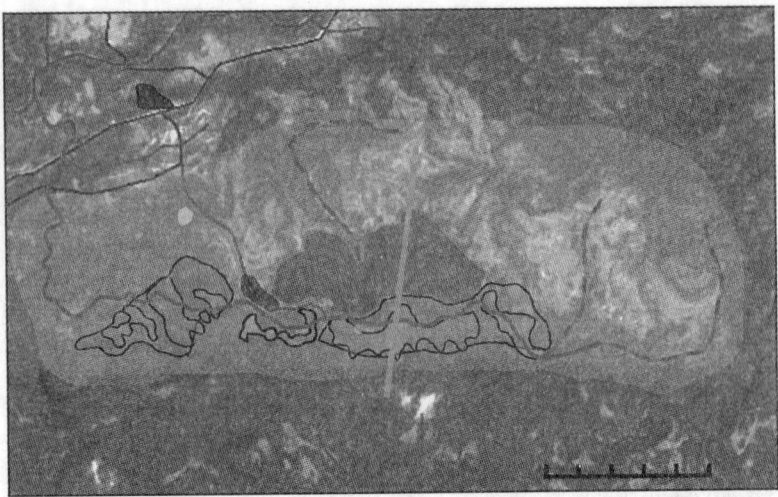

(Disclaimer: This image has been redrawn by the author (Kimihiko Hyakumura) and is not to scale. It does not represent any authentic national or international boundaries and is used for illustrative purposes only.)

**Figure 14.9:**
*Cross-sectional view of land and forest use in Village K*

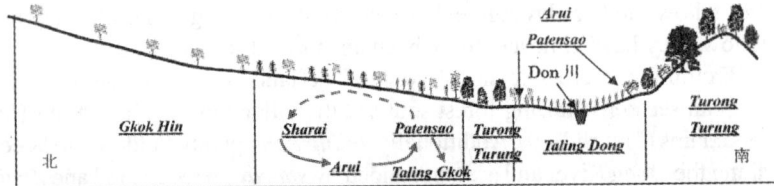

*Source:* Author.

## Discussion

The form of land use for agriculture in Village K has changed drastically in the past 40 years. At the time that the village was established, swidden fields accounted for all agricultural land, but after life on the plains, the villagers started introducing paddy fields, which were then expanded with external assistance. The government's later policies to halt swidden agriculture then accelerated the move towards paddies. Thus, in Village K, the land-use shift from swidden to paddy fields was advanced by three major factors: (1) desire to introduce paddy agriculture in a village on the plains, (2) technical support from Vietnam and (3) banning swidden agriculture because of a government policy. The first was a self-motivated factor of the villagers, but the second and third factors could be described as external pressure or influence. The majority of villagers in Village K actually desired to shift towards paddy farming, so they made the shift in land while incorporating external factors. One could say that they had a strong desire to shift to paddies, to the extent that approximately 90% of the households eventually owned paddy fields, even though the residential area is located in the hills and only a limited amount of land is suited to this type of agriculture. Even where they are today, some families are attempting to clear dry dypterocarp forests to convert to paddy fields.

On the other hand, only 40% of the households are engaged in swidden agriculture, a significant decrease since the village was established. As stated above, this arrangement can mean that the importance of swidden fields has declined compared to paddy fields, as a result of the shift to paddy farming. However, although it has declined, swidden farming is likely to continue in the future. Almost 80% of level A households continue to be engaged in swidden agriculture, which continues to be a source of rice to complement paddy fields. Many level C households see swidden agriculture as an important means of securing a supply of rice. The land that is most suited for paddy farming in Village K is almost completely occupied; thus, the need to continue swidden agriculture will remain in the future.

Additionally, NTFPs are an important food safety net for villagers from Village K. Rice shortages occurred during the years when this study was conducted, and most households, including level A households, which have considerable amounts of paddy and swidden land, were using NTFPs as a rice alternative; thus, NTFPs were serving as famine foods. During normal years, half of the households had to use yam potatoes, including households in level C that produced only a limited amount of rice; thus, this food was important for the households. Bamboo shoots, which serve as a major side dish during the rainy season, were also an important food source for Village K.[6] These NTFPs are gathered mainly

---

[6] During the author's field study to the village in the midst of a rainy season, bamboo shoots were a supplement to the villagers' meals.

from dry dypterocarp forests. Even if the dry dypterocarp forests are not as significant as agricultural land, they are in fact important for securing the livelihoods of the villagers.

Swidden agriculture in Village K is not used in the dense *turung* and *turong* forests; it is conducted by clearing land in secondary forests such as *patensao*.[7,8] In other words, these are not the dense forests that concern the government. Granted, forest resources will be temporarily lost if secondary forests are cleared for swidden agriculture use, but within 6–15 years the secondary forest will regenerate itself, depending on the natural re-generation of the land. Additionally, after being used for swidden agriculture, villagers continue to own the site for future use in a rotatable manner as swidden agriculture land. In other words, swidden fallow land is used as a part of the swidden farming process in their agricultural system. Swidden fallow forest might not appear to be cultivated and might not appear to have any use; however, although the government could observe it as worthless empty fallow land, it is actually a potential swidden field to grow rice in the future. Additionally, the total area of the swidden fields is gradually declining. In other words, the swidden agriculture of Village K suggests that the pressure on forest resources is gradually decreasing.

As far as one can tell by observing the current land and forest uses for swidden agriculture, it is difficult to conclude that the forests will dramatically decline in the foreseeable future. Even if swidden agriculture continues, the pressure on the land is expected to be low. Furthermore, the villagers have customary forest use that displays many sustainable aspects. Because the Lao government's forest sector cannot provide sufficient funds or personnel for forest management, it is most practical to seek ways to take advantage of the local people's customary land and forest use, for the benefit of forest management.

The beginning of this chapter states that, among the three strategies in terms of the local people, the *adjustment strategy* applies best to this case study. In tropical Asian countries where forest protection is urgently needed, policies must be designed to mitigate the negative impacts on local people that arise from the establishment of a protected area and must promote smooth transitions. However, as stated above, the local

---

[7] Many researchers have already noted cases in which fallow secondary forests (bush fallow) were preferable compared to fallow forests (the clearing of primary forest) for local people (Sato, 1999).

[8] Takeda (2001) and others have described cases of local people's sustainable use of second-growth vegetation.

people of Village K were not subject to a strict management structure in the context of the protected area management policies. This freedom is not because the facilitation of the aid organisations had ensured that the local people had the rights to the land and forest use. Village K was not a pilot village for priority support from aid organisations, and it was the provincial and district agriculture and forestry offices (PAFO and DAFO) that were engaged in specific matters that related to the protected areas. In this context, there had been some 'slippage' by local government bodies in terms of addressing forest resource use in protected areas to secure the livelihoods of local people (Hyakumura, 2010). Thus, one could say that this arrangement was actually close to an eclectic strategy.

At the same time, private sector companies and investors, who are external stakeholders that are separate from the policy implementation agencies, were not interested in forest resources in the area or they were not able to become interested. A major reason for this lack of interest was that the territory that included Village K was part of a protected area. The area was not targeted for economic development; thus, it could not be subject to intervention by external stakeholders. Furthermore, the limited access to the village was likely another factor in the absence of intervention by external stakeholders. Where the road access is good, there have been other cases in which rubber plantations and other forms of economic development have been introduced, even though the territory belongs to a protected area (Schipani, 2007).

Based on this case, below is an attempt to examine graduated membership and the commitment principle, which are important design principles of collaborative governance presented in Chapter 1. The first is graduated membership. Stakeholders involved in forest resource management in Village K include the villagers, the local people in nearby villages, DAFO and PAFO of the local governments and finally DOF of the central government. Aid organisations are excluded here because they had withdrawn from protected area management programmes already in 2002. As indicted above, the local people of Village K could be described as being the core members of forest management. Among those core members, it was the influential 'class' in economic levels A and B in the village that possessed the greatest decision-making authority. Villagers in level C, of whom the majority depended on the use of NTFPs and swidden agriculture, were separate from the influential class and had only limited influence on village decisions about forest use. Additionally, the gathering of NTFPs near the village boundaries created a resource relationship with neighbouring villagers, but this arrangement was not a major factor that influenced the livelihoods of Village K individuals. At the same time, the involvement of government agencies, the official management bodies of protected areas, is relatively small. Government agencies have limited personnel and budgets, making it difficult for them to take a hands-on approach to protected areas as a whole. Their specific commitments were limited to such aspects as awareness-raising about protected areas and the posting of signs. In fact, when the government started to engage in efforts to raise awareness about protected areas starting in 1998, it issued orders to halt swidden agriculture and logging, but because of the 'slippage' of government agencies, people were able to continue swidden cultivation as before, without any major problems. These activities occurred because (a) swidden agriculture by villagers was declining, and (b) the reduction in forest cover due to swidden agriculture was only temporary; thus, the pressure on forest resources was not very serious. Next, we consider the commitment principle. In reality, it is the local people of Village K who have the greatest authority on forest management. Levels A and B (which include the influential class) in Village K hold considerable authority, while the right to speak out is relatively low for level C, which is more dependent on the forest resources. At the same time, as stated above, the involvement of government bodies is relatively low.

Thus, if we assess using the design principle of collaborative governance, we can see that, practically speaking, the local people have the decision-making authority for forest resource management. It appears that an application of the commitment principle is not feasible because it is inevitable that the class that is most dependent on the forest resources would have low authority, given the power relationships within the village. However, under collaborative governance, some consideration for them is necessary, such as providing authorities to forest-dependent people to participate in discussions among the villagers, to make the commitment principle applicable.

At the same time, one might expect regulations to be strictly imposed on protected areas, but the local people were not specifically subjected to such rules. The reason is that the forest management policies in the protected area were not applied very stringently due to the local people's dependence on the land for a livelihood (Hyakumura, 2010), and the influence of the external stakeholders was prevented because of restrictions on their use of forest resources in protected areas, leaving the local people to continue being the core members of forest management. Thus, this case study fits under category I, which is characterised as a high degree of autonomy and a high dependency of local people on a forest, out of the cases shown in Chapter 1. In recent years, however, economic development and poverty reduction programmes have been proceeding at a rapid pace nationwide in Laos. It is possible that their influence will reach Village K in the not-too-distant future. Village K, admittedly, is located within a protected area; thus, economic development activities would not normally be advanced here due to regulations. However, if there are changes in external factors, such as improved access to the village, the influence of external stakeholders could reach this village, and the potential exists for a relative reduction in their self-determination in the use of these forest resources.

# References

Department of Forestry (DOF) of Laos (2005a). *Forestry Strategy to the Year 2020 of Lao PDR.* Vientiane, Laos.
———— (2005b). *Report on the Assessment of Forest Cover and Land Use During 1992– 2002.* Vientiane, Laos.
———— (2007). *Indicators for Monitoring of Forestry Sector Performance.* Vientiane, Laos.
FAO (2010). *Global Forest Resource Assessment 2010.* Rome: Food and Agriculture Organization.
Fujita, S. (2012). Site inspection of Stra Enso Corporation plantations in Laos. *Japanese Journal of International Forest and Forestry,* 36(83), 26–30 (in Japanese).
Grandin, B. (1988). *Wealth Ranking in Smallholder Communities: A Field Manual of Intermediate Technology Publications.* London: Intermediate Technology Publications.
Hyakumura, K. (2001). Issues of Conservation Forest Management Policy in Lao P.D.R.: Toward effective policies that reflect the real local conditions. *Forest Economy,* 54(12), 22–33 (in Japanese).
———— (2002). Forest use in southern Laos: Famine plants and forest taboos, *f Shinrin Kagaku,* 36, 76–78 (in Japanese).
———— (2003). Forest and land use in Laos. *Tropical Ecology Newsletter,* No. 52, 7–9 (in Japanese).
———— (2010). "Slippage" in the implementation of forest policy by local officials: A case study of a protected area management in Lao PDR. *Small-scale Forestry,* 9(3), 349–367.
Hyakumura, K. & Scheyvens, H. (2012). Financing REDD-plus: A review of options and challenges. In S. Managi (ed.), *The Economics of Biodiversity and Ecosystem Services* (pp. 148–163). Milton Park, Oxon: Routledge.
IPCC (2007). *Fourth Assessment Report: Climate Change 2007* (104 p.). Geneva: Intergovernmental Panel on Climate Change.
Lao-Swedish Forestry Programme (LSFP) (2001). *Fact Sheets on National Bio-diversity Conservation Areas (NBCAs) in Lao P.D.R.* Vientiane, Lao PDR.

NAFRI (2009). *Rubber Development in Lao PDR: Why Is Rubber Booming?* Presentation materials for the ASEAN Rubber Conference 17–18 July 2009, Vientiane, Laos.

Salter, R.E. & Phanthavong, B. (1989. *Needs to Priorities for a Protected Area System in Lao PDR.* Forest Resource Conservation Project, Lao Swedish Forestry Programme. Vientiane, Lao PDR, 29 pp.

Sato, R. (1999). Current trends in studies of swidden agriculture in the tropics: Ecological aspects and historical contexts. *Japanese Journal of Human Geography*, 51(4), 47–67 (in Japanese with English abstract).

Schipani, S. (2007). Ecotourism as an alternative to upland rubber cultivation in the Nam Ha national protected area, Luang Namtha. *Juth Pakai*, 8, 5–17.

Seki, Y. Ko Ko & Furukawa, N. (2009). *Forest Restoration in China: Beyond Socialism and Belief in the Market.* Tokyo: Ochanomizu Shobo (in Japanese).

Sigaty, T. (2003). *Legal Framework of Forestry Sector for Forestry Strategies to the Year 2020 Lao PDR* (123 pp.). Lao PDR: Vientiane.

Takeda, S. (2001). Benzoin production in swidden fallow fields in northern Laos, Noko no gijutsu to bunka. *Journal of Cultivation Technologies and Culture*, 24, 1–18 (in Japanese).

World Resources Institute (WRI) (2005). *Millennium Ecosystem Assessment—Ecosystems and Human Well-being: Synthesis.* Washington, DC: Island Press.

Pollard, D. (2010). *Living PLANET Report 2010: Biodiversity, Biocapacity and Development.* Gland Switzerland: World Wide Fund for Nature.

# East Asia

# SECTION IV

# East Asia

# 15

# China: Mechanism Design of Community Co-management in Forest Governance

*Huilan Wei, Ting Zhu, Haiyun Chen and Ganesh P. Shivakoti**

## Introduction

In China, the fate of forest resources is closely linked with modern Chinese history. Since 1949, the establishment of the People's Republic of China, the central Chinese government has instituted a series of policies and laws to manage and protect forest resources. Achievements and failures have both occurred in the long history of the development of these policies and laws. Since 1978, the Chinese government concentrated almost all of its energy on developing its market economy, and the resulting extensive economic growth has been remarkable. However, as a result of this rapid economic growth, the whole country has also paid a huge cost in natural resources. The over-exploration and destruction of natural resources have seriously impacted sustainable economic development.

In the 21st century, as a strategy of natural resource protection, forest resources have attracted special attention from the central government.

* Many thanks go to the colleagues in Lanzhou University, Asian Institute of Technology and Tongji University. The same thanks are also due to Professor Inoue Makoto in Tokyo University, who provided us huge support and help from the beginning until now. In addition, this study is supported by the National Social Science Fund of China (14CGL031), supported by the Program for Young Excellent Talents in Tongji University, and sponsored by the Shanghai Pujiang Program (14PJC055).

To protect forest resources, afforestation and territory greening were defined as a common duty of management organisations. A series of forest programmes conducted by the central government have been successfully implemented or are currently being implemented, including the Three-North shelterbelt programme in 1978, Agroforestry in Plain Areas in 1987, Conservation Forest on Upper and Middle Reaches of Yangtse River in 1989, Afforestation of Taihang Mountains in 1990, Coastal Windbreak Forest in 1991, Huai River and Tai Lake Conservation Forest in 1995 and so on. All these forest resource protection programmes have made China the world's most extensive plantation state of afforestation. Between 2000 and 2005, the growth of the forest areas was over 4,000,000 ha, which accounts for almost half of the forest growth areas in all of Asia. On 6 September 2011, the First Conference of Forestry Ministry of The Asia-Pacific Economic Cooperation (APEC) was held in Beijing; Hu Jintao, chairman of the Chinese government, attended this conference and presented an important strategic policy, namely strengthening the regional cooperation and achieving green growth; this informed the public that the Chinese government planned to pay more attention to natural resource management and environmental conservation to provide a better foundation for economic development. Forest resource management will be one of the most important topics in the process of decision making in the Chinese government in the future.

Although a series of policies have been implemented, some shortages and problems in forest resource management could not be solved immediately (Chen, Shivakoti, Zhu, & Maddox, 2012). Relevant laws and regulations have emphasised on the duties of forest resource usage restriction, with little or no consideration of the impact of these laws and regulations on the livelihoods of local residents. The conflict between livelihood development and forest protection has deteriorated the relationship between local communities and management departments. To solve livelihood problems and reduce tensions, the government gradually adopted policies to improve the livelihood of local peoples, including reducing agricultural taxes and providing food subsidies. These measures did not work as effectively as expected, leading to the development and gradual implementation of community-based co-management (CBCM) in some natural forest reserves. The application of CBCM has satisfied the requirement of participatory democracy and linkages among different constituencies, levels of governments and economic sectors (James, 2008; Stephen, 2006; Zhu, Shivakoti, Chen, & Maddox, 2012).

In this chapter, we focus mainly on the current status of forest resources; the policy changes China's forest resource management has experienced; the history, background and application of CBCM in China; and the operation status of the local CBCM mechanism, which was analysed through a case study. At the end of this paper, feasible recommendations will be provided for CBCM improvement and forest resource management perfection in future.

## Current Status of Forest Resources in China

China has approximately $150 \times 10^6$ ha of forest, covering 21.2% of its land surface. This forested land is concentrated in three of five regions (Table 15.1)—the Northeast (including Heilongjiang, Jilin and the eastern part of Inner Mongolia), the Southwest (including Sichuan, Yunnan and Tibet) and the South. Together, the Central North and the vast Northwest account for only 12.6% of the country's forest land (Sen, Cornelis, & Bill, 2004).

The overall majority of forest land (58%) is collectively owned and managed by the rural collectives (townships, administrative villages and natural villages). The collective forests are generally located in south and central China. The remaining forest land (42%) is operated and owned by the central and local state (province, prefecture, county and state forest companies) (Liu, 2001; Xu, Tao, & Amacher, 2004). The state forestry area is concentrated in the north-central and northeast region of China bordering Russia and Mongolia, which are largely composed of plantations and bamboo forests (Table 15.2) (Peter, 2006; Sen et al., 2004).

**Table 15.1:**
*China's forest resources, by region*

| Region | Land area | | Forest resources | | | |
|---|---|---|---|---|---|---|
| | *Area (10⁶ ha)* | *% of total* | *Area (10⁶ ha)* | *% of total* | *Timber volume (10⁶ m³)* | *% of total* |
| Northeast | 194.8 | 20.3 | 43.9 | 27.6 | 3340.3 | 29.6 |
| Central north | 69.1 | 7.2 | 9.6 | 6.1 | 191.8 | 1.7 |
| Northwest | 309.1 | 32.2 | 10.3 | 6.5 | 767.3 | 6.8 |
| South | 152.0 | 15.8 | 57.9 | 36.4 | 2020.8 | 17.9 |
| Southwest | 235.4 | 24.5 | 37.1 | 23.4 | 4946.5 | 43.9 |
| Overall | 960.3 | 100.0 | 158.9 | 100.0 | 11266.6 | 100.0 |

**Sources:** SFA (1999, 2001, 2002) and Sen et al. (2004).

**Table 15.2:**
*China's forest resources, by ownership*

| | State owned | | Collectively owned | | Total |
|---|---|---|---|---|---|
| Category | Amount | % | Amount | % | Amount |
| Area ($10^6$ ha) | 62.0 | 48 | 67.2 | 52 | 129.2 |
| Volume ($10^6$ m³) | 7124.2 | 71 | 2961.5 | 29 | 10085 |
| Of which plantations area ($10^6$ ha) | 7.7 | 26 | 21.4 | 74 | 29.1 |
| Volume ($10^6$ m³) | 378.3 | 37 | 634.7 | 63 | 1013.0 |
| Economic forests ($10^6$ ha) | 1.6 | 8 | 18.6 | 92 | 20.2 |
| Bamboo forests ($10^6$ ha) | 0.3 | 7 | 3.9 | 93 | 4.2 |
| Total forested area ($10^6$ ha) | 63.9 | 42 | 89.7 | 58 | 153.6 |

**Sources:** SFA (1999, 2001, 2002) and Sen et al. (2004).

# Review of Forest Policy Management in China

## Evolvement of Forest Policy in China

The destiny of forest resources is closely connected with the history of development in China. From the 1940s to the 1950s, the Chinese government implemented relevant policies that mainly focused on planting trees and forest production. From the 1960s to the 1970s, due to some political campaigns, such as 'cultural revolution', the forest resources suffered from overwhelming destruction and forest management was almost paralysed. Beginning in 1978, when policies of reform were initiated, stable policies and laws for forest resource management gradually developed. Table 15.3 shows a detailed list of the significant issues in the history of China's forest resource management.

## CBCM Development in China

After 1978, the Chinese government created a series of laws and policies to manage forest resources, such as 'Forest and wildlife natural reserves management approaches in China' (1985) and 'China Natural Reserve Ordinance' (1994), etc. However, these laws and policies only considered the negative influence that local communities have on natural resources, with little or no consideration to the impact of these

**Table 15.3:**
*Evolvement of forest resource management in China*

| Time | Details |
|------|---------|
| 1947–1949 | China proclaimed the 'Outline of the China Land Law', it was clearly regulated that forest would from then on fall under the direct administration of the government (Peter, 2006) |
| 1949–1956 | China government carried out a forest policy characterised by tree planting on barren lands and timber harvesting in major forest regions (Richardson, 1990; SFA, 2001; Sen et al., 2004) |
| 1957–1958 | Collectivisation proceeded quickly in this stage because of the establishment of the People's Commune System in which higher participation in forestry production by villagers was observed, but the efficiency of afforestation was in low level (Cannon & Jenkins, 1990) |
| 1962–1976 | Because of readjustment of policies and measures to manage land and forest resources, forest industries began to grow rapidly in the 1960s. However, the 'Cultural Revolution' resulted in a severe lessening of afforestation throughout China and a weakening of forest management systems. State forestry farms and community forest regions suffered from serious hardship until 1976 (Yajie, William, Gordon, & Liping, 1997; Zhang, 1989) |
| 1978–1989 | China government began to implement the policy of reform and opening, and most of previous policies were modified greatly. 'Decision on Some Issues concerning Forest Protection and Forestry Development' was issued in March 1981, which clarified a series of institutions, including the assessment and registration of forest resources, the distribution of use rights to 'family land', and the establishment of the forestry contract responsibility system under 'responsibility hills'. The forest legislation officially entered into force on 1 January 1984. The legislation provided the legal basis for the Ministry of Forestry to formulate relevant policies (Richardson, 1990; Sen et al., 2004) |
| 1990s | As a result of a series of reforms in the administrative hierarchy, state-owned forest companies became increasingly autonomous and the central forestry authorities loosened their control over corporate decisions pertaining to the production of forest products. The State Forestry Administration (SFA), became the new forestry authority since 1998, and its functions are to protect existing natural forests, and preserve biodiversity and a range of forest-related values and attributes, as well as engage in afforestation of barren hills (Peter, 2006; SFA, 1999) |
| 2000 to until now | Chinese forest policy has changed a lot in many relevant fields and aspects since the 21st century, especially in property rights of state-owned and collective-owned forests, legislation of forest management and conservation, forest industries, etc. (Cao, Chen, Shankman, Wang, Wang, & Zhang, 2011; Jintao, Ran, & Gregory, 2004; Xu et al., 2004) |

*Source:* Authors.

laws and policies on the livelihood of local community residents. This compulsory forest protection deprived local residents of some development rights and led to significant social and economic problems (Wei & Zhang, 2006). For example, after the establishment of natural reserves, the income of local community residents obtained from forest resources was obviously reduced because the protection policies restricted their use of forest resources. With the expansion of China's policy reform in the 1990s, local communities were no longer satisfied with the reduced incomes caused by the forest protection policies. The desire to improve local livelihood threatened the sustainability of local natural resources and sometimes resulted in systematic deforestation.

The conflict between livelihood development and forest protection has deteriorated the relationship between local communities and government. To solve livelihood problems and reduce tensions, the government gradually started to introduce CBCM as an alternative management model to natural reserve management. CBCM is a people-centred, community-oriented and partnership-based management model (Robert & Rebecca, 2006). It focuses on community, and as a participatory management model, horizontal (across the community) and vertical (with external to the community organisations and institutions) links are necessary. CBCM in China involves different stakeholders, including the government, local community, non-government organisations (NGOs) and other external groups. According to the different assignments in such management organisations, different stakeholders have different rights and responsibilities. Here, we need to emphasise two points. One is that in the process of CBCM, the leading role and position of the Chinese government cannot be replaced by any other stakeholders, so the partnership between government and community cannot be equal and fair. Government plays the important roles of institution maker, management supervisor and original organiser of CBCM. The other point is that CBCM focuses on the function of community. Instead of previous 'top-down' management models, CBCM gives more attention to the application of indigenous management knowledge and the autonomous management ability of local residents, with the purpose of bestowing local residents with the responsibility of forest resource protection. Moreover, CBCM in China is always connected with natural reserve management because most CBCM projects and activities are applied and implemented in the hotspots of natural resource concentrative areas, and most natural reserves are also plotted out in the same areas.

The application of CBCM was accompanied by the implementation of a series of funded projects that are supported by the Chinese government and foreign NGOs, including the Global Environment Facility (GEF), World Wide Fund (WWF), World Bank (WB) and some other international organisations, with the main purposes of conserving China's biodiversity, promoting socio-economic development in nature reserves and their surrounding communities and sustainably using China's natural resources (Yang, Jin, & Wang, 2008). Since August 1995, under the specific direction and guidance of the WB and China's State Forestry Administration, China's 10 national nature reserves from five provinces implemented the 'GEF China Nature Reserves Management Project', which has a duration of six years; one of the most important goals of this project was to establish CBCM (DWAP-SFA, 2002). In 1998, with funding from the Dutch government, the 'China–Dutch Cooperation in Forest Conservation and Community Development Project' was launched in Yunnan Province; the project involves four prefectures and cities in Yunnan Province of six nature reserves, and the construction of CBCM is one of the most important project components. During October 2002 to October 2008, GEF has implemented the 'Sustainable Forestry Development Project', which has chosen CBCM as the sub-project of 'Protect District Management' and involves 13 national nature reserves from seven provinces. In addition, WWF has implemented many conservation and development projects in Baima Snow Mountain and Zhongdian in Yunnan Province, Qinling area in Shanxi Province, wetlands in middle reaches of the Yangtze River of Sichuan Province, Tibet and so on. One of the main objectives of all of these projects is to implement and extend the CBCM approach (Li & Zuo, 2006).

## Case Study of CBCM Institution Arrangement and Mechanism Operation

To obtain a better understanding of China's CBCM, we chose Baishuijiang National Nature Reserve (BNNR) as a typical case to elaborate the institutional arrangement and mechanisms of operation of CBCM at a local and practical level.

### Study Area

It was key to select an appropriate study area to support our research, so according to the research objective and characteristics, we selected BNNR as it is one of the most important giant panda habitats in the world and is characterised by a rare biodiversity gene pool because it is situated in the transition zone from sub-tropical to warm temperatures in China. However, it is also a conflict hotspot between livelihood development and forest resource and biodiversity protection. In these years, a series of CBCM projects were applied in this area; we also performed research and experiments in this area in the past years. Importantly, we were able to obtain the support of the local government and academic institutes for our scientific research (Chen et al., 2012, 2013a, 2013b). All of these factors led to us selecting BNNR for our research platform.

The BNNR is located in Wenxian County, south of Gansu Province in China. This protection zone is in the shape of a ribbon 110 km long from the east to the west and 20 km wide from the south to the north, covering an area of 223,671 ha. It was established in 1978 and is mainly used for the protection of giant pandas, other rare wildlife and the ecosystem. In November 2000, the BNNR was formally included in the International Man and Biosphere Reserve Network. The management's responsibility of the BNNR is vested in the Baishuijiang National Natural Reserve Management Bureau (BNNRMB), which falls under the jurisdiction of the State Forestry Administration (Chinese central government). Based on management purposes, the BNNR is divided into three sub-areas: testing area (100,310 ha), buffer area (26,032 ha) and core area (97,329 ha). Most local people live in the testing area, less in buffer area, and no one is allowed to live in the core areas. Productive and commercial activities in buffer areas are also prohibited. Limited agriculture and subsistence hunting is permitted only in the testing area (Chen et al., 2012; Zhang & Wu, 1995; Zhu et al., 2012).

## Institutional Arrangement of CBCM in Study Area

As more and more CBCM projects supported by foreign NGOs have been implemented in China with positive effects, the leaders of BNNRMB used CBCM to mitigate conflict between resource protectors and local residents in the BNNR. Instead of using a government-guided management model, a new type of co-management model based on local communities was formed along with a series of CBCM committees. CBCM committee members, who are selected from local villagers, officers of the BNNRMB, researchers from academic institution and NGOs, share the responsibility and rights for the management of forest resources. As key committee members, villager representatives play leading roles as organisers, managers and supervisors of local community and forest resource management. All the other co-managers from outside are 'co-operators' and take charge of technology and funding. CBCM committees hold villager conferences regularly to resolve conflicts among villagers and gather public opinions. The strongest demands of the villagers are designed as a development project, and the CBCM committee seeks wider community support.

The CBCM approach was applied in the BNNR in 2003 as a potential instrument for improving natural resource management by attempting to conform to public opinion. Local CBCM organisations have developed throughout China that are associated with these projects and increasing numbers of local villages actively participate in CBCM projects. However, currently, CBCM mainly focuses on short-term projects, and CBCM has not been formally admitted by law. In fact, although the coordinated ideal between development and protection has been reflected in relevant laws and regulations for China's natural reserves, the management pattern of CBCM is not protected by constitutional rules. Fortunately, the contents of CBCM have been elaborated in some regional or provincial regulations, especially in some natural reserve management regulations.

The seventh item of the 'Management Regulations in Gansu Baishuijiang National Natural Reserve, China', which was constituted in 2000, stated that: Longnan municipal government should organise Wen County and BNNRMB to establish coordinated organisation that is mainly responsible for dealing with major issues in the reserve. Meanwhile, the BNNRMB should set up joint prevention and protection organisations, namely grass-roots CBCM organisations, with the local community and local government. These organisations should take charge of setting up protection regulations, expanding public protection education, assigning responsibilities to different areas and performing protection duties.

The 14th item stated that: The BNNRMB should establish a unified layout of the suitable lands for afforestation in the testing area of the reserve; support local community residents to develop economic forests, firewood forests and timber forests; and encourage residents to cultivate medical products and provide relevant technical services for residents. In regard to fee-paid labour services and tourism services for the natural reserve construction and forest resource protection, the BNNRMB should coordinate with relevant management departments at first and then give priority to local community residents for employment in these areas. Both local government and the BNNRMB should provide relevant support for local development projects in the reserve, such as constructing roads, small scale hydropower stations and marsh gas pools, developing a household breeding industry, applying solar energy, replacing firewood stoves with energy-saving stoves, etc. Although most of the existing regulations and policies are in theory supportive of CBCM, there is inadequate legal documentation to ensure the rights of the local community and the long-term mechanism of CBCM. Therefore, in practice, improvement is still required for CBCM institution construction.

## Operational Mechanism of CBCM in Study Area

Based on our survey of the study area, there are four main types of operational mechanisms in the application process of CBCM, namely the institutionalisation of the primary job responsibility of the CBCM committee members, the election mechanism for CBCM committee members, the CBCM management process and the CBCM project application mechanism.

### The Primary Job Responsibility of CBCM Committee Members

The institutionalisation of the primary job responsibilities for CBCM committee members is one of the basic requirements of efficient human resource allocation and needs to accommodate the interests of all parties; it is a concrete manifestation of the essence of condominium. Currently, the management personnel of CBCM committees in the study area are mainly composed of a competent leader from the local government, a leader from the protection management bureau, the project coordinator, representatives of other interest groups and representatives of the villagers. The local management organisations have established and regulated the primary responsibilities of committee members according to duty division described in Table 15.4.

### The Election Mechanism for CBCM Committee Members

The representatives of the villagers, women and the vulnerable group (collectively called community representatives in the following text) are selected from the community residents to be the spokesmen of the common wishes and interests of the community. Because of their special role in the community, the transparent election process for community representatives becomes one of the chief premises of the CBCM committee,

**Table 15.4:**
*Duty division table for CBCM committee members*

| Committee members (number) | | Duty division |
| --- | --- | --- |
| Director (1) | | To take charge of the overall project implementation, to harmonise the relationship among CBCM committee members, to be responsible for coordinating the project content, to host CBCM committee meeting, to supervise the operation of CBCM project and the regular management work, and to summarise the work experiences |
| Assistant director (2) | Leader from protection management bureau | As a representative of protection management bureau, to propose constructive ideas for forest resource protection and to supervise resource protection status, to coordinate the conflict between development project implementation and forest patrol, to improve the natural resource management in local community |
| | Leader from local government | As a representative of local government, to provide governmental support and assistance for CBCM, to supervise the process of CBCM projects and to ensure project actualisation |
| Representative of villagers (1) | | As a representative of villagers, to express their wills, to communicate with villagers and officials, to lead villagers to participate in CBCM projects and to fulfil CBCM committee task |
| Representative of women (1) | | As a representative of women, to express their wills and aspiration, to summon women for study and training, to encourage women taking part into CBCM projects and to fulfil the CBCM committee task |
| Representative of vulnerable group (1) | | As a representative of vulnerable group, to express their wills, to join in the implementation of CBCM project, to encourage vulnerable households improving livelihood and joining protection work, and to fulfil the task from CBCM committee |
| Representative of other interest groups (1) | | As a representative of relative forest enterprise, environmental organisations and scientific researchers, to join CBCM project planning, to provide suggestions, financial support or technical guidance to CBCM project, to encourage more NGOs and other external organisations to join the work of local CBCM committee |
| Project coordinator and accountant (1) | | As a project coordinator and an accountant, to submit work plan of CBCM committee to superior administrative department and to pass on the instructions from superior department in time, to coordinate the relationship among relative administrative departments, NGOs, external organisations and villagers, to report problems to superior department for solution, and to take charge of financial work for CBCM committee |

*Source:* Authors.

and it is also the most important way for local villagers to take part in community resource management. At present, communities usually use an open voting method in the villager meeting for community representative elections. The detailed steps of this open voting method are described here. First, one village meeting is held by the village leader. In that meeting, a staff member of the BNNRMB briefly introduces the responsibilities of the community representatives, the basic organisation construction of the CBCM committee and the main procedures for community representative elections. Then, community representative candidates are nominated by the villagers or through self-recommendations. As soon as the candidates are nominated, they are divided into three groups according to the different types of community representatives (the representatives of villagers, women and the vulnerable group), and these candidates give their own campaign speeches one by one. After the campaign speeches, there are three rounds of voting by the villagers for the election candidates. As per the rules of voting, each villager can vote once in each round of voting, and the voting is based on secret ballots. Villagers who join the meeting have the right to vote, those who do not join the meeting are treated as waivers. Villagers over the legal age of 18 years, regardless of nationality, gender and education level, own the right of election into public service and the right to vote. The voting results are determined by the staff of the BNNRMB. The villagers who obtain the highest number of votes in each round of voting become the community representatives. A single villager can only be in charge of one position. This open and transparent electoral mechanism ensures community resident participation in grass-roots condominium rights. As for the other committee members, such as the director and the vice directors, they are authorised or elected by the superior administrative department or organisations.

## The CBCM Management Process

The CBCM work does not simply include the election mechanism for the committee members and the establishment of a management committee. The more important work is the standardisation and institutionalisation of the CBCM process. Local communities implement their management work abiding by a three-stage project implementation order:

1. *The primary stage:* In this stage, the committee members are selected and trained, a co-management organisation is established and the duties of the committee members are specified.

2.  *The planning stage:* The work in this stage would be completed by the local CBCM committee. The main task is to devise a community sustainable economic development plan, namely Community Resource Management Plan (CRMP) according to the resource distribution and the social development condition of the local community. Committee members should collect relative information through discussions with villagers or household survey questionnaires by making use of participatory rural appraisal (Participation Rural Appraisal). After a comprehensive survey on the status of the community economy, farm production, living conditions, utilisation of natural resources, community organisations and traditional culture, the committee members would perform a statistical data analysis, draw a community organisation chart and natural resources distribution chart, identify the main conflict between natural resource utilisation and protection, and determine the community economic development level.

3.  *CBCM project approval and implementation stages:* The committee members would devise the CBCM project according to the local status, and the project would be submitted to the superior administrative department for approval. If approved, the committee would prepare the project participation agreement for the community residents and organise the local community residents to join the project, sign the relative agreement and take part in the technical training for the project. Moreover, the CBCM committee would take charge of project supervision and evaluation. However, if the project was rejected, the committee members would re-enact CRMP and prepare for the next round of project applications.

## CBCM Project Application Mechanism

The implementation of the CBCM project is the core work of the CBCM committee. It facilitates the openness of the CBCM information system, thereby, helping to reduce CBCM cost. Moreover, it ensures that the CBCM mechanism operates effectively. The local community residents abided by the following CBCM project application process in this study area:

1.  *Confirm the project:* Community residents confirm their selection of development projects based on the comprehensive understanding of their own economic status, education level, development ability and the integrated natural environmental conditions, such

as the area location, the climate characteristics and the resource distribution.

2. *Apply co-management project:* After the confirmation of the project, the households with the desire to participate submit a written form of project application. They must clearly indicate the project name, the detailed project loan, technology or materials necessary, and the expected project application.

3. *Verify project application:* After receiving the application materials, the CBCM committee members verify these project applications with a comprehensive review of the applicants' credit standing, development capacity, project feasibility and other factors. The applications that do not meet the requirements of the project are rejected by the CBCM committee, and reasons for the rejection or suggestions for improvement are also given to the applicants.

4. *Review project application in superior administrative department:* The verified project applications are submitted to the BNNRMB for further review.

5. *Obtain approval:* The relative administrative department approves the project applications, starts to implement the project and allocates the loan funds.

6. *Sign project agreement:* The applicants with project approval sign a joint management agreement with the local CBCM committee and receive project assistance from the committee.

7. *Repay the project assistance:* Project participants repay the project assistance in accordance with the repayment schedule after they receive the benefit of the project.

8. *Re-apply with a new or the same project:* The households that repay the project assistance on time can apply again for a new or the same project in the next term.

## Management Strategy and Guidelines Analysis of CBCM in China

### Management Strategy

Inoue concluded that there are three strategies for sustainable forest use and management, namely the resistance, adjustment and eclectic strategies (Inoue, 2001). China's CBCM, from what we have discussed above, can be classified as an eclectic strategy, which is identified as a collaborative governance of natural resources. In fact, CBCM in China emphasises four main principles: people-centred, community-based, resource-oriented and partnership-based. CBCM implementation emphasises the positive participation of different stakeholders. Government usually plays a coordinated role. Community residents participate in the process of forest resource management, and other stakeholders, such as NGOs, provide relevant information, technology, funding and training, for sustainable forest resource management and livelihood improvement. Partnerships are established among different stakeholders.

## Guidelines Analysis

To overcome the ignorance of minorities with less political power participating in collaborative governance, Inoue (2011) further proposed three vital guidelines, 'graduated membership', 'commitment principle' and 'trust building' to generate conditions for collaborative governance. We will analyse these three guidelines in terms of the application of CBCM in China.

### Graduated Membership of Executive Management Body

For this guideline, 'open-minded localism' (local community residents provide resources and an open environment to outsiders) is thought to be a required factor in collaborative governance. Based on 'open-minded localism', some of the local people act as core members (first-class members) and cooperate with other graduated members (second- and third-class members) (Inoue, 2011). Based on the analysis in Table 15.4 on CBCM committee members in China, the CBCM committee members, who are also 'the first-class members', include local community residents, government officials, representatives of NGOs and research institutes, etc. CBCM committees are not just composed of limited local community residents but also include other stakeholders, and they have the right to establish management rules and to help manage CBCM projects. In addition to members of the CBCM committee, other local community residents and stakeholders should be classified into the 'second- and third-class members'. It is necessary to note that this classification is mainly based on the roles people play in the implementation of CBCM.

## Commitment Principle for Decision Making

According to the description of the three vital guidelines, the 'principle of involvement' emphasises the factor of 'speaking' and the 'commitment principle' focuses on the factor of 'decision making' in CBCM implementation. More importantly, the 'commitment principle' emphasises that the authority of stakeholders should correspond to their degree of commitment to relevant activities. Furthermore, whether a decision is considered to be fair depends on whether the decision-making process was considered to be legitimate (Inoue, 2011). In the section of operational mechanism of CBCM in study area, we systematically analysed the election mechanism and the management process of the CBCM community and uncovered three key principles: 'fairness', 'equity' and 'openness'. The operation of the CBCM committee should be fair and not have any preference for special groups or individuals. All community residents should have an equal opportunity to participate in CBCM activities, and all election mechanisms, management processes, activity implementations and benefit distributions should be open for all the stakeholders. Although we emphasise 'fairness', 'equity' and 'openness', this does not mean that every stakeholder possesses the same rights and responsibilities. Members of the CBCM committee and other people who contribute to the CBCM development should have much more important roles in the relevant activities of the CBCM implementations, and they should have many more rights for speaking and decision making.

## Trust-building with Outsiders

According to the guidelines of Inoue, trust-building with outsiders is a precondition for 'graduated membership' and the 'commitment principle' (Inoue, 2011). In our study area, through signing a bilateral contract between the management department and local community, the BNNRMB (the entrusting party) bestowed part of the forest resource management rights on the local communities (the agent party) to reduce the government supervision and management cost and also to encourage local residents to join CBCM. According to the items of the contract, on the one hand, the BNNRMB is in charge of the protection and development strategy in the contracted area for the local community, organising external experts in relevant research institutes to evaluate and monitor the conservation activities, helping local communities to seek development projects funded by the government or NGOs, and providing relevant support of funding, technology and policy. On the other hand, as the consignee of the protection agreement, local communities should establish patrol teams for forest protection and set up protection institutions and management regulations. The main responsibility of patrol teams is to routinely check for any illegal behaviour that may damage or impact wildlife habitats, such as logging, poaching, mining and collection of non-timber forest products, and to report the relevant illegal incidents to BNNRMB (the entrusting party) in a timely manner. This allows local communities (the agent party) to fulfil their obligation of protection and monitoring.

It is especially important to note that the protected regions in the agreement are state-owned forests in terms of property rights in China, so local communities do not have the right to benefit from forest resources of specific regions. Community residents can receive remuneration to improve their livelihood conditions through the participation and management agreement or through participation in various types of CBCM projects. As for the necessary use of firewood, forest products, herbs and other forest resources for livelihood development, local community residents can capture them in the collective forests or private hilly forest land, which is permitted by the government.

## Governance Bottlenecks

Although CBCM systems have been established in the BNNR, they are still in the early stages. There have been a series of problems and gaps identified during the operation process of CBCM, including the shortage of policy and law support, the existence of trust crisis and the handicap of decentralisation (Chen et al., 2012; Zhu et al., 2012).

## Shortage of Policy and Law Support

Although CBCM has been implemented in some natural reserves and has obtained the permission of the local government to some degree, there are still no formal laws or policy documents to grant CBCM a legal status. This will impact the long-term development of CBCM. Because CBCM projects or activities usually involve complicated and comprehensive topics, such as livelihood of the local community residents, forest resource management and biodiversity conservation, integrated management is required. Thus, there are issues with CBCM projects not only related to different levels of government and local communities but also related to enterprises, researchers and NGOs. Such co-management platforms require laws and policies for background support to ensure such management systems operate in the long term. Therefore, it is quite necessary that a relevant law and policy system be instituted.

### The Existence of Trust Crisis between Community Residents and CBCM Committee

The trust crisis is mainly due to the low management capability and the main body of CBCM leadership being concentrated in the CBCM committee. The critical reasons for the trust crisis are 'difference between saying and doing', 'the promises of CBCM projects not being realised' and 'community residents cannot get real benefits from CBCM projects'. Thus, the distrust between community residents and CBCM committees will further this trust crisis. Therefore, strengthening the leadership in the CBCM committee is a key factor in the process of solving the trust crisis for the further development of CBCM.

### The Handicap of Decentralisation

Despite that the application of CBCM has been permitted by the government to some degree, some government officials are antagonistic towards the devolution of forest resource management to communities. Such a decentralisation management model has threatened their power, and they cannot totally trust the capacity of local communities to make the right decisions for resource protection. From the perspective of public rights, there is no incentive for officials to alter their role from a supervisor to a coordinator or ministrant. Paternalistic attitudes and punitive relations towards community resource users are deeply ingrained in many government resource management officials. However, local community residents are the main body in forest resource management and biodiversity conservation. They live in communities for generations and understand clearly the distribution, quantity, structure, quality and problems of local forest resources. They understand their own development constraints, potential and opportunities more than outsiders. At the same time, they also have a certain capacity to solve problems of development.

Therefore, decentralisation will be an inevitable trend in the future, and both the concept and behaviour transformation are badly needed for forest resource managers.

## The Handicap of Local Institutional Development

A fundamental challenge for CBCM projects remains the problem of developing appropriate institutional mechanisms at the local level to facilitate local control and management of natural resources. Hardin's 'The Tragedy of the Commons' in 1968, Ostrom's 'Revisiting the Commons' in 1999 (Ostrom, Burger, Field, Norgaard, & Policansky, 1999), and many other publications have discussed the institutional problems of natural resource management. Typically, however, institutional development has tended to create new and formal institutions and to ignore the existing remnants of traditional or indigenous resource management institutions. As a result of this, most evolving CBCM activities and projects are designed by CBCM committees and subcommittees, without much regard to the suggestions and comments advanced by the local elite. This tends to alienate traditional authority and undermine the CBCM initiatives.

## The Negative Impact of CBCM Projects

Different stakeholders participate in the process of CBCM project implementation with different interesting targets. Government departments work with the hope of strengthening management and conservation, generating community resident participation with residents who believe that CBCM projects can provide real benefits to improve their livelihood conditions, and causing NGOs to expect valuable project experiences and reports. At the beginning of each CBCM project, everything is fresh to community residents, and they have a positive attitude towards participating in CBCM projects under the catalysing of NGOs and the support of the local government. Unfortunately, some of activities are waylaid in the formal phases and are never implemented. NGOs leave after obtaining their valuable reports, but the community members still live in the area of the failed CBCM project, and their lives return to their previous status after the implementation of the project.

Therefore, how to maintain the sustainable development of CBCM projects is a significant topic for future management.

Obviously, there is no perfect management model in this world. Human beings are always busy with searching for, addressing and improving problems. That is how people understand the world. Similarly, CBCM is not an all-purpose management method for forest resource management. All the problems mentioned above are study directions for the future, such as how to improve CBCM laws and relevant policies, how to address the trust crisis and set up an incentive mechanism in the process of CBCM development, how to integrate indigenous cultures into the institutional arrangement during the process of CBCM, how to encourage active participation of local community residents in forest resource management and biodiversity conservation, and how to improve CBCM projects to be efficient in the long run.

# References

Cannon, T. & Jenkins, A. (1990). *The Geography of Contempoary China: The Impact of Den Xiaoping's Decade*. London: Routledge.

Cao, S.X., Chen, L., Shankman, D., Wang, C.M., Wang, X.B., & Zhang, H. (2011). Excessive reliance on afforestation in China's arid and semi-arid regions: Lessons in ecological restoration. *Earth Science Reviews*, 104, 240–245.

Chen, H.Y., Shivakoti, P.G., Zhu, T., & Maddox, D. (2012). Livelihood sustainability and community based co-management of forest resources in China: Changes and improvement. *Environment Management*, 49, 219–228.

Chen, H.Y., Zhu, T., Krott, M., Calvo, J.F., Shivakoti P.G., & Inoue, M. (2013a). Measurement and evaluation of livelihood assets in sustainable forest commons governance. *Land Use Policy*, 30, 908–914.

Chen, H.Y., Zhu, T., Krott, M., & Maddox, D. (2013b). Community forestry management and livelihood development: Integration of governance, project design and community participation. *Regional Environmental Change*, 13, 67–75.

Hardin, G. (1968). The tragedy of the commons. *Science*, 162, 1243–1248.

Inoue, M. (2011). *Prototype Design Guidelines for 'Collaborative Governance' of Natural Resources*. India: IASC.

James, S.G. (2008). Key principles of community-based natural resource management: A synthesis and interpretation of identified effective approaches for managing the commons. *Environmental Management*, 45, 52–66.

Li, X.Y. & Zuo, T. (2006). *Literature review of community resources co-management in China's natural reserves. Co-management: From Conflict to Cooperation*. Beijing, China: Social Sciences Academic Press.

Jintao, X., Ran, T., & Gregory, S.A. (2004). An empirical analysis of China's state owned forests forest. *Policy and Economics*, 6, 379–390.

Liu, D. (2001). Tenure and management of non-state forests in China since 1950: A historical review. *Environmental History*, 6, 247–249.

Ostrom, E., Burger, J., Field, C.B., Norgaard, R.B., & Policansky, D. (1999). Revisiting the commons: Local lessons, global challenges. *Science*, 284: 278–282.

Peter, H. (2006). Credibility of institutions: Forestry, social conflict and titling in China. *Land Use Policy*, 23, 588–603.

Richardson, S.D. (1990). *Forests and Forestry in China: Changing Patterns of Resource Development*. Washington, DC: Island Press.

Robert, S.P. & Rebecca, R.G. (2006). *Fishery Co-management: A Practical Handbook*. Ottawa, Canada: International Development Research Centre.

Sen, W., Cornelis, G., & Bill, W. (2004). Mosaic of reform: Forest policy in post-1978 China. *Forest Policy and Economics*, 6, 71–83.

SFA (State Forest Administration) (1999). *Forestry Development in China 1949–1999*. Beijing: China Forestry Press.

———— (2001). *China Forestry Development Report 2000*. Beijing, China: China Forestry Press.

———— (2002). *China's Forestry Development Report 2001*. Beijing, China: China Forestry Press.

Stephen, T. (2006). *Communities, Livelihoods, and Natural Resources Action Research and Policy Change in Asia*. Ottawa: International Development Research Centre.

Wei, H.L. & Zhang, K.R. (2006). *Theory and Method of Comprehensive Effectiveness Evaluation in Natural Reserves: Gansu Baishuijiang National Nature Reserve*. Beijing: China Science Press.

DWAP-SFA (Department of wild animals and plant protection in the state forestry administration) (2002). *Natural Reserve Community Co-management*. Beijing: Chinese Forestry Press.

Xu, J.T., Tao, R., & Amacher, G.S. (2004). An empirical analysis of China's state-owned forests. *Forest Policy and Economics*, 6, 379–390.

Yajie, S., William, B.J., Gordon, G., & Liping, G. (1997). New organizational strategy for managing the forests of southeast China: The share-holding integrated forestry tenure (SHIFT) system. *Forest Ecology and Management*, 91, 183–194.

Yang, J., Jin, L., & Wang, L. (2008). Co-management in community from the perspective of development intervention. *Rural Economy*, 10, 42–45.

Zhang, J.G. (1989). *Theory and Practice of Ecological Forestry—A Research into the Way of Forestry Development*. Nanping: Fujian Forestry College Press.

Zhang, K.R. & Wu, G.H. (1995). *Comprehensive Scientific Investigation Report in Gansu Baishuijiang National Nature Reserve*. Lanzhou: Gansu Science and Technology Press.

Zhu, T., Shivakoti, G.P., Chen, H.Y., & Maddox, D. (2012). A survey-based evaluation of community-based co-management of forest resources: A case study of Baishuijiang National Natural Reserve in China. *Environment Development and Sustainability*, 14, 197–220.

# 16

# Korea: Conditions for Sustainability of Traditional Village Woods *Maeulsoop*

*Koo Ja-Choon, Kweon Deogkyu and Youn Yeo-Chang\**

## Introduction

As the importance of governance in natural resources management has been increasingly recognised, forest commons, an example of common pool resources, have been attracting attention. Forest commons are forests used in common by a large number of heterogeneous users, with defined boundaries of the forest and its user group and legally enforceable property rights to forest benefit streams (Chhatre & Agrawal, 2008, 2009; Schlager & Ostrom, 1992). Village woods in South Korea were previously formed based on shared cultural and religious backgrounds and have been managed by the communities (Kim & Jang, 1994). Given the definition of forest commons provided above, village woods can be classified as forest commons. Today, village woods in South Korea are called Maeulsoop. Stories about the formation of village woods are recorded in old texts, including Dongkukyisangkukjip, Samguksagi, Samgukyusa, Yangchonjip, Dongkukyeojiseungram, Sejongsilok and Mokminsimseo, as well as old maps.

\* This research was supported by the National Research Foundation of Korea (Grant Number 32A-2010-1-B00161). The authors wish to thank the anonymous journal reviewers for their advice and valuable comments.

The colonial government of Korea published 'Chosun's Limsu' in 1938 (Tokumisu, 1938), which listed 189 village woods. Through literature reviews and field studies, Forest for Life (2007) revealed that approximately 71.4% of village woods have been lost due to increasing demand for land and timber, as well as natural disasters. Even today, there is a possibility of loss and degradation of village woods. Realising the magnitude of the problem, since 2003, the Korea Forest Service and local governments have made an effort to conserve the remaining village woods and to encourage non-government organisations to restore village woods that have disappeared or degraded.

To design policy alternatives for conservation, local peoples' attributes towards village woods and communities that could determine village woods' quality and sustainability need to be identified. The quality of village woods should be evaluated in terms of sustainability as well as their perceived utility to the villagers. The core of mainstream sustainability thinking has coalesced into the integration of three dimensions: environmental, social and economic aspects (Adams, 2006). Thus, ecological quality and socio-economic quality must be considered together, and several studies have examined this relationship in greater detail. Adams et al. (2004) and Wunder (2001) showed that there is a diametrically opposing relationship between human livelihood and biodiversity conservation (Persha, Agrawal, & Chhatre, 2011). On the other hand, Naughton-Treves (2005) demonstrated a shift in favour of protected areas wherein resource use of local communities is allowed. Furthermore, Agrawal and Chhatre (2006) found that the higher utility of forests could result in better forest conditions because villagers are more likely to protect their forests for securing its utility. Persha, Fischer, Chhatre, Agrawal, and Benson (2010) also proved that tree species' richness and livelihoods could go together. They argued that the trade-off and synergy between conservation of forest commons and their utility should be considered simultaneously in decision making for sustainability of forest commons. Recently, Chhatre and Agrawal (2009) and Persha et al. (2011) analysed the trade-offs and synergy between ecological quality and the quality of socio-economic characteristics of forests simultaneously using a quadrant model as seen in Figure 16.1. With reference to the research approach used by Persha et al. (2011), we evaluate the sustainability of village woods.

This chapter aims to analyse the relationship between the sustainability and attributes of village woods with an emphasis on ownership and self-governance. Ostrom (1999) defined a self-governed forest as one in

**Figure 16.1:**
*Quadrant of the dependent variable*

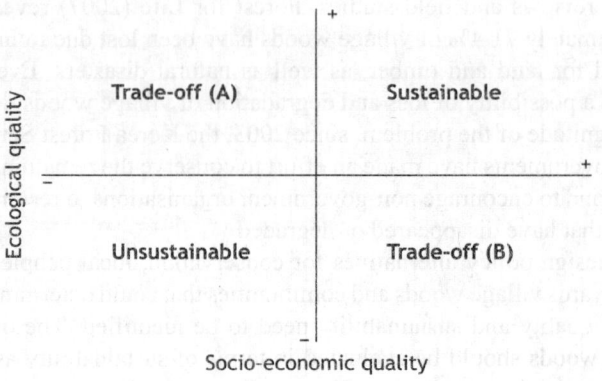

Source: Authors.

which actors who are major users of the forest and involved over time in making and adapting rules within collective-choice arenas regarding the inclusion or exclusion of participants, appropriation strategies, obligations of participants, monitoring, sanctioning and conflict resolution. It is generally taken for granted that there are positive relationships between ownership and the right of the local people to manage and use the commons; however, ownership and the right for commons in South Korea were separated after the new land ownership registration system was enforced during the Japanese colonisation in the early 20th century (Kim, 2008). Thus, the two rights should be considered separately. Based on this background, we formulated two hypotheses as follows. First, if the ownership is given to the community, village woods are likely to be conserved. As Jodha (1986) and Poffenberger (1990) showed, the loss of ownership combined with substantial increases in population, greater commercialisation of forest products and technological change that made rapid deforestation possible have increasingly threatened forests in all parts of the world. Second, if the community manages the village woods by themselves without external assistance, village woods are more likely to be conserved. Deforestation and degradation of forest resources are most likely to occur in open-access forests where effective governance has not been established (Ostrom, 1999). This chapter begins by describing the methodology and material. In the following sections, the survey results and estimated results are presented. Finally, the results are interpreted and discussed and concluding remarks are provided.

## Material and Methods

The discrete choice model is applied to the analysis of sustainability of village woods, which can be categorised into four groups as seen in Figure 16.1. The categorised value for sustainability of village woods assumed to be the outcome resulting from a series of community choice is statistically related to the attributes of the village woods and community as the dependent variable. Logit and probit are most commonly used models in the discrete choice analysis. These models are expanded according to the characteristics of the dependent variable (e.g., multinomial logit/probit, conditional logit, nested logit and mixed logit). Chhatre and Agrawal (2009) applied a multinomial logit model without ordering the categories of the states concerning the forests and their socio-economic qualities, and labelled the four categories as follows: sustainable forest commons, deferred forest commons, overused forest commons and unsustainable forest commons. On the other hand, Persha et al. (2011) selected the ordered logit model, assuming that there is an order among categories but no difference between the two trade-off categories.

To make the quadrant of sustainability, it is very important to select appropriate indicators to represent ecological quality and socio-economic quality. Various indicators that measure the condition or health of forest were identified (Woodall et al., 2011). Ochoa-Gaona et al. (2010) developed an ecological index based on qualitative and semi-qualitative data (e.g., canopy height, number of strata, tree cover, dominant trees and number of tree species, as well as the management status of and physical damages to the forest). Unfortunately, these indicators are not applicable in this study because village woods are entirely different from a typical forest. Kim and Jang (1994) stated that communities planted indigenous species in village woods. Consequently, a significant number of village woods consist of a couple of species in a similar age group. Korean red pine (*Pinus densiflora*) is the most common in village woods. Understory vegetation is not well developed and stand density is low compared to a typical forest because communities have managed village woods for special purposes such as leisure activities. Therefore, we had to select alternative indicators. Crown width has been used as an indicator for the health of individual trees (USDA Forest Service, 1999), but this indicator is affected by age and stand density. Thus, we think it is more appropriate to select the ratio between crown width and diameter at breast height (DBH) as an indicator of ecological healthiness. Many previous studies show that there is a positive correlation between crown width and DBH (Bechtold, 2003, 2004; Gringrich, 1964; Paulo, Stein, & Tome, 2002; Smith & Gibbs, 1970; Zhang, Ke, Quackenbush, & Zhang, 2010). Moreover, Bonner (1964) proved that stand density did not influence the correlation between crown width and DBH for the lodge pole pine (*Pinus contorta*). Zarnoch, Bechtold, and Stolte (2004) argued that an individual tree approach can be used to detect a forest health problem when trees in stand are not competitive.

Subsistence livelihood, used by Persha et al. (2011) as a socio-economic indicator, could not be used in this study because communities no longer collect timber and non-timber products from the village woods, meaning that communities no longer depend on village woods for substantive livelihood. Davidson-Hunt et al. (2001) categorised tourism and education based on natural and cultural heritage, biodiversity conservation, healing ceremonies, recreation and water quality as non-consumptive, non-timber forest products. Various uses of village woods such as protection from wind and flood, places for festivals and relaxation, and recreation sites for outsiders and *dang-san* (commemorative religious rites) were also reported in previous studies (Kim & Jang, 1994; Tokumisu, 1938). Thus, these non-timber forest utilities can serve as alternative indicators representing socio-economic benefits valued by communities (Kim & Jang, 1994; Park, 2006; Shin, 2009; Yoon, Kim, Jang, & Kim, 1998). However, the number of forest services recognised by contemporary communities cannot be used as indicators because village woods were formed for different purposes and the utilities valued by communities have changed. Finally, we decided to use the change in the number of forest services over time, specifically between the 1970s and the present, as a socio-economic indicator. We postulated that the more services of village woods were retained and developed, the more benefit communities would enjoy from village woods. The forest services of village woods are categorised as physical, mental and cultural, and recreational (Table 16.1).

It is difficult to judge that there is *a priori* order between trade-off (A) and trade-off (B). Salafsky and Wollenberg (2000) also defined trade-off (A) as a protected area that does not accommodate community activities for livelihood. Considering these factors, we selected the multinomial logit as an analytical framework. The multinomial logit is a generalised logistic regression that allows more than two discrete categories as dependent variables. For $M$ possible categories, the probability that respondent $i$ selects $m$th category is as shown in equation 1. $\beta_k$ is the regression coefficient associated with the $k$th category. $X_i$ is the vector of

**Table 16.1:**
*Forest utility functions of village woods in South Korea*

| Main category | Sub category |
| --- | --- |
| Physical functions | Protection from wind |
| | River bank |
| | Timber and non-timber product |
| Mental and cultural functions | Dang-san (place of commemorative rite for the god) |
| | Place for village festivals |
| Recreational functions | Recreation site for villagers |
| | Recreation site for outsiders (visitors) |

*Source:* Authors.

explanatory variables describing respondent *i*. In equation 1, the baseline category is *M*. Any category can be selected as the baseline category.

$$\Pr(Y_i = m) = \frac{\exp(\beta_m X_i)}{1 + \sum_{k=1}^{M-1} \exp(\beta_k X_i)} \tag{1}$$

Ownership and self-governance were selected as independent variables. Ownership and self-governance are dummy variables. Ownership scored the value 1 if the village woods are owned by the community as a whole and 0 otherwise. In South Korea, an autonomous village organisation that determines the major issues of the village has been established in nearly all villages. In the past, some villagers established a special organisation—called Song-gye—for managing pine trees. Shin (2009) stated that communities that formed Song-gye made a rule, established a fund and participated in forest management activities Therefore, it is natural to regard all village woods that have been managed by an organisation as self-governed assets; however, matters pertaining to the management of village woods are given low priority. Consequently, there are many village woods that are managed solely by the community but some that are also managed by local governments. The variable of self-governance scored 1 only if the village woods were managed solely by the community without any assistance from external authorities and 0 otherwise. Inoue (2011) suggested that there are three possible strategies by the local people may respond to external influences; resistance, adjustment and eclectic strategies. Thus, the self-governance corresponds to the resistance strategy. Forest size and land price were also included as control variables because these factors can influence the condition of the forest (Bottazzi, Cattaneo, Rocha, & Rist, 2013; Kim & Lee, 1992).

## Study Site

According to Forest for Life (2007), there are currently 37 village woods remaining in South Korea. However, most of the village woods listed in the report are located along the banks of rivers or the sea. Thus, the village woods identified by Forest for Life cannot represent the entire body of existing village woods. Youn et al. (2011) used literature reviews and field surveys to identify 144 village woods around Mt. Jiri in five counties (Namwon, Gurye, Hadong, Sanchung and Hamyang) (Figure 16.2). To restrict the spatial variance in observations, the study site was limited to one county, Namwon City (Figure 16.3).

**Figure 16.2:**
*Location of the Mt. Jiri*

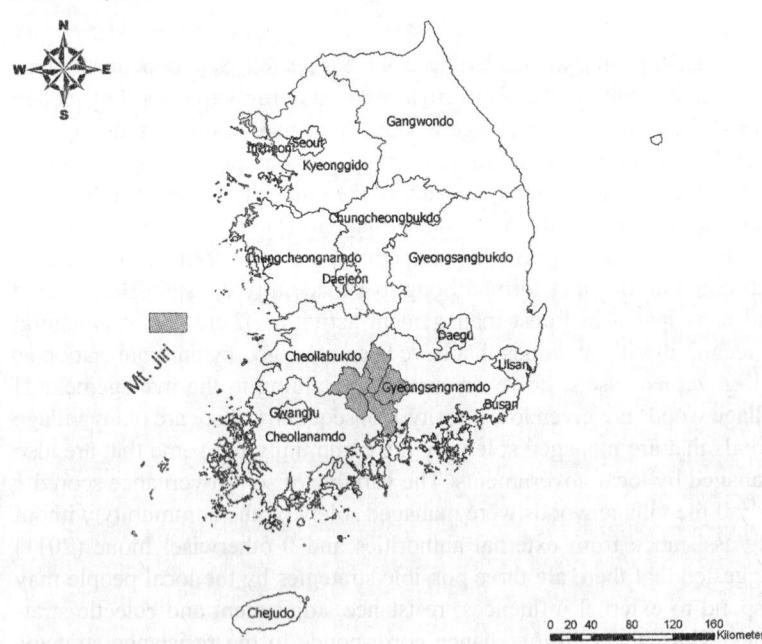

*Source:* Authors.

**Figure 16.3:**
*Village woods in Namwon City*

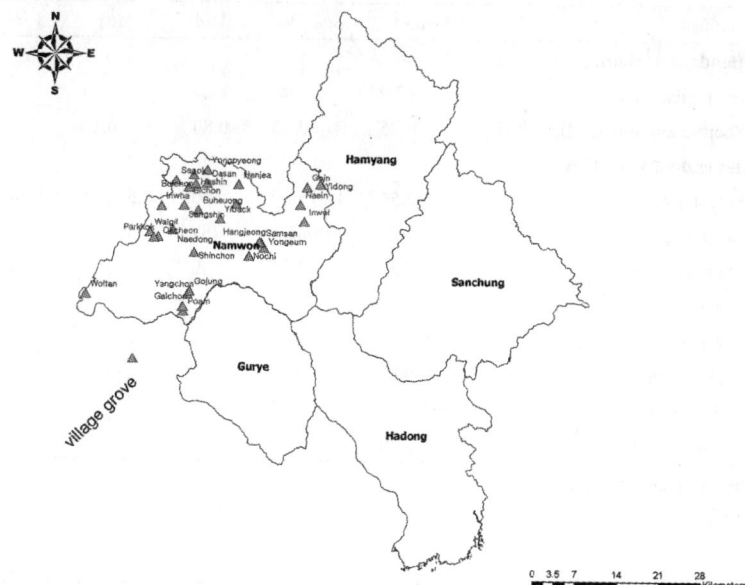

*Source:* Authors.

# Results

## Survey Results

All data were collected from a field survey in 2011. We visited all villages maintaining village woods in Namwon City. The ratio between crown width and DBH of all trees in village woods was measured directly. Information on the number of woods utility functions, ownership and self-governance were collected through in-depth interviews with current and former village representatives. Data on woodlot size and land price were collected from the website (http://www.onnara.go.kr/index.jsp) managed by Ministry of Land, Infrastructure and Transportation.

Table 16.2 describes the basic statistics of the variables. The number of forest utility functions decreased by 25% compared to the 1970s. Sixteen village woods (53%) are owned by the communities, while 14 village woods (43%) are managed by the communities without external

**Table 16.2:**
*Basic statistics of variables*

| Variables | Mean | Std. Dev. | Min | Max | VIF |
|---|---|---|---|---|---|
| **Dependent variables** | | | | | |
| Ecological quality | 17.9 | 3.08 | 12.1 | 25.11 | – |
| Socio-economic quality (%) | –0.25 | 0.33 | –0.80 | 0.50 | – |
| **Independent variables** | | | | | |
| Forest size (m²) | 2,062.56 | 1,915.39 | 500 | 8,555.76 | 1.36 |
| Land price (USD) | 7,256.7 | 7,081.1 | 288 | 30,237 | 1.48 |
| Ownership | 0.53 | 0.51 | 0 | 1 | 1.15 |
| *Village's own: 1* | *16 (53.33%)* | | | | |
| *Otherwise: 0* | *14 (46.67%)* | | | | |
| Self-governance | 0.43 | 0.50 | 0 | 1 | 1.04 |
| *Presence: 1* | *13 (43.33%)* | | | | |
| *Absence: 0* | *17 (56.67%)* | | | | |

VIF, variance inflation factor.
**Source:** Authors.

assistance. The multinomial logit model is a type of generalised regression model. Thus, multi-collinearity between independent variables should be checked. The last column of Table 16.2 shows the variance inflation factor (VIF), which quantifies the severity of multi-collinearity in regression analysis. Chatterjee and Price (1977) and Kennedy (1992) explained that there is a serious multi-collinearity if VIF exceeds 10. Furthermore, Judge, Hill, Griffiths, Lutkepohl, and Lee (1982) fixed 5 as a threshold to determine multi-collinearity. All independent and control variables were included in the probability function (equation 1) because all VIF values were smaller than 5.

The level of sustainability represented by quadrants a, b, c and d (shown in Figure 16.4) is treated as the dependent variable representing: (a) those belonging to the 'sustainable' category of above-average ecological and socio-economic quality; (b) those belonging to the 'unsustainable' category of below average ecological and socio-economic quality; and (c and d) those woods that were above average on one dimension but below average on the other. Among categories (c) and (d), case (c), high ecological quality and low socio-economic quality, was labelled as 'trade-off (A)', while (d), low ecological quality and high socio-economic quality, was labelled as 'trade-off (B)'; 9 (30%),

**Figure 16.4:**
*Four groups of the dependent variable*

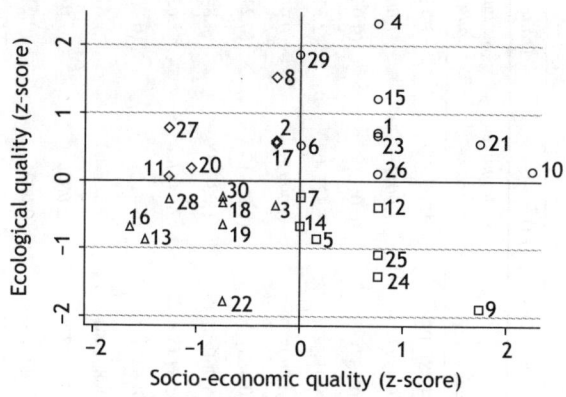

| number | village | number | village | number | village |
|--------|---------|--------|---------|--------|---------|
| 1 | Galchon | 11 | Buheuong | 21 | Yongpyeong |
| 2 | Goin | 12 | bichon | 22 | Woltan |
| 3 | Gojung | 13 | Sagok | 23 | Yidong |
| 4 | Inwol | 14 | Samsan | 24 | Yiback |
| 5 | Naedong | 15 | Sangshin | 25 | Inwha |
| 6 | Naein | 16 | Shinchon | 26 | Jungcheol |
| 7 | Nochi | 17 | Yangchon | 27 | Poam |
| 8 | Dasan | 18 | Okcheon | 28 | Hasin |
| 9 | Parkkok | 19 | Walgil | 29 | Hanjea |
| 10 | Bulchon | 20 | Yongeum | 30 | Hangjeong |

*Source:* Authors.

8 (26.7%), 6 (20%) and 7 (23.33%) were categorised as the sustainable group, trade-off (A) group, unsustainable group and trade-off (B) group, respectively.

## Estimated Results

The data were analysed using STATA 11.1. Table 16.3 shows the estimated results of the multinomial logit model. The goodness of fit can be measured by likelihood ratio index (LRI). Hensher and Johnson (1981) commented that LRI values between 0.2 and 0.4 are considered to be

**Table 16.3:**
*Estimated results of the multinomial logit*

| Baseline category | Unsustainable outcome | | | Trade-off (A) | | Trade-off (B) |
|---|---|---|---|---|---|---|
| Outcome category | Trade-off (A) | Trade-off (B) | Sustainable outcome | Trade-off (B) | Sustainable outcome | Sustainable outcome |
| Forest size (m²) | −2.95E-05[a] | −5.08E-04 | 1.04E-03** | −4.78E-04 | 1.07E-03* | 1.54E-03* |
| | 0.9999[b] | 0.9995 | 1.0010 | 0.9995 | 1.0011 | 1.0015 |
| | (0.946)[c] | (0.518) | (0.032) | (0.567) | (0.050) | (0.068) |
| Land price (USD) | −3.28E-05 | −1.34E-04 | −3.65E-04** | −1.01E-04 | −3.32E-04* | −2.31E-04 |
| | 0.9999 | 0.9999 | 0.9996 | 0.9997 | 0.9997 | 0.9998 |
| | (0.732) | (0.344) | (0.035) | (0.483) | (0.054) | (0.213) |
| Ownership | −5.97E-01 | −1.66E-03 | 4.82E-01 | 5.96E-01 | 1.08E+00 | 4.84E-01 |
| | 0.5502 | 0.9983 | 1.6198 | 1.8145 | 2.9441 | 1.6225 |
| | (0.648) | (0.999) | (0.717) | (0.637) | (0.412) | (0.704) |
| Self-governance | 2.57E+00* | 1.69E+00 | 3.31E+00** | −8.80E-01 | 7.32E-01 | 1.61E+00 |
| | 13.1282 | 5.4444 | 27.3069 | 0.4147 | 2.0800 | 5.0156 |
| | (0.065) | (0.231) | (0.033) | (0.479) | (0.602) | (0.218) |

| Intercept | −6.13E-01 | 1.04E+00 | −1.38E+00 | 1.66E+00 | −7.71E-01 | −2.43E+00 |
|---|---|---|---|---|---|---|
| | (0.665) | (0.478) | (0.395) | (0.292) | (0.649) | (0.167) |
| No. of observation | 30 | | | | | |
| Log-likelihood | −31.708 | | | | | |
| Pseudo $R^2$ | 0.2314 | | | | | |
| Prob > chi$^2$ | 0.0864 | | | | | |

Significance levels: *10%, **5%, ***1%.

[a]Estimated coefficients.

[b]The numbers in italic represents the ratio of the probability of choosing one outcome category over the probability of choosing the baseline category.

[c]The value in bracket is $p$ value (P>|z|).

**Source:** Authors.

**Table 16. 4:**
*Small-Hsiao (1985) tests for IIA assumption*

| Omitted category | $\chi 2$ | d.f. | $P>\chi 2$ | Evidence of $H_0{}^*$ |
|---|---|---|---|---|
| Trade-off (A) | 3.276 | 10 | 0.974 | For $H_0$ |
| Trade-off (B) | 1.809 | 10 | 0.998 | For $H_0$ |
| Sustainable | 4.102 | 10 | 0.943 | For $H_0$ |

*$H_0$, Odds (category-J vs. category-K) are independent of other alternative.
**Source:** Authors.

extremely good fits. The LRI of this model is 0.2314. Therefore, the estimated results are statistically significant.

In case of the multinomial logit, the assumption of independence of irrelevant alternative (IIA) must be met to ensure an unbiased and consistent parameter. A null hypothesis that the probability of choosing a certain outcome is independent of the probabilities of choosing another outcome was formulated. Small and Hsiao's (1985) method was applied to test the null hypothesis for each possible omitted category. As a result, the null hypothesis was supported by all the three cases (Table 16.4).

An estimated coefficient in multinomial logit should be interpreted as the impact of independent variables on the result of the category outcome over the baseline category chosen a priori. A one-unit increase in the variable of self-governance (here, the change from absence to presence of self-governance) is associated with a 3.31 increase in the relative log odds of belonging to the sustainable category over the unsustainable category. We converted estimated coefficients into the relative risk ratio, i.e., the ratio of the probability of the category outcome being selected to the probability of the baseline category being selected. The relative risk ratio for a one-unit increase in the value of self-governance is 27.31 for belonging to the sustainable category over the unsustainable category. However, ownership has little impact on the probability of the outcome category to be selected for all cases.

## Discussion and Conclusion

We started this research with a question: what kind of policy reform can make forests more sustainable in terms of ecological and socio-economic quality. In this study, we attempted to answer this question by analysing the relationship between sustainability and attributes of village woods in South Korea with an emphasis on the ownership structure and self-governance. The result of our analysis with cases in the City of Namwon, South Korea, support the hypothesis that self-governance is an important element for sustaining the village woods, whereas the hypothesis referring to property ownership was not supported by the cases studied. One possible explanation for the deviations from the expectation results of the role of property ownership can be referred to the history of community management of natural resources. Shin (2009) stated that the *song-gye* autonomous organisations secured the right to use the forests for their subsistence regardless of ownership. Although the modern land ownership system deprived communities of their ownership rights by statutory laws, they have been using and managing village woods because they did not recognise the concept of the modern ownership system in the case of village woods (Kim, 2008). The villagers of communities studied seem to treat the property ownership and governance of village woods as a separate matter (Figure 16.5). Sixteen village woods (62.5%) owned by communities as a whole are being managed by the communities

**Figure 16.5:**
*Relationship between ownership and self-governance*

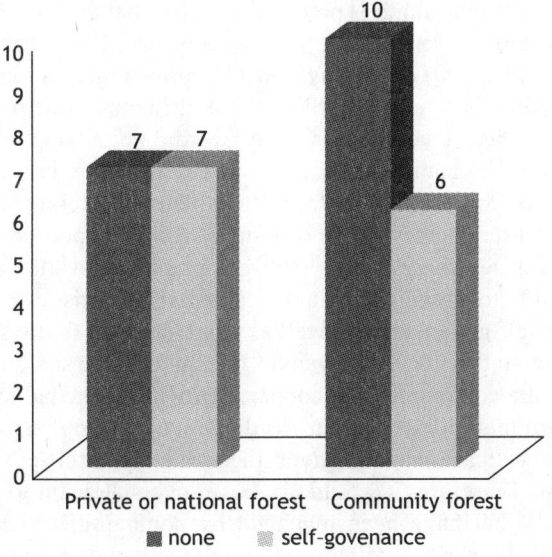

*Source:* Authors.

themselves without outsiders' involvement. On the other hand, a half of the village woods not owned by the community as a whole are being managed by the communities.

Several studies emphasise on the importance of self-governance in common pool resource management (Ostrom, 1999). The results of this study also provide empirical evidence supporting previous studies by Chhatre and Agrawal (2009) and Persha et al. (2011). Proceeding from what has been described above, it is possible to conclude that the sustainability of village woods is influenced more heavily by the power of self-governance than by property ownership. This conclusion is virtually identical to the findings of Chhatre and Agrawal (2009), who showed that forest ownership has little influence on the probability of forest use being categorised as sustainable or unsustainable. Thus, to make village woods sustainable in Korea, policy should strive to support communities in managing village woods on their own. This seems difficult to achieve, however, due to hollowing out of rural areas, the ageing population and the increased proportion of outsiders in rural area. Ostrom (1999), through a review of previous studies (Baland & Platteau, 2000; McKean, 1992; Ostrom, 1990, 1992a, 1992b; Schlager, 1990; Tang, 1992; Wade, 1994), summarised the attributes of the resources and users that enhance the likelihood of users self-organising to avoid the social losses associated with open-access, common-pool resources as follows: (1) feasible improvement, (2) indicator, (3) predictability, (4) spatial extent, (5) salience, (6) common understanding, (7) discount rate, (8) distribution of interests, (9) trust, (10) autonomy and (11) prior organisation experience. Thus, there is a need to utilise these attributes to help communities organise themselves. In this study, we did not attempt to collect information on the characteristics of resources and users in such detail that the data could provide clues for us to explain whether the resources situations and user characteristics of village woods support community efforts to organise village woods' self-governance. Nevertheless, it is usually possible to assess the boundaries and characteristics of village woods under self-governance, as well as to predict their future state.

We also found that 10 village woods, one out of three sites, are being managed by the communities in cooperation with municipal authority. We can regard that villagers of these village woods adopted the eclectic strategy. This case was not given the value 1 for self-governance in estimation. Therefore, it would not be an overstatement to say that village woods that have been managed by communities themselves were more likely to be sustainable than village woods managed with external assistance. In other words, resistance strategy is more suitable than eclectic strategy in South Korean cases. The village woods are still well understood and appreciated by most villagers due to their intangible services, which are fairly evenly distributed among villagers regardless of their social status. As rural societies in South Korea has become more connected to outside markets and governments, the trust among village members has been deteriorated and the decision-making process has become less autonomous. Recently, retired urban people have begun to settle in rural villages for a second life after retirement, but the resulting heterogeneity of community members has become a factor weakening the trust among villagers. Because of the high opportunity cost of capital during the economic development of the nation, many village woods were converted to other land uses; for example, some village woodlands were converted to public school grounds and roads. Such social changes have made it difficult for villagers to organise self-governance in village woods management in Korea. However, recent researches have emphasised on the importance of governance with government in natural resources management (Lawrence, De Vreese, Johnston, Konijnendijk van den Bosch, & Sanesi, 2013). Thus, trust building is an important element of the collaborative management of village woods in Korea, especially between land owners and community members. The commitment principle suggested by Inoue (2011) could be applied to make the village woods governed collectively with the participation of villagers in Korea. The results of this study do provide a clue to this end, since the self-governance by community members is an important factor for sustainable village woods management. The commitment of community members to the management of their village woods inherited from their ancestors can make self-governance of village woods stronger. But how to make them committed to village woods management remains unanswered in this study.

There are some problems remaining in terms of selecting the dependent variable representing the ecological quality of village woods. We have already made clear that the ratio of DBH and crown width cannot represent complete ecological quality. Zarnoch et al. (2004) insisted that the volume, surface area and productivity of the crown should be considered altogether. However, considering the specificity of the village woods in South Korea, it is not easy to add or replace these with another ecological indicator. Therefore, further studies comparing the results from possible indicators such as soil are needed. The dependent variable was classified based on the average value in this chapter. However, the average value cannot be regarded as an absolute value determining the sustainability. When observations are added, the average value will shift, and consequently, the groups of dependent variable will also change. Given these limitations, it seems more reasonable to apply the thresholds, if they can be identified, determining the sustainability rather than the average value. Thus, further studies are needed to determine an appropriate threshold.

# References

Adams, W.M. (2006). *The Future of Sustainability: Re-thinking Environment and Development in the Twenty-first Century.* Gland, Switzerland: The World Conservation Union.

Adams, W.M., et al. (2004). Biodiversity conservation and the eradication of poverty. *Science*, 306,1146–1149.

Agrawal, A. & Chhatre, A. (2006). Explaining success on the commons: Community forest governance in the Indian Himalaya. *World Development*, 34, 149–166.

Baland, J.-M. & Platteau, J.-P. (2000). *Halting Degradation of Natural Resources. Is There a Role for Rural Communities?* Oxford, UK: Oxford University Press.

Bechtold, W.A. (2003). Crown-diameter prediction models for 87 species of stand-grown trees in the eastern United States. *Southern Journal of Applied Forestry*, 27, 269–278.

————— (2004). Largest-crown-width prediction models for 53 species in the western United States. *Western Journal of Applied Forestry*, 19, 245–251.

Bonner, G.M. (1964). The influence of stand density on the correlation of stem diameter with crown width and height for lodgepole pine. *The Forestry Chronicle*, 40, 347–349.

Bottazzi, P., Cattaneo, A., Rocha, D.C., & Rist, S. (2013). Assessing sustainable forest management under REDD+: A community-based labour perspective. *Ecological Economics*, 93,94–103.

Chatterjee, S. & Price, B. (1977). *Regression Analysis by Example.* New Jersey: John Wiley & Sons.

Chhatre, A. & Agrawal, A. (2008). Forest commons and local enforcement. *Proceedings of the National Academy of Sciences*, 105, 13286–13291.

————— (2009). Trade-offs and synergies between carbon storage and livelihood benefits from forest commons. *Proceedings of the National Academy of Sciences of the United States of America*, 106, 17667–17670.

Davidson-Hunt, I.J., Duchesne, L.C., & Zasada, J.C. (2001). Non-timber forest products: Local livelihoods and integrated forest management. In Conference Proceedings of Forest Communities in the Third Millennium: Linking Research, Business, and Policy Toward a Sustainable Non-Timber Forest Product Sector, 29 May 2001.

Gringrich, S.F. (1964). Criteria for measuring stocking in forest stands. *Proceedings of Society of American Foresters*, 23,198–201.

Hensher, D.A. & Johnson, L.W. (1981). *Applied Discrete-choice Modelling.* London/New York: Croom Helm, Wiley.

Inoue, M. (2011). *Prototype Design Guidelines for 'Collaborative Governance' of Natural Resources.* In 13th Biennial Conference of the International Association for the Study of the Commons, 10–14 January 2011.

Jodha, N.S. (1986). Common property resources and rural poor in dry regions of India. *Economic and Political Weekly*, 21, 13.

Judge, G.G., Hill, R.C., Griffiths, W.E., Lutkepohl, H., & Lee, T.C. (1982). *Introduction to the Theory and Practice of Econometrics*. New Jersey: John Wiley & Sons.

Kennedy, P.A. (1992). *Guide to econometrics* (3 edn.). Massachusetts: MIT Press.

Kim, H.-D. & Lee, K.-H. (1992). *A Study of the Relation between Land-price and Land Development*. Annual Conference of the Architectural Institute of Korea, pp. 255–260.

Kim, H.B. & Jang, D.S. (1994). *Maeul-soup*. Seoul: Youlhwadang.

Kim, S.W. (2008). The role of the Chosun Governor Headquarter in the formative process of the modern Woodland Ownership System. *Social Science Research Review*, 24, 23–40.

Lawrence, A., De Vreese, R., Johnston, M., Konijnendijk van den Bosch, C.C., & Sanesi, G. (2013). Urban forest governance: Towards a framework for comparing approaches. *Urban Forestry & Urban Greening*, 12, 464–473.

Forest for Life (2007). *Limsu of Chosun Dynasyty*. Seoul: Geobook.

McKean, M.A. (1992). Management of traditional common lands(Iriaichi) in Japan. In D.W. Bromley (ed.), *Making the Commons Works:Theory, Practice, and Policy* (pp. 63–98). Oakland, CA: ICS Press.

Naughton-Treves, L. (2005). *The Role of Protected Areas in Conserving Biodiversity and Sustaining Local Livelihoods* (pp. 219–252). Palo Alto: Annual Reviews.

Ochoa-Gaona, S., Kampichler, C., de Jong, B.H.J., Hernández, S., Geissen, V., & Huerta, E. (2010). A multi-criterion index for the evaluation of local tropical forest conditions in Mexico. *Forest Ecology and Management*, 260, 618–627.

Ostrom, E. (1990). *Governing the Commons: The Evolution of Institutions for Collective Action (Political Economy of Institutions and Decisions)*. Cambridge, UK: Cambridge University Press.

——— (1992a). *Crafting Institutions for Self-governing Irrigation Systems* (2 edn.). Michigan, USA: ICS Press.

——— (1992b). The rudiments of a theory of the origins, survival, and performance of common property institutions. In D.W. Bromley, D. Feeny, M.A. McKean, P. Peters, J.L. Gilles, R.J. Oakerson, C.F. Runge, & J.T. Thomson (eds), *Making the Commons Work: Theory, Practice and Policy* (pp. 293–318). San Francisco, California: Institute for Contemporary Studies Press.

——— (1999). *Self-governance and Forest Resources*. Bogor: CIFOR.

Park, B.W. (2006). Maeul-soup and culture. *Review of Korean Studies*, 33, 195–232.

Paulo, M.J., Stein, A., & Tome, M. (2002). A spatial statistical analysis of cork oak competition in two Portuguese silvopastoral systems. *Canadian Journal of Forest Research-Revue Canadienne De Recherche Forestiere*, 32, 1893–1903.

Persha, L., Agrawal, A., & Chhatre, A. (2011). Social and ecological synergy: Local rule-making, forest livelihoods, and biodiversity conservation. *Science*, 331, 1606–1608.

Persha, L, Fischer, H., Chhatre, A., Agrawal, A., & Benson, C. (2010). Biodiversity conservation and livelihoods in human-dominated landscapes: Forest commons in South Asia. *Biological Conservation*, 143, 2918–2925.

Poffenberger, M. (1990). *Keepers of the Forest: Land Management Alternatives in Southeast Asia*. Quezon City, Philippines: Kumarian Pr Inc.

Salafsky, N. & Wollenberg, E. (2000). Linking livelihoods and conservation: A conceptual framework and scale for assessing the integration of human needs and biodiversity. *World Development*, 28, 1421–1438.

Schlager, E. (1990). *Model Specification and Policy Analysis: The Governance of Coastal Fisheries*. Indiana: Indiana University.

Schlager, E. & Ostrom, E. (1992). Property-rights regimes and natural resources: A conceptual analysis. *Land Economics*, 68, 249–262.

Shin, D.H. (2009). The fate of the commons and sustainable ancient wisdom of Korea. *Comparative Korean Studies*, 17, 7–31.

Small, K.A. & Hsiao, C. (1985). Multinomial logit specification tests. *International Economic Review*, 26, 619–627.

Smith, H.C. & Gibbs, C.B. (1970). A guide to sugarbush stocking. Based on the crown diameter/D.b.h. relationship of open-grown sugar maples. *USDA Forest Service Research Paper* 8.

Tang, S.Y. (1992). *Institutions and Collective Action: Self-Governance in Irrigation*. California: ICS Press.

Tokumisu, N. (1938). *Chosun's Imsu*. Hanseong (Seoul), Korea: Japanese Governors-Genenal of Korea.

USDA Forest Service. (1999). Forest health monitoring. Field methods guide. U.S. Forest Service, National Forest Health Monitoring Program, Research Triangle Park, North Carolina, USA.

Wade, R. (1994). *Village Republics: Economic Conditions for Collective Action in South India*. California: ICS Press.

Woodall, C.W., Amacher, M.C., Bechtold, W.A., Coulston, J.W., Jovan, S., Perry, C.H., Randolph, K.C., Schulz, B.K., Smith, G.C., Tkacz, B., & Will-Wolf, S. (2011). Status and future of the forest health indicators program of the USA. *Environmental Monitoring and Assessment*, 177, 419–436.

Wunder, S. (2001). Poverty alleviation and tropical forests—what scope for synergies? *World Development*, 29, 1817–1833.

Yoon, Y.W., Kim, H.B., Jang, D.S., & Kim, J.T. (1998). A study on the landscape and utility of the windbreak forest as an element of a settlement near the east sea region in Kangwon province. *Journal of Korean Institute of Traditional Landscape Architecture*, 16, 59–81.

Youn, Y.C., Kweon, D., Lee, E.H., Youn, S.J., Choi, S.H., & Yun, J.O. (2011). Ownership, management, functions, and using status of Village Groves in Korea—the case of municipalities around Mt. Jiri. In Summer Meeting of the Korean Forest Society.

Zarnoch, S.J., Bechtold, W.A., & Stolte, K.W. (2004). Using crown condition variables as indicators of forest health. *Canadian Journal of Forest Research-Revue Canadienne De Recherche Forestiere*, 34, 1057–1070.

Zhang, W.H., Ke, Y.H., Quackenbush, L.J., & Zhang, L.J. (2010). Using error-in-variable regression to predict tree diameter and crown width from remotely sensed imagery. *Canadian Journal of Forest Research-Revue Canadienne De Recherche Forestiere*, 40, 1095–1108.

# 17

# Japan I: Histories of 'Property Wards (Zaisanku)' and 'School Forests (Gakkou-rin)'

*Haruo Saito and Taro Takemoto*

## Introduction

After the Edo shogunate era ended and Japan became a united country in 1868, it has experienced modernisation in various ways due to the impact of Western ideology and its social system. The forested areas did not escape this modernisation as the governors worked to remove the barrier created by the customs carried over from the earlier period.

However, there still exist many forests that are substantially managed by the local communities regardless of the spread of modernisation throughout the country. According to Inoue (2011), these forests that have adopted, though not necessarily intentionally, a localisation strategy (resistance strategy) are referred to as commons. It is notable, however, that the local community can maintain its independence over the management of the commons given the tide of the times. The cases we address in this chapter are those in which the forests have been managed according to local customs for decades or even hundreds of years. The first case is the property ward forest studied by Saito, and the other is the case of the school forest studied by Takemoto.

In these cases, the 'outsider', which is generally the local government, confronts the local commons, which results in collaborative governance of the simplest structure that comprise the local community and a local government administrator. This chapter discusses the process that the local community undergoes to maintain its management of the commons and abstracts implications for the future by considering the history of the commons from the perspective of collaborative governance.

First, we briefly illustrate the policies towards forest commons in Japan. Then, we explain the reality of localisation based on the two aforementioned cases from the perspective of collaborative governance. Finally, we discuss the implications on collaborative governance as derived from the commons that have adopted the localisation strategy.

## Japanese Policy over Forest Commons

### Traditional Commons and Modern National Government in Japan

Typical forest commons in Japan are called *iriai*. The traditional structure and system of the forest commons were introduced to the world by McKean (1982) as a self-governed system approximately 30 years ago. It is estimated that the *iriai* originated in the medieval age as various historical documents show that the *iriai* were already highly developed throughout Japan during the Edo period (1603–1868).

Since the beginning of the Meiji period, successive governments had been implementing policies that effectively negated the existence of communal forests, thus necessitating *iriai* to be subject to a wave of restrictive policies (Table 17.1). Initially, the forest commons faced difficulties consolidating their ownership when the Public-Private Ownership Separation Project (1873–1881), which the Meiji government established as its financial base by collecting land taxes, was enacted. Under this policy, as many forest commons were unable to prove their communal ownership, they were registered as national or individual lands.

Soon after, the forest lands were regarded as not only a viable tax source but also a significant source of national wealth by the government. From this perspective, the *iriai* were deemed undesirable because they were likely to be used as meadows and coppices to meet the local people's daily needs. The government believed that planting conifer trees and harvesting the land was a preferable use of the forest land as doing so would produce timber, which, in turn, would contribute to the

**Table 17.1:**
*Four major policies which dam* iriai *forests Japan*

| National policy | Period | Background/aim | Influence to the commons |
|---|---|---|---|
| Public–Private Ownership Separation Project | 1873–1881 | To clarify the basis of land tax, land ownership was needed to be registered | Communal lands which could not prove Its ownership was registered as nationally owned land |
| Town and Village Municipal Law | 1889 | To govern local society effectively, marginal local government system was established and each municipality was required a certain scale | A lot of communal lands converted into municipality-owned landSome hamlets could establish their 'property ward' based on their communal property |
| Hamlet-owned Forest Land Integration Policy | 1910–1939 | To make forest lands to contribute national wealth and strength, it was easy for the national government to control forests if forests were owned by municipality | A lot of communal lands converted into municipality-owned land |
| Commons Forest Modernisation Law | 1966 till now | To enhance economic productivity of forest lands, informal way of land owing was needed to be corrected | A lot of communal lands converted into corporation-owned, personally owned or municipality-owned land |

*Source:* Authors.

industrialisation and economic growth of the area. This government transformation of the forest land served as the background to the government policies regarding *iriai* after the adoption of the Public–Private Ownership Separation Project (Table 17.1). Among these policies, the one having the greatest impact on the cases investigated in this chapter is the second one, which addresses amalgamation.

## Municipal Amalgamation and Forest Commons

### Government's Policy on Local Administration and People's Protest

The national government had the following political motives when enforcing the Town and Village Municipal Act: to reinforce the modernised administrative foundation of towns and villages as a basic local government body through the amalgamation of former villages in existence since the Edo period, and accordingly, to integrate the lands owned by former villages into the fundamental property of new municipalities, regarding them as administrative assets. This means that the ownership of the village commons was transferred to the new municipalities formed after the amalgamation. As a result, the policy created a great deal of backlash from villagers, and there was fear that the amalgamation implemented by the government would not progress as expected. The communal forest was a primary source for the supply of daily commodities that were indispensable for the subsistence of the farming and mountain villages at that time. Indeed, it was an important lifeline for the people in these villages. In response to the resistance mounted by the villagers, the government had little choice but to compromise, which was as follows.

(a) *Old Custom Right:* The Town and Village Municipal Act stipulated the right to use municipal tenure based on old customs. This right is called the 'old custom right'. In this case, the ownership of the communal lands goes to the municipality, and the municipality allows the hamlet (a village in the former age, a hamlet in the new municipality) old custom rights to the land. However, the right can be modified or diminished by a decision of the town or village assembly, not by the hamlet itself.

(b) *Property Ward:* The act also added another article stating that 'in the case of a part of a village or a town (i.e., hamlet) owning a property or a building structure, the village or town can set up a ward assembly or a ward general meeting for conducting administrative work regarding such property or building structure' (summarised by the author). The 'ward' as stated herein refers to the body that is treated as a property ward today.

These two articles were appended to the Local Autonomy Law, which was enacted in 1947, almost without change. It was then that the term 'property ward' appeared for the first time in statutory law.

*Further Amalgamation*

When the Town and Village Municipal Act was enforced, a great number of villages were merged, reducing the number of villages from 71,000 to approximately 15,000. This event is known as 'the great amalgamation of Meiji'.

After World War II, the Local Autonomy Law was enacted as a successor to the Town and Village Municipal Act. In this new system, municipalities were acquired to administrate more tasks, such as establishing and managing junior high schools. To operate these tasks effectively, it was recognised by the government that enlarging the scale of municipalities was necessary. Hence, the Municipality Amalgamation Promoting Act was enforced in 1953. The aim of this policy was to ensure that all municipalities had more than 8,000 residents. As a result of this policy, the number of municipalities was reduced to 3,500. This merger is known as 'the great amalgamation of Showa'. Although adherence to the old custom right is unclear, it is believed that the stipulation of property wards worked well as over 2,000 property wards were established during this great amalgamation (Izumi et al., 2011).

Further, a municipality amalgamation policy was recently enacted to improve the financial status of municipalities. Based on the Special Mergers Law (1995), amalgamations made earnest progress after the year 2000 when the number of municipalities was reduced to 1,700.

## Local Commons Suffering from Merged Municipality: A Case Study of Property Wards (Zaisanku) in Toyota City, Japan

As stated before, the property ward system was created as a result of the compromise between the national government and the commons users throughout Japan. The central government of Japan was willing to construct a local administrative system that would enable it to sufficiently control local society as the central government wanted to integrate some small villages into a larger municipality that would include the village's property, especially the communal forests. The villagers so vehemently protested this policy that the central government was in danger of failing to integrate the small villages. Therefore, to successfully integrate the villages, the government developed the property ward system, which allows local units smaller than municipalities (property wards) to be established and to possess their own communal property.

This system has an absorbing feature from the standpoint of governance as it has two co-existing centres of authority. The one is that of the local community, and the other of the municipality to which the property ward belongs. In other words, this system provides a form of collaborative

governance. As such, it is the simplest form in which a local administrator may participate in the governance of a communal forest as an
outsider. In the Local Autonomy Law, it was stipulated that the administrator of the property ward is the mayor of the municipality. Therefore,
the latter's authority is stronger than that of the local community, which
actually is contrary to the commitment principle. This particular feature
of the property ward system is investigated in the following case study.

## Outline of the Field

The Inabu district, located adjacent to Gifu prefecture and Nagano prefecture, is a mountainous region with a slightly cold climate (Figure
17.1). The central part of the district has an elevation of approximately
500 m. Cultivable land areas such as flatlands and gentle slopes are
scarce, demonstrated by the fact that only 3.4% of the land is cultivated.
Forests cover 85.8% of the land, and the number of forest owners is

**Figure 17.1:**
*Location of 13 property wards of Inabu district in Toyota City*

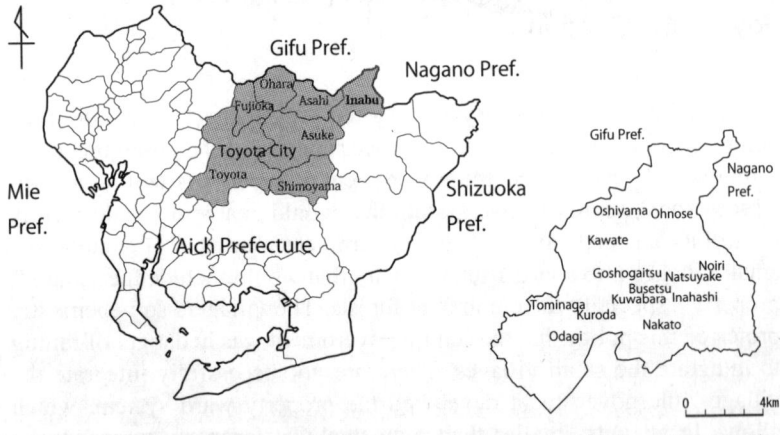

Aich Prefecture and location of Toyota City and Inabu district    Inabu district and location of 13 property wards

*Source:* The map was created by the author (Haruo Saito), by using National Land
Numerical Information of Japan (http://nlftp.mlit.go.jp/ksj-e/index.html).
(Disclaimer: This image has been redrawn by the author (Haruo Saito) and is not to scale.
It does not represent any authentic national or international boundaries and is used for
illustrative purposes only.)

547 per 371 farm households, thus indicating the district is dependent on the mountains and forests.

In the Inabu district, there are 13 property wards corresponding to 13 existing community districts (Inahashi, Ohnose, Oshiyama, Odagi, Kawate, Kuroda, Kuwabara, Goshogaitsu, Tominaga, Nakatoh, Natsuyake, Noiri and Busetsu). These community districts were once independent villages. After they were merged into two municipalities (Inahashi village and Busetsu village), they each established their own property wards, enabling them to possess and manage their own property, especially the forest lands. In 1940, the two villages merged and became the town of Inabu, though the 13 property wards were retained.

In 2005, the long-standing administrative system of the town of Inabu underwent a significant change. It was merged into the already large Toyota City together with Asuke, Asahi, Fujioka town, Shimoyama and Ohara village.

The Inabu district is positioned as a minority group in Toyota City, which has a population of 420,000. Accordingly, with a population of 2,800, the population ratio of the Inabu district to the total population of Toyota City is 0.67%. The population density of the district is 28.5/km², which is the lowest of those of the former towns and villages before the merger (Table 17.2).

**Table 17.2:**
*District-by-district comparison of Toyota City*

| | Population | Population density (per sq. km) | Area (sq. km) | Ratio of forest (%) | Forest area per capita (ha) |
|---|---|---|---|---|---|
| Current Toyota City | 424,128 | 461.8 | 918.47 | 68.6 | 0.15 |
| Districts before merger | | | | | |
| Toyota | 379,312 | 1307.5 | 290.11 | 35.6 | 0.03 |
| Fujioka | 19,922 | 303.8 | 65.58 | 73.2 | 0.24 |
| Ohara | 4,305 | 57.8 | 74.54 | 82.8 | 1.43 |
| Asuke | 9,095 | 47.1 | 193.27 | 86.4 | 1.84 |
| Shimoyama | 5,369 | 47.0 | 114.18 | 85.8 | 1.82 |
| Asahi | 3,312 | 40.3 | 82.16 | 82.1 | 2.04 |
| **Inabu** | **2,813** | **28.5** | **98.63** | **85.8** | **3.01** |

*Sources:* The statistical information of Toyota City (June 2009) and 2000 World Census of Agriculture and Forestry in Japan.

## Thirteen Property Wards as Local Commons

### Existing Community Districts and Property Wards

The community district, as an administrative body, follows the principles of the former villages of the Edo period and is, therefore, a different entity from the property ward, which is a special local public entity. However, both entities have been managed for the same purpose (local autonomy) as they have been recognised by the residents of the Inabu district as identical units. This is because of the process by which current property ward-owned properties have been used and managed as communal forests since the Edo period. In the following section, we examine this process and the management status of the commons in the town of Inabu before the town merged with Toyota City. The outline of the 13 property wards of the Inabu district is shown in Table 17.3.

### Changes in the Form of Use and Management: Elaboration of the Commons System

During the Edo period, when the *sou-yama* (*sou* means village or rural community, and *yama* means forest) is recorded to have been implemented, communal forests were used mainly for mowing grass to feed the cattle and for logging to produce charcoal (Inabu Education Board, 2000). At the end of the Edo period, in the area designated as *tomey-ama*, the individual use of the forest by users with commons rights began to be regulated to prevent excessive logging and a communal tree planting project was implemented. This type of utilisation was created to save bankrupt farmers after they had experienced recurring crises, including the Great Famine of the Tempo Period (Hiramatsu, 1929; Tokoro, 1970).

After the Edo period, optimal utilisation of communal forests became a focus of the villagers, depending on the socio-economic situation, even when the communal forests were owned by property wards. As the authority of the property wards was respected by the local governmental administrators of the town of Inabu, each property ward could develop and use the forest freely (Table 17.3).

(i) *Allotment utilisation:* In the allotment utilisation (called *wariyama*, *wari* means divide) type, a property ward-owned forest is divided and allotted to individual members of the ward, a practice similar to providing private ownership within a defined period. When the contract

# Table 17.3:
Outline of the 13 property wards of the Inabu district

| Name of property ward | Number of households | Type of property | Area (ha) | Administrative organisation | Type of usage | | | Purpose of (former) mokutekirin | Regulations |
|---|---|---|---|---|---|---|---|---|---|
| | | | | | Direct | Wariyama (allotted) | Lease | | |
| Inahashi | 134 | forest, grassland, residential, cemetery, field, wasteland, roads | 621.3 | Assembly | yes | yes | yes | youth association, fire defence, school, hospital | Regarding forestation, and wariyama |
| Ohnose | 80 | forest, grassland, residential, cemetery, field, wasteland, roads | 176.7 | Assembly | yes | yes | yes | youth association, fire defence, school | Vary depending on each kumi |
| Oshiyama | 48 | forest, grassland, residential, cemetery, rice field, wasteland, roads, channels | 173.8 | Assembly | yes | yes | yes | youth association, school | Regarding wariyama |
| Odagi | 93 | forest, grassland, residential, cemetery, wasteland, roads | 260.7 | Assembly | yes | yes | yes | youth association, fire defence, shrine, aiding retired warriors, electric pole | not inevestigated |
| Kawate | 48 | forest, grassland, residential, cemetery, rice field, field, wasteland | 130.3 | Assembly | yes | yes | no | youth association, temple, school, repairing channels | Regarding superficies |
| Kuroda | 104 | forest, residential, cemetery, wasteland | 133.7 | Assembly | yes | yes | yes | youth association, school, lady's society, shrine | Regarding wariyama |

(Table 17.3 Continued)

(Table 17.3 Continued)

| Name of property ward | Number of households | Type of property | Area (ha) | Administrative organisation | Type of usage Direct | Wariyama (allotted) | Lease | Purpose of (former) mokutekirin | Regulations |
|---|---|---|---|---|---|---|---|---|---|
| Kuwabara | 126 | forest, grassland, residential, cemetery, wasteland | *65.63 | Assembly | yes | yes | yes | fire defence, youth association | Regarding wariyama |
| Goshogaitsu | 109 | forest, grassland, residential, cemetery, field, wasteland | **55.14 | Assembly | yes | yes | no | not inevestigated | not inevestigated |
| Nakatoh | 31 | forest, grassland, residential, cemetery, wasteland | 141.2 | Assembly | yes | yes | yes | bridge | Regarding wariyama and bunshurin (shared forest) |
| Natsuyake | 91 | forest, grassland, residential, cemetery, wasteland, roads, channels, spring site | 114.6 | Assembly | yes | yes | yes | youth association, lady's society | Regarding wariyama |
| Noiri | 61 | forest, grassland, residential, cemetery, field, wasteland, roads | 238 | Assembly | yes | yes | no | youth association, fire defence, lady's association | The byelaw of the property ward |
| Busetsu | 121 | forest, grassland, residential, cemetery, field, wasteland, roads | **62.40 | Assembly | yes | yes | yes | not inevestigated | not inevestigated |
| Tominaga | 9 | forest, residential, cemetery, wasteland | 18.41 | General meeting | yes | yes | yes | youth association, fire defence, shrine | Regarding wariyama, etc. |

* including the land communally owned by Goshogaitsu and Busetsu Property Wards.

** including the land communally owned by Kuwabara Property Ward.

**Source:** This table was prepared according to Izumi et al. (2009), materials (as of 27 July 2009) and interviews of Inabu Municipal Branch of Toyota City.

expires, *warikae* (re-allotment) is conducted. In case a user moves out of the property, the right of use is returned to the property ward. Users who planted trees, which happened in most cases, return the allotted land after cutting down the trees, or, if not, they receive a stipulated amount of money in compensation for the trees. According to interviews with elderly local residents, the system was borne out of local wisdom that deemed it wrong to allow land to flow outside the property ward.

(ii) *Purpose-based forest:* The earnings from logging were used not only as a revenue source for operating the entire property ward but also for specific purposes depending on the needs of each ward, such as the building of schools. The forested sites intended for specific purposes, collectively called *mokuteki-rin* (purpose-based forest), include the *seinen-yama* (forest for young men's association), the *shohboh-yama* (forest for firefighting), the *iryoh-yama* (forest for medical facilities), the *gaku-rin* (forest for school), etc. depending on the purpose.

These forested sites are managed and maintained through the collaborative work (*oyaku*) of the residents.

(iii) *Land lease use:* Furthermore because the high-speed economic growth period of Japan, a type of utilisation in which land lease revenue is earned under a land lease contract has significantly increased. As a starting measure, local volunteers in the Inabu district encouraged the Municipal Education Board of Nagoya City to build an open-air education centre. This decision was also a countermeasure against the decline of forestry due to the import liberalisation of foreign lumber. Presently, in a time when forestry revenue is not expected to rise due to the unusual slump in lumber prices, land lease revenue has become a valuable revenue source.

Most recently, a project in which the property ward land will be opened as housing space so that urban dwellers can immigrate to these areas is going to be tried. This new approach, due to the rapid decline of the population in the area, is intended to stimulate growth and activity in the community.

### Enhanced Common Benefit in the Property Ward

As noted in the previous section, the utilisation type of property ward-owned forest in the Inabu district has been changing, as necessary, over time. The changes were the result of attempts to address the interests of the individual residents in the community district (private benefit), while simultaneously enhancing the benefits of the entire district (common benefit). The utilisation of the mountains and forests with the clear aim to benefit the public is evidenced through the communal tree planting programme implemented at the end of the Edo period.

In the beginning, communal tree planting was conducted in preparation for famines and as a social security for unforeseeable circumstances, but since the Meiji period, the focus has been on the enhancement of the common benefit through positive asset accumulation.

The *mokuteki-rin* attracts our interest as it reflects the historical background and circumstances unique to each district. The *gaku-rin* is a forest that serves as a revenue source for building schools and purchasing school equipment (Takemoto, 2009). When a school district straddled several property wards, a section of one property ward was designated as the *gaku-rin*, and accordingly, communal tree planting was conducted by all the property wards that comprised the school district. Another unique *mokuteki-rin* was the *hashi-yama* in the Nakatoh district, where a river runs through and divides the district. One forest area was designated as a *hashi-yama*, which provided both lumber and source of revenue for building three bridges that enabled the crossing of the river from both sides.

*Seinen-yama* and *shohboh-yama* existed in many property wards. The *seinen-yama* was a revenue source for conducting a festival where the main activity was the *seinendan* (young men's association). The *shohboh-yama* was a revenue source for purchasing fire pumps. These collaborative activities, which each property ward conducted on its own, were part of the public service that was assumed by the municipality. As a result, the *mokuteki-rin* completed its original role and is now positioned as a fundamental asset of each property ward.

Earnings from forested sites managed directly by each ward became a source of general revenue for the autonomous administration of the ward. Under the current circumstances, when forestry revenue is not expected, land lease revenue makes up a substantial portion of the property ward's total revenue, and the utilisation fee of the *warichi* becomes an important revenue source for the autonomous administration of the ward. Some property wards owe their major financial source to the administration subsidy from Inabu, but the revenue which the property ward independently earns is contributed to the entire management of the district, including expenses for various activities in the district as well as a contribution to the development of the local infrastructure, as these are resources for public benefit (Table 17.4).

**Table 17.4:**
*Items of general expenditure in community districts*

---

I. Conference Cost

    1) General meeting 2) Account settlement 3) Handover process 4) Board meeting

II. Security-assurance Cost

    1) Fire-fighting 2) Health and sanitation

III. Subsidy for the Activities of Various Organisations

    1) Sports clubs 2) Women's Association 3) Elderly People's Association 4) Children's Association

IV. Ceremonial Cost

    1) District festivals 2) Mountain God festival 3) Mountain products Festival

V. Rewards of Officers

    1) Ward Head 2) Deputy Ward Head 3) Accounting 4) Kumi Leader 5) Assembly Members 6) Public Health 7) Agriculture 8) Shrine Parishioner Representative 9) Forestry 10) Sports 11) Sports Support 12) Local History 13) Drainage 14) Residents Representative

VI. Forest Management Cost

    1) Forestation 2) Forest Roads 3) Communal Work

VII. User Fee for Infrastructure Development and Improvement

    1) Roads within the district 2) Street Lights

VIII. Facilities Expenses

    1) Land rental fee 2) Expenses for utilities 3) Management costs 4) Equipment costs 5) Shrine repair costs

IX. Other Costs

    1) Clerical costs 2) Forest cooperatives levy funds

---

*Source:* Authors.

## Change of Governance over Property Ward Forests

### The Property Ward and the Administration of the Town of Inabu

Inabu was in the position of administrating the property ward under the Local Autonomy Law. In most former property wards, an ordinance regarding the establishment of the ward was not enacted nor was a decision-making body established (Izumi, Saito, Asai, & Yamashita, 2011; Watanabe, 1974). In contrast, a management structure and system for establishing a property ward was well prepared in Inabu after the Local Autonomy Law came into effect. An ordinance regarding the property ward was stipulated in 1949. As a decision-making body, a general meeting was established in the Tominaga Property Ward, while an assembly was established in the 12 other property wards in accordance with Article 295 of the Local Autonomy Law (The Assembly or the General Meeting). The procedure for collectively putting the land ownership in the name of the property ward was also established, as previously mentioned.

Such emphasis on the formation of various operating systems only indicates one of Inabu's positive responses towards property wards. The operational details of property wards were, in substance, inextricably linked to those of existing community districts. The town of Inabu permitted the utilisation of property wards through their autonomous management, which was a great convenience for the wards.

Each community district claimed subsidies necessary for its autonomous management, as shown in Table 17.4, and the town disbursed the required amount as a transfer to the district. When a property ward offered a site within its district for the building of a high school to enrich the education of the area, Inabu assigned the site to be town-owned as an expedient measure to clear regulations for land registration. In this case, the town assumed the ownership rights of the property ward and became a nominal owner for the convenience of the property ward. Furthermore, when disbursing subsidies to each district, Inabu accepted that small and relatively poor wards should receive subsidies from property wards with abundant revenues and thus passed them on to those wards with lower revenues. Through the arrangement of a coordinator, the town of Inabu was very careful not to disrupt the administration of the property ward, even in those districts with less financial capability.

During the period of the former town of Inabu, property wards, whose framework has been formally established, virtually acted as part of an autonomous administration of the community districts (villages in the Edo period), and thereby property ward-owned resources retained their status as commons owned by existing community districts.

### Discussion of Amalgamation: Advocating the Maintenance of the Existing System

Since the amalgamation came to an issue in 2002, a discussion on how to treat a property ward after amalgamation occurred, though the discussion seemed to be underdeveloped. It was reported to the assembly and the residents that the management of the property wards under the town of Inabu would be conducted as had conventionally been done in the past even when the management moved to Toyota City after amalgamation.

*Toyota City's Intervention*

It was in 2005, immediately after the amalgamation, that the property wards were managed 'as had conventionally been done'. In 2006, the General Administration Division of Toyota City required every property ward to cease expenditures/disbursements to their local communities. The allegation of Toyota City was that the disbursement as a subsidy from the property ward to the community district would disintegrate the uniformity of the single municipality, which was stipulated to by law. As expenditures for local communities are paid through the city's administrative agency, the city administrators regarded that local communities in the Inabu district could obtain subsidies other than the regular subsidy from the city, which would violate the law.

As it turns out, the commitment regarding the property ward was withdrawn, in effect, unilaterally by Toyota City. The 13 property wards of the Inabu district, which completely lost discretion, faced difficulties in changing residents' communal use and management of forests as well as their lifestyle and autonomous management centring on forestry.

*Resistant Movement by Inabu Residents*

This withdrawal not only hindered the conventional forest management system but also started to affect the management of the property wards in various ways. The property wards held discussions with Toyota City in pursuit of flexible management of the property ward (restoration to its original state). Meanwhile, counsellors of Toyota City elected by the Inabu district and the advisor of the property ward chairperson's liaison committee of the Inabu district visited Aichi prefecture, the Ministry of Internal Affairs and Communications and Toyota City for intermediary negotiations to seek ways to reach a resolution to the problem. In 2008, the Property Ward's Problems Liaison Committee was established, and its resolution requiring the restoration of the conventional management system of the property ward was presented to Toyota City three times. In addition, two workshops were held in 2009 to help property ward dwellers understand the features of the institution and the concept of local autonomy. Various researchers, including the author and city officers were involved in these workshops.

Although the city had not intended to change its policy despite the resistance, Toyota City finally agreed to create an ordinance that would legitimise the property wards to manage themselves. As a result, the Ordinance for Supporting the Local Development of Property Wards was enforced in 2011, which officially permits management of the property wards. It is stipulated that 'the City mayor shall take care of any affairs of property ward such as historical background, feature of the institution and social function of property ward' (Article 3-2), thus reflecting the achievement of the resistant activities.

## Discussion from the Perspective of Forest Governance

### Significance of a Localisation Strategy

The property ward system itself, which was created as a compromise policy, may be regarded as a fruit achieved through the resistant strategy of the local communities against the idea of an outsider. However, it was also the result of the compromise by the local community that the property ward be subject to the municipality's rule of law. That is to say, the institutional design of the property ward system opposes the commitment principle proposed by Inoue (2011). However, as shown by this case study, many property wards have been managed as commons on the basis of the unrestricted intentions of the local communities (Watanabe, 1974).

We now consider the factors related to maintaining the independence of the local community in this case study. In the age of the former town of Inabu, the property ward was part of the local communities. Therefore, all dwellers in the town were also property ward residents who understood that the property ward essentially worked for their welfare. Naturally, town officers, who were also property ward residents, understood this relationship. That is, substantial homogeneity was retained between the property wards and the town that administrates the property wards.

Thus, it was significant for the local communities to seek local autonomy and adopt a localisation strategy by exploiting the institutional framework of the property ward. The existence of purpose-based forest reflects this significance. Although the former town of Inabu was a highly homogeneous area whose local communities all stood on mountainous area with little flatland, there were a number of differences among the 13 communities. For example, the aim of purpose-based forests was different according to each ward, as the purposes were created to reflect the surroundings and demands of each ward, thus again indicating the significance of the forests to the local community. Although the role of the purpose-based forests is now to provide funds for the property ward's general budget, it is still important that the forest commons be managed locally, as the ways of using the funds differ depending on the needs of each ward.

Further, the uses of property ward forests have changed through the ages. The property ward forest of today is used as a resource to attract new residents to the area for the purpose of re-introducing the tradition of the *iriai*, which had been the exclusive purpose of the forests for ages. It is important to note that the decision making executed by a small unit may contribute to the flexibility regarding the utilisation of the forest according to the changes of the time and the demands of the community. Although further investigation is needed, we may say that traditional local commons do not remain static as they are subject to change depending on the changes in their surroundings and that by doing so, they may be able to sustain themselves as independent communities, while adhering to the localisation strategy.

*Possibility of 'Cooperative Governance'*

On the other hand, we must recognise that the ability of the local community to sustain itself is limited. With respect to this case, the limitations of the property wards in the Inabu district are the result of the Great Amalgamation of Heisei.

As a result of this amalgamation, the administrator of the property ward became Toyota City, which is a huge industrialised city. It was soon realised that property wards were being governed by outsiders who had no common or shared understandings with the people of the Inabu district regarding the self-governance of a forest in a mountain village and no knowledge of the property ward system. In other words, the heterogeneity between the mountain village dwellers and the city administrators was extremely significant as the outsiders had much greater power than the property ward dwellers regarding the institutional design and population balance. Hence, the resistant strategy of the property ward dwellers was ignored for five years.

The resistant strategy was developed through continuing dialogue and by emphasising on the legitimacy of property wards to be managed locally. In this resistance effort, not only property ward dwellers but also researchers, including the author, who are outsiders, were involved. Even so, the arguments presented by the property wards were not accepted, again reflecting the heterogeneity and the large gap of power between the administrator and its subject. Thus, it was evident that the conflict would not be resolved by consociating among the property wards. In addition, although many local communities enacted a solitary resistant strategy against the national or local government to regain their rights to the confiscated commons, in most cases the efforts were never or only rarely rewarded (e.g., Hojo, 2000; Kobayashi, 1968).

Property wards in the Inabu district are now managed much as they were before because Toyota City amended its policy and prepared an original guideline for the legitimate operating of property ward systems under the administration of the city. Accordingly, the current status of the property wards is that of a collaborative governance between municipal administrators and local communities. However, under the governance system, it is the municipal administrators who must be open-minded. We can further conclude that the guideline prepared by the city is the result of the trust that was established when the city made concessions to the property wards. In addition, as shown in a project whereby land is provided to new residents, the local community is not always exclusive as the property ward dwellers have sufficient potential to make decisions and take action.

The lessons from this case include the following. Firstly, it is important to pay attention to the differences due to heterogeneity and power between the local community and the partner in the collaborative governance. Secondly, it is important that the outsider be open-minded and work to establish a level of trust when there are great differences between the collaborative parties as it is assumed that the stronger party is the more legitimate and that the heterogeneous outsider will not make concessions to the local commons.

# Modern History of School Forests in Japan

## Introduction

In Japan, there are approximately 3,057 schools holding forests. These forests are called school forests. The total area of school forests is 20,106 ha (the number of primary schools is 1,859, and the area is 7,009 ha). This accounts for 7.8% of all primary, junior high and high schools (8.1% of all primary schools). However, most of the school forests are not currently being used, and the histories of these forests are almost forgotten.

This chapter first intends to clarify why school forests were introduced, how they spread and the role they played in the dual structure of hamlets and administrative villages, which the Meiji (1868–1912) local autonomous system produced. Second, it explains the functional and systematic changes in communal relations (Abiko, 1986) with respect to the dual structure of natural villages and administrative villages from the Taisho period (1912–1926) to the Showa pre-war period (1926–1945). And finally, it illustrates the communal relations with school forests from the Showa post-war period (1945–1989) to the present. This chapter aims (1) to explicate why Arbour Day and school forests were reinstated during the Showa post-war period; (2) to determine what the communities demand from school forests by observing the transfers of the property rights of school forests, accompanied by the consolidation of villages and towns since 1958; and (3) to explain why school forests declined in numbers and in quality after the consolidation and to report new trends regarding the school forests that are being established for environmental preservation or education.

## Establishment of School Forest during the Meiji Period

In the very early period (from 1872 to 1889), school paddies and school forests were introduced as a way to create school funds. Though most of the school forests were established on lands owned by hamlets, some of them were established on lands sold by the Meiji government (Chiba, 1962). During this period, the number of the school paddies was greater than the number of school forests. This was especially true in the Aomori prefecture (Inoue, 1940).

During the first period (from 1889 to 1894), because of the Town and Village Act of 1889, approximately 75,000 hamlets were consolidated into 15,000 administrative villages (Ohishi, 1991). As the new administrative villages had the responsibility to establish primary schools, hamlets lost the right to establish their schools, though they still had a significant amount of real estate that was being used as communal properties (Stevenson, 1991) known as *iriai*. This is because the Meiji government did not unify the real estate under compulsion when it consolidated the natural villages. In 1890, the school fund was made possible by the enactment of the Rural Education Act. This act permitted the establishment of school forests as a way to raise school funds. Because the administrative villages did not own the lands for themselves, they used the lands of the hamlets when they set up the school forests.

During the second period (from 1895 to 1896), with the School Arbour Day instruction led by Nobuaki Makino, the vice-minister of education, as a trigger, the establishment of school forests became widespread throughout the country (Sonobe, 1940). Arbour Day was first established in Nebraska, United States, in 1872 by Governor Morton in response to the restoration of devastated land. Then, in 1895, B.G. Northrup, the chairman of the Board of Education in Connecticut, United States, visited Japan and went on a lecture tour about School Arbour Day, which was developed from Morton's Arbour Day (Dai-nihon Sanrinkai, 1895; Schaufller, 1909). Makino was so affected by Northrup's lecture that as part of the top-down governance structure, he instituted Makino's School Arbour Day and tied it into school funding.

During the third period (1897–1903), the administrative villages still did not own enough lands for school forests. To solve this problem, the Meiji government started to sell unreserved national forest lands to the administrative villages. Meanwhile, three major acts related to forests were enacted, which prescribed the sale of unreserved national forest lands. Thus until 1906, the number and the area of school forests established by using unreserved national forest lands increased.

During the fourth period (1904–1905), the period of the Russo-Japanese War, the Ministry of Education further promoted the establishment of school forests with the forestry division under the Ministry of Agriculture and Commerce. The Division of Forest believed that school forests were important for propagating their afforesting policy, which is why school forests increased in number and in area covered during this period. At the peak of development, in 1906, 2,968 school forests (7,200 ha) were established by using unreserved national forest land. However, based on the case studies we conducted, we found that the landowners of many school forests were hamlets or individuals.

During the fifth period (1906–1914), the expenses for education continued to accumulate because compulsory education was extended from four years to six years. Because of these expenses, rural governments found themselves in financial difficulties. Therefore, in response, the Ministry of the Interior implemented a rural improvement project. The goal of this project was to make the administrative villages financially sound while simultaneously maintaining order in the hamlets. To accomplish this, the ministry approved the unification of the forests of hamlets to school districts that existed between the hamlets and the administrative villages. Additionally, the sale of unreserved national forest lands to administrative villages continued during this period.

Fundamentally, the school fund institution was an excuse by the Meiji government to gain the cooperation of hamlets. However, school forests in this institutional framework cause many hamlets to reorganise into a single community. As the result of this reorganisation, a community whose foundation was school forests was born.

## Change of School Forests in the Taisho and the Showa Pre-war Period

During the first period (from 1912 to 1928), when the modern education system was introduced, communal relations established school forests as self-governing entities. Until the commemorative forestation of the Taisho Imperial accession, school forests were established for the purpose of forestation and raising school funds (Division of Forest, 1916). In the communities during this period, we found several organisations such as a young men's association, a local fire brigade, an alumni association and a local credit cooperative. However, as to forestation and management of school forests, hamlets and school districts continued to play active roles. When the government began spending money on compulsory education in 1918 and on forestation in 1920, the purpose of school forests became relatively unimportant.

During the second period (from 1929 to 1937), the Great Depression in 1929 resulted in Japan's rural communities to suffer extreme poverty as well. In this context, Dai-nihon Sanrinkai, the association for Japan's forests, introduced Forest Loving Day in 1934, which was celebrated with a commemorative tree planting, a practice implemented in the colonial Korean peninsula since 1911 (Sonobe, 1934). Otosaku Saito, the chief of the forestry division of the Government-General of the Chosen, introduced the commemorative tree planting with a reference to Makino's School Arbour Day and succeeded in turning the devastated mountains into green forested areas (Saito, 1935).

During the third period (from 1938 to 1945), under the national mobilisation system, which was started in 1938, the central government endorsed Forest Loving Day and recognised school forests forestation as a practical strategy to infiltrate the Emperor system (Division of Forest, 1938). The young men's associations, which were set up as self-governing functional organisations, were promoted as administration-governed organisations when the government considered the afforested school forests as national resources during World War II (Association of Imperial Flood Control Committee, 1941). To change such communal relations, there was a hometown patriotism logic that converted the spirit of a loving hometown into one of patriotism (Maruyama, 1944). In other words, tree planting and school forests served as intermediaries between a loving hometown and national patriotism.

## Restructuring of School Forests in the Showa Post-war Period and the Present

The communal relations around school forests, which were born in the Meiji period (1868–1912) or the early Taisho period (1912–1926), can be called property communal relations. However, when the national mobilisation system was started, they changed to patriotism communal relations.

There were three actors responsible for renewing Forest Loving Day and school forests. The actors were the general headquarters (GHQ)/ Supreme Commander for the Allied Powers (SCAP), the bureaucrats and the communities. The GHQ/SCAP considered that Forest Loving Day and school forests could alleviate some of the social anxiety that existed immediately after World War II. The Japanese nation felt uneasy because she had lost the absolute Emperor. The owners of forests also felt uneasy because of the possibility of emancipation of the forest. However, making the Emperor the ceremonial master on Forest Loving Day alleviated the first anxiety, and the school forestation along with the other forestation movements alleviated the second. It was actually the bureaucrats who proposed the resumption of Forest Loving Day and school forests to the GHQ/SCAP (Kokudo ryokka suisin iinkai, 1970) because they wanted to maintain their system and their organisation. Therefore, they emphasised that Forest Loving Day be first introduced by the United States as B.G. Northrup's School Arbour Day and that a tree-planting campaign should have nothing to do with the war campaign. On the other hand, the communities needed funding to build their new school houses, especially for junior high school, which was introduced under the new school system in 1947. During the post-war rehabilitation, Forest Loving Day and school forests encouraged communities both physically and mentally. Thus, patriotism and communal relations were strengthened autonomously.

Accompanied by the consolidation of villages and towns in 1958, the property rights of school forests owned by municipalities usually went to new municipalities or new property wards (*Zaisan-ku*) that were established to manage former municipalities' properties, such as forests (Kawashima et al., 1959–1968). However, the property rights of some school forests went to other entities. For example, Matsuo ward in Nagano prefecture set up a new property ward just to keep its school forest (Matuo sonsi hensan iinnkai, 1982). Takase school forest in Oita prefecture was established as a hamlet forest before the Takase administrative village forest was established (Takase seisan shinrin kumiai, 1990). As a symbol of administrative village consolidation, the school forest has moved to the Takase forest producer's association along with other village forest properties. The property rights of the Aihara school forest in the Tokyo metropolitan prefecture established in the Meiji period finally went to a legally incorporated foundation (Kyoyuchi, 1978). Residents of Aihara regarded school forests as a symbol of public welfare. Concluding from these examples, patriotism communal relations insisted that the municipalities manage their school forests for the public welfare. Thus, the municipalities prevented their school forests from unifying into new municipalities or returning to property communal relations by fitting them into other legal frameworks, such as incorporated foundations. Accordingly, it became a system of public welfare relations.

Japan became independent from the GHQ/SCAP in 1951, and the second school afforestation plan was then implemented. However, school forests had lost their role as school funders because subsidies for education and forest management were already provided to new municipalities. Therefore, only a peaceful image of tree planting was left. Most of school forests faded into municipality fund ordinances from the mid-1960s to the mid-1970s. With the increasing interest in the destruction of nature since the 1970s, school forests were regarded as fields of environmental preservation and education. From the late 1990s, municipalities, prefectures and the state started new institutions to establish school forests for environmental education.

## Conclusion

The change in the communal relations around school forests was first generated when several hamlets using a new modern primary school cooperated to retain the funds for the new modern school after the village consolidation in 1890. The role of this relation, which we called property communal relation and which existed mainly from the late Meiji era to the early Taisho era, weakened when the administrative villages improved their financial conditions. However, property communal relation changed patriotism communal relation, which brought out patriotism from the spirit of a loving hometown through school forests afforestation for the war commemoration and the celebration of the Emperor's anniversary. When post-war rehabilitation was begun, relations after World War II—patriotism communal relations—became stronger. However, because of the Showa village consolidation, the relation was forced to separate community sentiment from community property. Some communities recovered their original function, some were annexed by the administration, and some created public communal relations to maintain the function of school forests.

Going through such history, the ownership of school forests today is quite varied. Not only is there the original *iriai* organisation, there is the property ward (*Zaisan-ku*), the incorporated foundation, and the community-based forest productive cooperation as well as the public bodies such as the local and national governments. However, unless the owner and the user do not correspond with one another, school forests are conventionally managed and used by the local people.

Since the Meiji restoration, Japan has modernised ownership and utilisation of the forests. The use of school forests has changed from fodder and firewood to school funding and then to environmental education. Ownership has also changed. However, the community around school forests has sustained the use of natural resources for school and children under the consensus of the local residents. What lessons can be learned from these unique school forests?

Inoue, focusing on the relationship between the commons and the outsiders, suggests three vital design guidelines for the collaborative governance of natural resources along with Agrawal's category of commons factors: graduated membership for group characteristics, commitment principle for institutional arrangements and trust building for the external environment.

As for the communal relations of school forests, in the beginning, the hamlet, using the *iriai*, was considered to change its area and resource usage in the course of accepting the modern education system in the local area. Therefore, because of the principle of subsidiarity under the Meiji local autonomy system, we find a very strong influence of outsiders, and because the objective is to maintain modern school facilities that are open to everybody, we observe open-minded localism. On the other hand, management organisation of school forests is under the control of the hamlet. Namely, *iriai* members are given priority over other villagers in the selection of the management organisation of school forests, which may confirm the existence of graduated membership.

Although school forests are managed mainly by *iriai* members in the property communal relation stage, in regard to the patriotism communal relation stage, school personnel are the members of the organisation that manage the forests, the planting of trees by the students, and the selection to membership, which is also decided in detail. On the other hand, some school forests are owned by public entities, such as the administrative village or the *Zaisan-ku*, a management organisation that actually retained its power. Commons seem to establish connections conventionally with administrations that are generous with their money but sparing with their advice. It may be an exaggeration to call it a commitment principle, but in the process of organisational change from a natural village to a communal relation around a school forest, we can determine control of the membership. Today, the number of actors engaging in school forest management is increasing because the objectives of school forest are shifting to environmental education. Inoue's commitment principle might be more important if new members, such as non-profit organisation (NPOs), affected the organisation in the near future.

As for trust building, communal relations around school forest have been, in some cases, organised for more than a century. Surveying cases throughout Japan, modernisation causes the majority of the school forest be annexed to administration property, usually through village consolidation. Commons have maintained their organisational body, though it is vulnerable due to trust building relations with the external environment. Built by a compromise of natural resource management between hamlets and administrative villages under the influence of consolidation, communal relations and trust itself, with respect to school forest may have survived or remained by accident or by effort of the residents.

Along with Inoue's suggestion, I applied three guidelines to the school forests case: graduated membership, commitment principle and trust building. Presenting first some discussion points, I find it important to determine what factor of the commons promotes change to a new organisation or maintains the original organisation. Though school forests present a case whereby commons accept change to new organisations, some returned to the hamlet and some were annexed to administration. In view of this, it is posited that many factors of the commons are characterised not by the nature of changing to new organisation or maintaining the traditional one, but by the nature of flexibility or discretion to select an organisation. Second, though the communal relation around school forests modernised ownership and utilisation, the conventional and local way of consensus building was maintained. However, because of the need for trust building with outsiders, communal relations are vulnerable to external circumstance change such as village consolidation.

Inoue's three vital design guidelines for the collaborative governance of natural resources tries to develop conventional and local trust building by implementing rational rules such as graduated membership and commitment principle. While these guidelines reduce vulnerability to external circumstance changes, conventional and local diversity of commons would be lost and there could be no return of a consensus building of commons model. This may be compensation for the strategy of the pincer attack by globalisation and localisation.

## Implications from Two Case Studies

One can say that in local commons, such as the *iriai*, the welfare of and benefits to the community and its members are the first priority, whereas resource utilisation policies can be changed through member assembly.

However, in the case studies, the life of the internal community had been exhausted and the community people needed to cooperate with external actors due to modernisation. In the school forests (*Gakkorin*) case, a common benefit between communities that used the same primary school was generated and expanded not only to communities but to municipalities and to the state. In the case of property ward (*Zaisan-ku*), because the community and the city government benefited in common, such as the preservation of water-shed management and welfare services, they could cooperate with each other.

From the case studies of *Gakkorin* and *Zaisan-ku*, we learned that along with modernisation, the community found its way to retain the *iriai* by evaluating the common benefits with the outsider. However, to change community benefit to common benefit with an outsider, the community had to clarify the benefit and accept the commitment of the outsider. It is conceivable that, at the same time, discretionary power to change the rules by members of the *irai* assembly was lost.

Being confronted with globalisation, implications from the case studies suggest that the community must clarify common benefits with possible commitment members to enable firms, administration, NPOs and scholars to commit to the commons. However, paradoxically speaking, such commitment tends to restrict the flexibility and discretion that is originally associated with the commons.

When selecting common benefits, we must carefully consider the strategy for the continuance of the *iriai* in the age of globalisation, in view of the successful *Gakkorin* and *Zaisan-ku* cases in which communities have lasted and maintained the balance of the commons and the external commitment. Though we can consider common benefits as education, welfare, environment and so on, which are relevant to both community and outsiders, such benefits could have broken commons if the external commitment is too strong and could have excluded outsiders if the external commitment is too weak. It is desirable to take the time to establish relations between the community and the outsiders when selecting common benefits.

# References

Abiko, R. (1986). Farm land revolution and community. *Rekishi Hyouron*, 435, 32–45 (in Japanese).

Association of Imperial Flood Control Committee (1941). *School forest* (53pp.). Tokyo: Association of Imperial Flood Control Committee (in Japanese).

Chiba, M. (1962). *Study of School District* (489pp.). Tokyo: Keiso-Shobo (in Japanese).

Dai-nihon Sanrinkai (1895). Arbor day in school. *Dai-nihon Sanrinkai Kaihou*, 150, 1–20 (in Japanese).

Division of Forest (1916). *The Commemorative Forestation for Taisho Imperial Accession*. Tokyo: Division of Forest (in Japanese).

———— (1938). *School Forests of Primary Schools* (131pp.). Tokyo: Division of Forest (in Japanese).

Hiramatsu, H. (1929). Communal forest at Inahashi of Sanshu region. *Journal of Japanese History of Economics*, 2, 59–72 (in Japanese).

Hojo, H. (2000). *Law Sociology of Iriai* (537pp.). Tokyo: Ochanomizu Shobou (in Japanese).

Inabu Education Board (2000). *History of Inabu Town*. Inabu: Inabu Town (in Japanese).

Inoue, M. (2011). *Prototype Design Guidelines for 'Collaborative Governance' of Natural Resource*. Presented at 13th Biennial Conference of the International Association for the Study of the Commons, 12pp., Hyderabad, India.

Inoue, Y. (1940). Research of school paddies as commons. *Kagami-kenkyu-houkoku*, 49, 1–62 (in Japanese).

Izumi, R., Saito, H., Asai, M., & Yamashita, U. (2011). *Commons and Local Autonomy: Past, Present and Future of Property Wards* (233pp.). Tokyo: J-FIC (in Japanese).

Kawashima, T., Ushiomi, T., & Watanabe, Y. (1959–1968). *Dissolution of Communal Property Rights I–III*. Tokyo: Iwanami-Shoten (in Japanese).

Kobayashi, M. (1968). *Study of Common Rights on National Lands* (557pp.). Tokyo: Tokyo Daigaku Shuppankai (in Japanese).

Kokudo ryokka suishin iinnkai (1970). *20 Years of Tree Planting* (363pp.). Tokyo: Kokudo ryokka suishin iinnkai (in Japanese).

Kyoyuchi, A.H.L. (1978). *History of Aihara Communal Property* (209pp.). Tokyo: Aihara Hozen Kai (in Japanese).

Maruyama, M. (1944). Pre-formation of nationalism. *Kokka-gakkai zassi*, 58(3–4), 321–364 (in Japanese).

Matuo sonsi hensan iinnkai (1982). *History of Matuo* (898pp.). Iida: Matuo sonsi hensan iinnkai (in Japanese).

McKean, M.A. (1982). The Japanese experience with scarcity: Management of traditional common lands. *Environmental Review*, 6(2), 63–91.

Ohishi, K. (ed.), (1991). *Administrative Villages in Modern Japan* (774pp.). Tokyo: Nihon keizai hyouron-sya (in Japanese).

Schaufller, R.H. (1909). *Arbor Day* (376pp.). New York: Moffat Yard and Company.

Saito, O. (1935). A secret history of commemorative tree planting in Korea. *Chousen Sanrin Kaiho*, 120, 11–18 (in Japanese).

Sonobe, I. (1934). Background of forest loving day. *Sanrin*, 616, 10 (in Japanese).

——— (1940). A secret history of Arbor day movement in Japan. *Sanrin*, 689, 2–13 (in Japanese).

Stevenson, G.G. (1991). *Common Property Economics: A General Theory and Land Use Applications* (256pp.). New York: Cambridge University Press.

Takase seisan shinrin kumiai (1990). *History of Corporative* (169pp.). Hita: Takase seisan shinrin kumiai (in Japanese).

Takemoto, T. (2009). *A Study of School Forest* (446pp.). Tokyo: Nobunkyo (in Japanese).

Tokoro, R. (1970). An institution of communal forest at Inahashi of Inabu Town, Kita Shitara County, Aichi Prefecture: Regarding with Inahashi Village and a gentleman farmer Mr. Furuhashi. *Bulletin of the Tokugawa Institute for the History of Forestry, the Fiscal Year of 1969* 113–139 (in Japanese).

Watanabe, Y. (1974). *Iriai and Property Ward*. Tokyo: Keiso Shobo.

# 18

# Japan II: Implications of the Commons for Endogenous Development of a Mountain Village

*Hironori Okuda and Makoto Inoue*

## Introduction: Essence of the Issue of Mountain Villages in Japan

### Population of the Mountain Village

Chiba (1976) defined a mountain village as originally being a community in which people who were not involved with rice cultivation but were engaged in hunting, slash-and-burn cultivation, woodworking and so on. Mountain villages were not favourable for rice cultivation and were usually very poor areas where people depended on forest resources. Thus, the population that mountain villages could support was restricted, and eventually these villages became sources for industrial labour. After World War II, mountain villages attracted jobless people and repatriates from overseas, and accordingly, their populations increased until around 1955, after which the population began to decline rapidly due to rapid growth of the Japanese economy (Figure 18.1).

The mountain village residents began migrating to cities in search of high and steady income. This migration was further promoted by the decline of charcoal production and the stagnation of earning from

**Figure 18.1:**
*Population trend in mountain villages (1920-2000)*

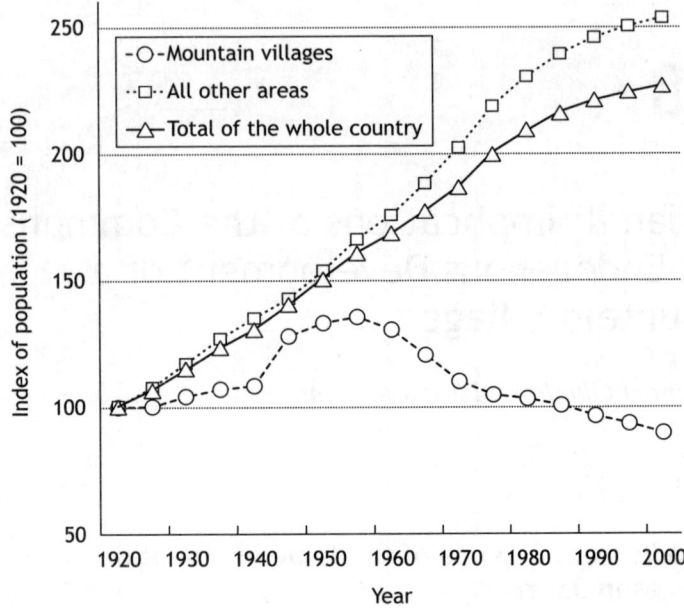

**Note:** A mountain village is a municipality where at least 75% of the area is forested.
**Source:** Population Census Data (The Ministry of Internal Affairs and Communications).

forestry activities. Mountain villages have played important roles in disaster prevention, water supply and environmental preservation, among other services. It was, however, more important to consider that mountain villages would be places where people can live like humans. To maintain the vitality of mountain villages, the Japanese government implemented the Mountain Village Development Act in 1965 to set targets and formulate plans for mountain village development. Based on this plan, local, industrial and social infrastructures, such as roads and water and sewer facilities, have been improved. However, mountain villages are still plagued by population reduction and ageing. Moreover, local agriculture and forestry have fallen into a depressed state due to globalisation, and new measures for attracting new industries have been mostly unsuccessful. Cohort analysis of mountain villages using population census data from the ministry of internal affairs

**Figure 18.2:**
*Population trend in mountain villages (1960-2020)*

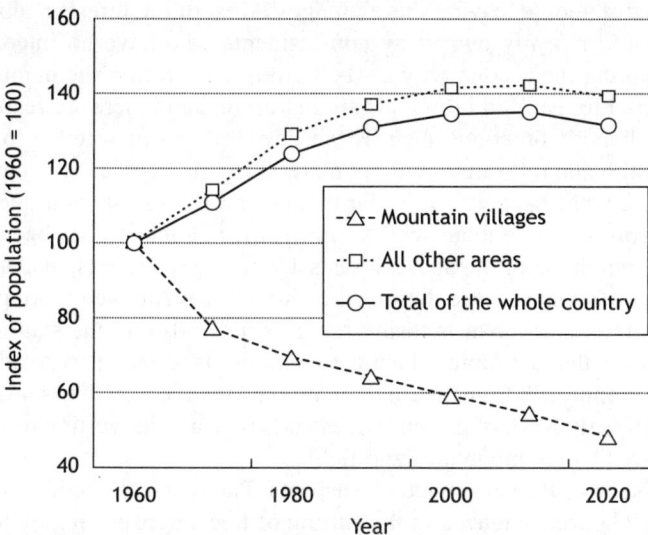

**Note:** Mountain villages are specified by the Mountain Village Development Act in 1965.
**Source:** Population Census Data (The Ministry of Internal Affairs and Communications).

and communications indicates that the total population of mountain villages, which was 7,645,000 in 1960, will decrease to 3,684,000 by 2020 (Figure 18.2).

## Issues on Promoting Mountain Villages

Do people really need to live in mountain villages? This is a question that often arises when discussing mountain village promotion. During the period of high economic growth, when there was migration from such villages to major cities, migrants were assimilated into cities even though they congregated among themselves. However with today's economic globalisation, factories have been moving overseas in search of cheap labour or have been automated, and thus the absorptive ability of cities has been waning. The workplaces have also decreased and cities are now full of part-time or temporary workers causing social conditions in the cities extremely unstable. At the same time, if we consider the mountain villages, there is a remarkable increase in the number of people leaving these villages. Most of the forests and farmlands are presently owned by non-residents who have no interest in maintaining the productivity of their lands. As a result, the number of neglected forests and farmlands are increasing, and there are fears that public benefit functions such as the development of water resources and health and relaxation will not materialise. As people leave farming villages, what happens to the farmlands and the forests is a question that nobody is able to answer. Recently, much attention has been paid to the purchase of Japanese forests by foreign interests, which has become a national issue that must be addressed. An even more pressing issue is that Japan is losing track of its lands and the state of its rural areas that are rapidly losing population. It is necessary to adopt a positive approach to maintain the nation's lands in excellent condition through daily work of growing trees and crops and harvesting products generated by the mountains and fields.

Ebisuno (1967) examined two aspects. The first is the fluid aspect in which labourers engaged in the cutting of trees, transporting of goods, production of charcoal, mining, etc. often emigrate out of rural communities. The second is the fixed aspect in which there are common customs and common local deities that are worshipped in the forests owned by the community who do not move from there. However, in recent years, the fixed fundamental social structure has begun to undergo a major upheaval, as exemplified by the many examples whereby successors have sold the fields or quit managing the fields, houses, forests, etc. that they inherited from their ancestors, and they have instead moved to other places. This means that compared with urban areas, the standard of living, which includes health care, education, etc., that had once been considered normal is now recognised as full of defects. As a result of the democratisation that occurred after World War II, the 'ie' system or patriarchal family system became a modernised family, and what should have been the protected estate of the ie became an individual's asset that was supposed to enrich the lives of everyone, thus liberating people from their ie.

However, according to Iguchi and Kitagawa (1985), one reason for the exodus of the population from the rural areas is that the decline in farm and forest production, such as charcoal-making, that started in the 1960s led to the deterioration of the living conditions. There were many cases where people had to leave the rural communities to maintain their cash income. Two of the most pressing rural issues today may be: (1) how to maintain the income of people who live in small communities and (2) how to provide opportunities to support the lives of farmers and foresters who live in those communities.

## Purpose of the Chapter

Mountain villages suffer from the depression of industries such as agriculture and forestry, the lack of public facilities for daily living and the fragility of the social organisation. As a result, social structures of many communities are on the verge of collapse. To revitalise these communities, it is necessary to form new identities in areas that hitherto had little economic attraction, to give local residents a clear image of the area in question, and to develop new methods for utilising local resources (Tachikawa, 1998).

Throughout the world, many developing countries are pursuing the same type of Western economic development that countries such as Japan and South Korea have achieved, while the countries that have achieved economic development have begun a scramble for resources. If every country on the globe pursues the Western-style monophyletic and extrinsic economic development, it will exacerbate the deterioration of the global environment and the depletion of resources. Thus, nations and communities are being forced into a situation where they will have to pursue polyphyletic and independent endogenous economic development.

At the Seventh Special Session of the United Nations general assembly, the Dag Hammarskjold Foundation (1975), in a report titled 'What Now?' proposed two new types of approaches to development: endogenous development and self-reliance. According to the report, 'If the development is the development of man as an individual and as a social being, aiming at his liberation and at his fulfilment, it cannot but stem from the inner core of each society'.

In the 'Deployment of the Endogenous Development Theory', Tsurumi (1996) stated that the processes and lifestyles that result from the achievement of the common goals of people in an area are automatically created by those people using local resources such as water, wood and minerals. At the same time, these processes and lifestyles are automatically created by new technologies in which traditional techniques are integrated with outside knowledge, technology and systems in accordance with the particular natural environment, cultural heritage and history of the area. This process of social change by which the goal is achieved is endogenous development. In this last case, some residents of mountain villages have been integrating traditional techniques with new technology to utilise local resources, such as water, wood and minerals, on a sustainable basis. This manifestation of Tsurumi's (1996) endogenous development has enhanced the quality of life in such villages. By understanding the state of activities and the networks of persons involved with this endogenous development process in the Tsukimoushi-machi district of Tono City, Iwate prefecture, the purpose of this study is to examine the background behind endogenous development using the commons theory and collaborative governance theory.

## Research Method

We conducted interviews with residents in the mountain community of Tsukimousi-machi district of Tono City, Iwate prefecture in January 2004, June 2005, November 2007 and June 2010.

## Commons and Endogenous Development in a Mountain Village in Iwate Prefecture

### Satoyama Supported in Mountain Villages

Kitao (2001) defines the concept of *satoyama* in Japanese as 'forests located near farms and mountain villages that have been used by farmers for their living'. In the Edo period, such *satoyama* were indispensable for people in mountain villages as places to collect manure, forage for oxen and horses, and provide materials and fuel use for home and lodges. People in mountain villages have wisely managed such areas by treating them as *iriai* and setting various types of limits on their exploitation. These common forests were an important element of rural life, and they served as a tool to control farmers. They were also a symbol for the solidarity of village communities. The *iriai* resembled the rigid local commons (Inoue, 1997) for which certain rules and rights regarding their management and exploitation (when, in which areas, with which tools and to what extent to gather resources) were imposed within the community.

However, after the Meiji restoration of 1868, the new government ignored such traditions and instead focused on securing tax income. Accordingly, the new government tried to impose the concept of private land ownership by selling or transferring the land to individuals. The Meiji government did not allow systems, such as the *iriai*, where the land was held in common, and instead the government ensured that the land belonged either to private owners or to public authority.

As wood resources ran short and the purpose of forest exploitation changed from fuel production to material production, the *iriai* were increasingly identified for the purpose of turning them into national forests and placing the forest resources under state control. In this way, by taking advantage of the local farmers' weak concept of land ownership, the government rapidly redefined land ownership regarding the *iriai* and converted them into national or private forests, thus making the exploitation of the *iriai* by local farmers impossible as a result of afforestation and reforestation aimed at the development of forest resources.

As a result, during the Meiji period, conflicts occurred successively between owners of large forests and those who sought to maintain the right of access to the *iriai*, and some cases, such as the Kotsunagi Affair in Ichinohe-cho, Iwate prefecture, continued until after World War II. The *iriai* that remained as village-owned forests became community-owned forests as a result of the introduction of the municipality system in 1889 and the subsequent merger of different villages. Many such forests became the property of local authorities because of the regularisation and integration process of community-owned forests that began in 1910.

With the enactment of the Town and Village Merger Acceleration Law in 1953, 40% of the forests owned by old-system municipalities were handed down to the new municipalities created by these mergers, but in regions where people did not want such succession of forest ownership, these forests became, once again, community-owned. It is estimated that approximately 2.2 Mha of such *iriai* existed nationwide in 1955. In 1966, a law was enacted aimed to establish rights regarding the use of the *iriai*. The use of the *iriai* was allotted to rights holders who had formed cooperatives that they funded based on their proportion of user rights. From 1966 to 2008, a total of 6,636 *iriai* equalling 574,175 ha were regularised. Today, what we consider as *iriai* consist of community-owned forests, forests owned by local authorities open to local residents and those owned by associations (Figure 18.3).

## Evolution of the Exploitation of the Iriai

In the Meiji and Taisho eras, wood production was mainly designed for fuel production. Saito (2003) estimates that in the first phase of the Meiji era between 36.5 and 73 million tons of fuel wood was produced for home cooking and heating, and it was necessary to cut down another 5,000 ha of forest for fuel wood every year for steel manufacturing. Kato (2003) notes that the production of charcoal, which was approximately 700,000 tons in the early Meiji era, increased to approximately 1.4 million tons in the late Meiji era and to 2.2 million tons in the late Taisho era. However, such exploitation of fuel wood and production of charcoal, which supported the economy in mountain villages decreased sharply following the fuel revolution. The production of fuel wood, which was 19.9 million cubic metres in 1955, is now approximately 0.1 million cubic metre.

On the other hand, the Japanese Income-Doubling Plan implemented in 1961 and the high economic growth of more than 10% a year in the late 1960s accelerated the demand for material wood, thereby tightening the supply and demand of building materials and pulp wood and resulting in continuous price increases. As a result, urgent measures were taken to increase the supply of wood, and in 1965, 50 million cubic metres of wood were put on the market. However, as the logged areas became farmland and the wood price declined because of the large-scale import of cheap foreign wood, deforestation abated and the production of domestic timber decreased to 17 million tons in 2005. Moreover, the Japanese forestry has long been stagnant, as evidenced by the total afforested area, which was at 405,000 ha in 1960 and shrank to 29,000 ha in 2005 (Figure 18.4).

Against this backdrop of the Japanese forest industry, the *iriai*, which were indispensable for people in mountain villages as sources of fertiliser, forage for oxen and horses, and materials and fuel for houses and lodges during the Edo period, were utilised in the Meiji and Taisho eras as a source of cash income through the production of charcoal and other forest products. After the rapid decline of such production following the fuel revolution, Japanese cedar and Japanese cypress were planted in

**Figure 18.3:**
*The change of* Iriai

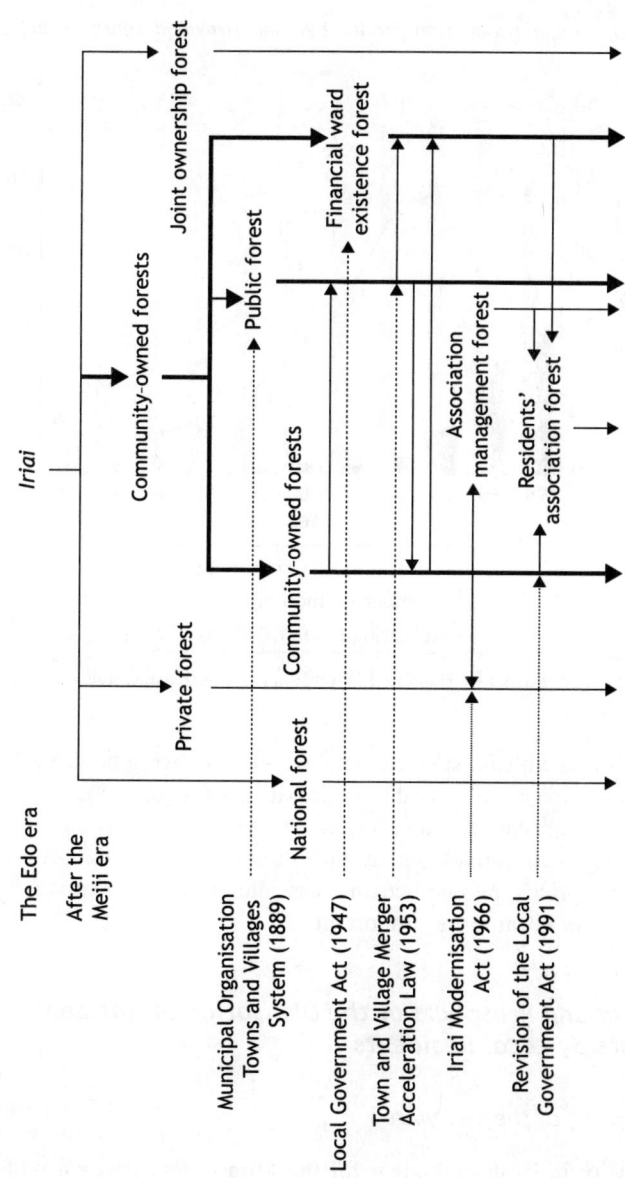

*Source:* Authors.

**Figure 18.4:**
*Trends of timber production for lumber and firewood/charcoal and planting area*

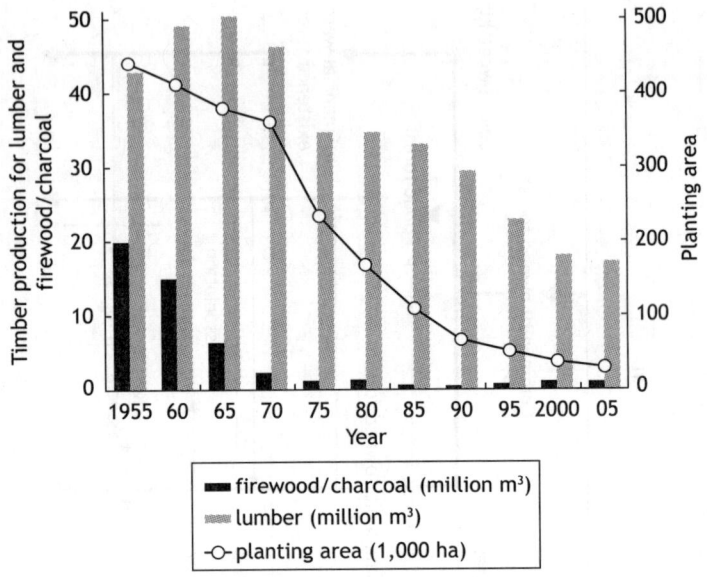

*Source:* Forestry Statistics Handbook, Wood supply-and-demand table.

these forests with assistance from the national government. Such planted areas were managed with the participation of local residents for several years after forestation. However, as the trees grew, it became unnecessary to manage these forests, and as a result of that and the long-term decline of timber prices, people became less interested in forest management, and the *iriai* became less important.

## History and Prospects of the Utilisation of National Forests by Local Residents

### Description of the Study Area

Tono City in Iwate is known for the 'Tono Story', which was written by a folklorist named Kunio Yanagida. This city, which is located in the Tono Basin, the largest basin in the Kitakami Highland in the southern Iwate prefecture, has prospered since ancient times as the castle town of the Tono Nambu Family. It is located approximately 60 kilometres southwest of Morioka City via Route 396 and has a strategic position as a transportation node between coastal and inland areas.

### Description of the National Forest Wood Production Cooperative of the Tono District (Kokuseikyo)

In the process during which former *iriai* were turned into national forests, people who were denied access to such national forests, especially in the Tohoku region, founded cooperatives for charcoal and other commodities and were able to acquire fuel wood and by-products through trade or at reduced prices. Many of these cooperatives broke up or were turned into business establishments for contracting plantations due to the conversion from self-supplied fertiliser to chemical fertiliser and the decline of charcoal production following the fuel revolution. However, in Iwate prefecture, where the charcoal industry has continued to play an important role, this type of cooperative (in the form of the National Forest Wood Production Cooperative, hereafter, Kokuseikyo) is still very active.

Kokuseikyo has organised people living in these local communities and has supported their lives by harvesting hardwood forests, by producing pulp materials and selling them to pulp companies and by functioning as agencies for dealing with timber production and the establishment of plantations in national forests. Such activities allowed local people to remain in the region or work in cities as seasonal workers. However, the activities of such cooperatives have been on the decline, as evidenced from the sales of timber, which reached 15,000 m³ in 1986 and decreased to 9,000 m³ in 2009, and from the contracted area for plantations, which was 1,200 ha in 1986 but has decreased to 300 ha in 2009 (Figure 18.5).

### Life in the Mountain Village of Tono

The village where we conducted the survey has a highly ageing society with nine households (32% of the 28 households) consisting of elderly people at 60 years of age or older. In this village, seven households have lost contact with Kokuseikyo because the former members have left due to their age, they have no successors or they work outside the village, while five households have members still working in Kokuseikyo. In the survey village, 12 households (43%) once had or still have contact with Kokuseikyo (Table 18.1). Thus, Kokuseikyo still plays an important role in maintaining the village. However, its activities and influence

**Figure 18.5:**
*Trends in the amount of contracted work of the Kokuseikyo*

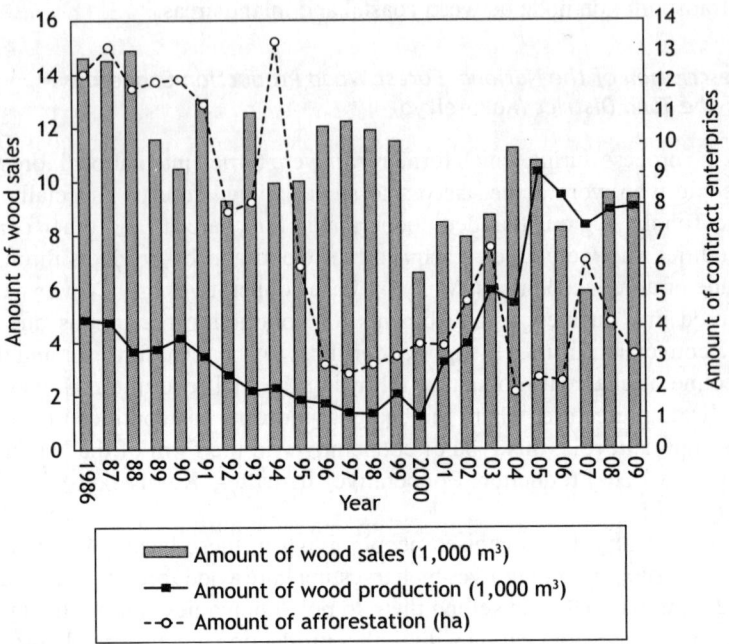

Amount of wood sales (1,000 m³)
—■— Amount of wood production (1,000 m³)
--o-- Amount of afforestation (ha)

*Source:* Authors.

**Table 18.1:**
*Relation between households in mountain communities in Tono city and Kokuseikyo membership*

| Community | Households consisting entirely of members 60 years old or older | Households with one or more members who are 59 years old or younger | Total |
|---|---|---|---|
| Oide | 5[4] | 4 | 9[4] |
| Oonodaira | 4 | 15[4] (4) | 19[4] (4) |
| Total | 9[4] | 19[4] (4) | 28[8] (4) |

**Note:** The number in parenthesis ( ) denotes the number of households with at least one Kokuseikyo member.
The number in brackets [ ] denotes the number of households with at least one previous Kokuseikyo member.
The result of interviews conducted in June, 2005.
*Source:* Authors.

have been shrinking every year, and considering the harsh situation sur-rounding the management of national forests, it has become necessary to consider how to obtain new revenue sources by using local resources.

The people of the village surveyed work cooperatively to grow shiitake mushrooms to supplement their incomes. In 1985, the Hayachine Coop-erative was founded under the direction of Kokuseikyo for the purpose of producing wood for cultivating shiitake mushrooms. This cooperative has a 65-year contract with the government for leasing national forest land to produce the wood (oak, *Quercus serrata*) for cultivating the shii-take mushrooms. The cooperative will make its first harvest 25 years after plantation, and it plans to make subsequent harvests twice every 20 years and to plant seedlings to replace the harvested trees. Thus far, 4,000 seedlings per hectare have been planted, and the process-sharing proportion is 30% for the government and 70% for producers. The estab-lishment of the plantation has been carried out systematically since 1986, and in 2006 the planted area reached 157 ha (Figure 18.6). In this way, the people in the mountain village in Tono borrow national forest land from the government to create a system that allows them to collectively make use of the land. Such activities will open the national forests to the local communities and will give the communities a chance to recover the right to use the forests.

**Figure 18.6:**
*Trend of the forest area contracted to produce wood for cultivating shiitake mushrooms*

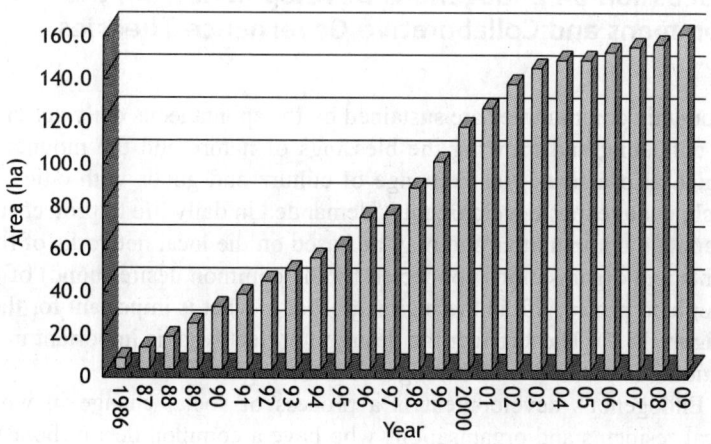

*Source:* Authors.

## Residents' Network for the Endogenous Development of the Local Economy

In Tsukimoushi-machi, Tono City, residents are growing *Q. serrata* trees in state forests to produce beds for growing shiitake mushrooms, which is the main product of the region, by obtaining the assistance and advice from the prefectural government, the manager of state forests and the national cooperative society. The network of the residents has caused a social change related to the growing of the *Q. serrata* trees in state forests and has led to the endogenous development of the local economy. What connects the network is the strong desire of the residents who want to live in the region. The forests of *Q. serrata* are indispensable and essential for the residents who live in the region and are, therefore, considered very necessary commons. The activity is an act of restoring the former commons of the *iriai*, for which the residents of today are unconcerned, and using it for producing beds for growing shiitake mushrooms. Not only have the residents played a major role in planning and designing the *Q. serrata* forest commons, but support from the outside, such as advice from state forest managers on where to grow the trees and grants from the prefectural government and national cooperative society has also been instrumental.

## Discussion on Endogenous Development from the Commons and Collaborative Governance Theories

Mountain communities are sustained by the spontaneous daily activities of collecting and growing the blessings of nature and the mountains, which also activate the exchange of culture and goods with cities. As such, endogenous development is demanded in daily life to protect such mountain communities and may be based on the local networks of residents and organisations connected by the common desire (bond) of the people to protect life in the region, as that is what is important for them (Figure 18.7). The stronger the desire to protect what is important in the region, the easier it is for endogenous development to occur.

Endogenous development is a process of social change in which local residents and organisations who have a common desire (bond) to protect what is important, ask outsiders for knowledge, technologies and

**Figure 18.7:**
*Structure of 'endogenous development' and 'the local commons'*

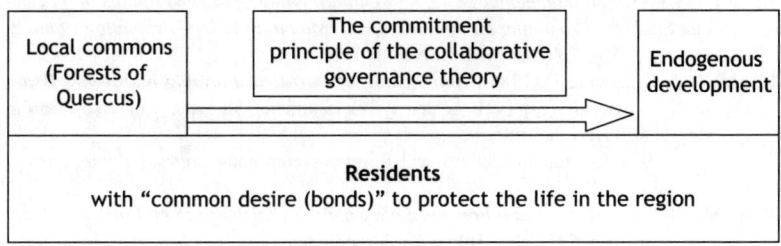

*Source:* Authors.

systems and allow the outsiders to be involved in the making of decisions and the planning and designing of activities based on the degree of their involvement in the commons (the commitment principle of the collaborative governance theory), which, in turn, activates local activities that protect, grow and use the commons. The activities to protect, grow and use the commons must be practised by the local entities and be open to outsiders, or to 'open-minded localism (of the collaborative governance theory)'. The collaborative governance theory explains the process of endogenous development. These activities, as described herein, could not lead to endogenous development unless outside entities supported them and were involved in the planning, designing and practising of them. To lead the mountain communities towards endogenous development, it is critical that outsiders, depending on the degree of their involvement in the commons (the 'commitment principle'), be involved in the planning and designing activities. It was predicted that the local residents and organisations would seek increased involvement of outsiders and request more knowledge, technologies and systems as the desire of the residents to protect what is important is weaker, and endogenous development is more difficult to achieve and sustain.

# References

Chiba (1976). *Ecology of a Mountain Village* (pp. 5–36). Japanese Folklore Lecture 1 (Economic Traditions). Tokyo: Asakura Publishing.

Dag Hammarskjold Foundation (1975). *The 1975 Dag Hammarskjold Report on Development and International Cooperation Was Prepared on the Occasion of the Seventh*

*Special Session of the United Nations General Assembly* (New York, 1 to 12 September), Uppsala, Sweden.

Ebisuno, M. (1967). *Fundamental Issues in Mountain Villages, in 1965 Annals of the Special Study for Promoting Mountain Villages, Mountain Village Promotion Committee* (pp. 1–75). Tokyo: Tokyo University Press.

Iguchi, T. & Kitagawa, I. (1985). *Entrenching of Part-time Farming and Aging Issues in Mountain Villages* (vol. 1, pp. 1–24). Shimane: Shimane University San'in Regional Research Center.

Inoue, M. (1997). Examination of tropical forests as commons. *Journal Environmental Sociology*, 3, 15–32.

Kato, M. (2003). *Forestry and a Forestry Policy until the Taisho Era: An Encyclopedia of the Forest* (pp. 525–528). Tokyo: Asakura Shoten.

Kitao, K. (2001). *Satoyama, the Forest and the Forestry Encyclopedia* (pp. 347–348). Tokyo: Maruzen.

Saito, M. (2003). *The History of the Forest Management: An Encyclopedia of the Forest* (pp. 407–413). Tokyo: Asakura Shoten.

Tachikawa, M. (1998). *Diversification of Residents' Consciousness and Farm Village Social Composition by the Increase in Outside Jobs and Non-Farm Houses.* Research Trends of Regional Problems in Hilly and Mountainous Areas, Yokendo, Tokyo, pp. 94–100.

Tsurumi, K. (1996). *Development of Endogenous Development Theory* (332pp.). Tokyo: Chikuma Shobo Publishing.

# Synthesis and the Way Forward

# 19

# Towards an Effective Policy for Forest Management in Asia

*Ganesh P. Shivakoti, Inoue Makoto, Juan M. Pulhin,*
*Shubhechchha Sharma, Edward L. Webb and Tapan K. Nath*

The 29 case studies from 14 countries analysed in this volume bring rich insight on the issues related to local community's dependency on forestry coupled with multi-level governance arrangements prescribed and implemented through several institutional arrangements. These chapters also highlight how diverse level of autonomy due to complex nature of multi-level governance as mediated by institutional arrangements and multi-scale linkages of social ecological systems have resulted in varied performance of forest condition as well as changes in the livelihood of the forest dependent communities (Mwangi & Wardell, 2012; Nagendra, 2007; Poteete, 2012). All of the cases also examined the application of prototype design guidelines as evaluative criteria for advancing forest health and community welfare. Several scholars have identified critical variables such as group size, heterogeneity, dynamic leadership, adequate funding and clear understanding of rights and responsibilities by the users that are important for the better and under performances of forest resources (Agrawal & Ribot, 1999; Ghate & Nagendra, 2005; Nagendra, 2007; Potete & Ostrom, 2004). But the challenge remains in analysing performance in order to suggest effective policies through better linkage and integration of local-level social ecological system (Ostrom, 2009). In earlier work, Shivakoti and Ostrom (2008) developed a framework for comparing complex case studies by defining basic elements of the framework and applied it to explain both successful and unsuccessful outcomes. In this chapter, we have expanded the framework by including additional elements of governance, forest conditions and rules, external environment, challenges of the resources system and prototype design guidelines. Before we present the comparative perspectives on the role of these elements on the performance of forest systems, we define each concept comprising the elements. Towards the end of the chapter, we have classified forests based on the level of dependency and autonomy and draw some policy implications for multi-level governance and collaborative management of forest resources.

## Governance

In order to depict overall governance issue reflecting support structure of a resource system, we examined the network mechanism within and outside systems assigning categories ranging from excellent to poor; presence or absence of legal pluralism; degree of reliable information availability; linkage with higher level of governance; and provision for social capital building. In addition, important elements of governance also include the process and mechanism of conflict management; property right situation; process of rule formation and power to modify the rules relating to forest management; and harvesting rules through a fair accountability and transparency mechanism. While social network mechanisms are directly linked to better information, linkage with the external resources and reciprocity; context-specific property rights situation, conflict management mechanisms and other rule formation process are the indictors of whether the governance mechanisms are equitable, fair and open in terms of its pluralism. Finally, good governance is reflected by management and harvest rules whether they are transparent and accountable to the users.

## Forest Conditions and Rules

The condition of the forest is one of the most important evidence for the result of good governance and management. There is a direct link between the condition of the forest and presence of rules in use for equitable distribution or its absence due to elite capture. For achieving equity in benefit-sharing from forest products whether the resource system is centrally governed by state or collaboratively between the state and local community or self-governed by the community themselves, monitoring plays an important role in the overall condition of the forest as well as its sustainability (Presha et al., 2011). An important indicator for the sustainability of the forest condition can be attributed to how resilient the resource system is in the recovery from initial failure. The system failure can occur due to overharvest either from the outside or by someone inside the system. In order to improve the physical condition of the forest, synergy and balance between physical improvement and societal rules and monitoring and harvesting are crucial. Only those systems where there is a balanced approach of managing forest condition and following rules over time can sustain in the long run.

## External Environment

For effective multi-level governance and management of forest resources in the present day of global collaborative mode, provision of conducive external environment is absolutely essential. We have included role of technology and its access either at low, medium or high level as one of the important elements of external environment. Level of articulation is another important element of external environment. How outside stakeholders and local community value the importance of their forest system and participate in governance and management of such resource translates directly into the official recognition and standing which paves the way for stewardship of the resource. To link with industry and market for the forest product, negotiating the compensation for carbon credits and other environmental services, the local community need external support including the state support. Finally, the modalities implemented for effective forest governance must be locally acceptable. All these elements comprise the evaluative criteria for external environment.

## Challenges of the Resources System

When we consider the forest system as a management unit, there are several challenges for its effective management. It is very important to examine the success or failure of the systems in terms of forest condition through an in-depth historical analysis, which should be the basis for new policy for effective governance. Community participation in maintaining the forest condition directly translates into the success of new governance mechanism. The other elements of challenge include the size of the resource systems, user group and heterogeneity. Particularly, the large forest system to be managed by a small group or large number of users managing a small-sized forest is an important challenge for managing the resources (Ostrom, 2009). Heterogeneity in the endowment of the community must be respected while crafting new governance policy. Variation in productivity within a forest system is another major challenge for equitable management policy. Similarly, at community level diverse poverty level of users and its level of importance in supporting livelihood are other two important challenges of forest resource system.

## Prototype Design Guidelines

Chapter 1 of this volume has discussed in detail the applicability of all eight design guidelines as suggested by Inoue (2009). In this chapter, three most important elements of these guidelines viz., *graduated membership, commitment principle and fair benefit sharing* are summarised at low, medium and high levels. As discussed in detail in Chapter 1, graduated membership is based on the representation of local people who act as core members and have the strongest authority and cooperate with other graduated members who have relatively weaker authority. Having a clear and graduated membership boundary implies exclusion of non-members. The commitment principle, on the other hand, is a principle for decision making in which the authority of stakeholders is recognised to an extent that corresponds to their degree of commitment to relevant activities. In terms of fair benefit distribution, we imply that benefit distribution is not necessarily equal, but is fair in accordance with cost bearing. Fair benefit distribution criteria is violated when there is fear of elite capture and the core members of the forest users group paying attention on the timber and other prime forest benefits rather than the poor household's dependence issues on non-timber forest products (NTFP) and other day-to-day livelihood needs from the forest.

## Comparative Perspectives on the Role of Multi-level Criteria in Managing Forest Systems

By using the framework outlined above, we have synthesised all 29 cases from 14 Asian countries in the following five tables. The results of the synthesis show that self-governance arrangement with or without outside assistance has worked better in maintaining good networks within and outside forest systems with several fora for legal avenues including reliability of information provided as presented in Table 19.1. The community-initiated forestry programmes of Bangladesh, Bhutan, Indonesia (collaborative governance forests in Kalimantan, Nagari systems in West Sumatra), Nepal, Philippines, Thailand and Vietnam (Forest Land Allocation programme) fall under better governance arrangement having better link with the higher level of governance. These social ecological systems have also above-average endowment of social capital and more secured property rights situation. The rules are decided either by the users themselves or jointly together with higher authorities. These arrangements have flexibility in modifying the rules through an accountable and transparent process. This could be taken as the reason for high accountability and transparency in community-initiated village common forest (VCF) in Bangladesh, community forest in Nepal and community forest in Bhutan.

Table 19.2 presents the comparative results of forest condition relating to sustainability associated with the rule following and enforcement mechanisms. It is interesting to note that although lower level of self-governance and autonomy were reported in the forest of Japan, Korea and Malaysia, the forest systems in these countries are good and stable. On the other hand, although collaborative and self-governed forestry systems in Indonesia, Sri Lanka and Vietnam have reflected better mechanisms (Table 19.1), these forestry systems are weak in addressing problem of elite capture, equity issues and ineffective monitoring mechanism, resulting in lower evidence of sustainability and poor recovery from initial failure. There is a similar level of positive relationship between the governance and forest conditions and rules in the forestry of Bhutan, Nepal and Philippines. These findings point out towards the importance of autonomy and dependence of resources for their long-term sustainability, which we will bring as an important issue towards the end of this chapter.

**Table 19.1:**
Diversity in governance arrangement for managing forest in Asia

| Country | Case | Networks | Legal pluralism | Reliable information | Links with high governance | Provision for social capital building | Conflict management | Property right situation | Rules formation | Power to modify rules | Forest management | Harvesting rules | Accountability and transparency |
|---|---|---|---|---|---|---|---|---|---|---|---|---|---|
| Bangladesh | Upland settlement project | Medium | Absent | Not available | High | Medium | Complex in solving | Poor-limited access and limited withdrawal | State | State | State in coordination | Restricted | Low |
| | Community initiated village common forests | Good | Present | Available among users | Medium | High | Easily solved, simple and indigenous | Strong | In coordination | In coordination | Users | Allowed | High |
| India | Joint forest management | Poor | Present | Not available | High | Medium | Medium | Medium-access and limited withdrawal, manage | State | State | In coordination | Limited | Low |
| Nepal | Centrally controlled national forestry | Poor | Present | Not available | Low | Low | Complex in solving | Poor-limited access and limited withdrawal | State | State | State | Restricted | Low |
| | Community forestry | Excellent | Present | Available to users | High | High | Easily solved | Very strong | In coordination, locally dominant | In coordination, users dominant | Users | Allowed | High |

| | | | | | | | | | | | | | |
|---|---|---|---|---|---|---|---|---|---|---|---|---|---|
| Sri Lanka | Community-based pro-poor leasehold forestry regime | Medium | Present | Less available | High | High | Complex in solving | Strong | DFO, in coordination, locally dominant | DFO and users | Users | Limited harvest | High |
| | Small grant programme for operations to promote tropical forests | Poor | Present | Less available | High | High | Difficult | Limited harvesting | State | State | In coordination | Limited | Poor |
| | Farmer's wood lot | Poor | Present | Less available | High | Medium | Medium | Limited harvesting | State | State | In coordination | Limited | Medium |
| Bhutan | Jigne Singye Wangchuck National Park | None | Present | Not available to users | Low | Low | Not given | Poor-limited access and limited withdrawal | State | State | State | Restricted | Medium |
| | Community forestry | Excellent | Present | Available to users | Medium | High | Not given | Very strong | In coordination, locally dominant | In coordination, locally dominant | Users | Allowed | High |
| Indonesia | Collaborative forest governance | Poor | Present | Available | Poor | Medium | Complex | Access, Withdraw | State dominant | State | In coordination | Limited to users | Low |
| | Hutan Kemasyarakatan/HKm | None | Present | Not reliable | Poor | Medium | Easily solved | Access, withdraw | Not given | Not given | Users | Limited | Low |
| | Hutan Tanaman Rakyat/HTR | None | Present | Not reliable | Medium | Medium | Easily solved | Access, withdraw | Not given | Not given | Users | Limited | Low |

(Table 19.1 Continued)

(Table 19.1 Continued)

| Country | Case | Networks | Legal pluralism | Reliable information | Links with high governance | Provision for social capital building | Conflict management | Property right situation | Rules formation | Power to modify rules | Forest management | Harvesting rules | Accountability and transparency |
|---|---|---|---|---|---|---|---|---|---|---|---|---|---|
| | Hutan Desa/HD | None | Present | Not reliable | Poor | Poor | Complex | Access, limited withdraw | Not given | Not given | Users | Limited | Low |
| | Co-management between state and Nagari system | Poor | Present | Less available | Medium | Medium | Complex | Access, limited withdraw, manage | In coordination, state dominant | State | Users | Limited to timber | Medium |
| | Self-governed Nagari system | Poor | Present | Available among users | Low | High | Locally, easily solved | Strong | Users | Users | Users | Allowed | High |
| Korea | Self-governed village woods | Poor | Absent | Not given | High | Not given | Easily solved | Strong | Coordination, users dominant | Users and state | Users | Allowed | High |
| Thailand | Self-initiated community forestry | Good | Absent | Available to users | High | High | Easily solve | Strong-not endorsed by law | Users in coordination with RFD | Users in coordination with RFD | Users | Allowed | High |
| | RFD supported/ coordinated community forestry | Good | Absent | Available to users | High | High | Complex and lingering | Medium-complex limited withdrawal, manage | State dominant | State dominant | In coordination | Limited withdrawal | High |
| China | Community-based co-management | Good | Absent | Available | High | Not given | Easily solved | Medium | In coordination | In coordination | Users | Allowed | High |

| | | | | | | | | | | | | |
|---|---|---|---|---|---|---|---|---|---|---|---|---|
| Laos | Village in a protected area | Poor | Absent | Not given | Medium | Medium | Complex | Strong because locals do not follow state rules | State, strongly opposed by locals | Locals are resistant to state rules | Users | Restricted but locals to not listen | Low |
| Japan | Property wards | Poor | Resolved | Available | High | Low | Locally | Strong | In coordination | In coordination | Locally | Locally | High |
| | School wards | Poor | Resolved | Available | High | Possible | Locally but complicate | Strong | In coordination | In coordination | Locally | Locally | High |
| Malaysia | Forest management initiated by DFO with little community input | Good | Absent | Not given | High | Medium | State, difficult | Medium | State | State | State | Limited | Low |
| | Forest management actively by local community | Medium | Absent | Not given | Low | Medium | Locally, easily solved | Medium | Local, in coordination | Locally, in coordination | Locally | Allowed | High |
| Vietnam | Community-based forest management (FLA) | Good | Present | Available | Low | High | Locally, easily solved | Strong | User, in coordination | User | Users | Allowed | High |
| | State forest (Before FLA) | None | Present | Less reliable | Low | Medium | State-complicated | Poor | State | State | State | Restricted | Poor |
| Philippines | Community-based forest management | Good | Present | Available | Medium | High | Locally but complicated | Strong | In coordination | State in coordination with user | People's organisation limited | Timber limited | High |

DFO: District Forest Officer; RFD: Royal Forest Department.

*Source:* Authors.

**Table 19.2:**
*Sustainability of forest conditions associated with rules' enforcement and rule following*

| S. no. | Country | Case | Monitoring | Forest condition | Equity | Elite capture | Recover from initial failure | Sustainable evidence |
|---|---|---|---|---|---|---|---|---|
| 1 | Bangladesh | Upland Settlement Project | Rarely, ineffective | Degraded | Less equitable | High | Poor | Low |
| | | Community-initiated village common forests | Locally, effective | Good | More equitable | Low | Good | Medium |
| 2 | India | Joint Forest Management | Centrally, ineffective | Good | Less equitable | High | Poor | Low |
| 3 | Nepal | Centrally controlled national forestry | Rarely | Degraded | Less equitable | Not given | Poor | Low |
| | | Community forestry | User and DFO | Good | More equitable | Low | High | High |
| | | Community-based pro-poor leasehold forestry regime | Users and DFO | Good | More equitable | Low | High | High |
| 4 | Sri Lanka | Small Grant Programme for operations to promote tropical forests | Never | | Less equitable | More | Poor | Poor |
| | | Farmer's wood lot | Never | | Less equitable | More | Medium | Medium |
| 5 | Bhutan | Jigne Singye Wangchuck National Park | State, effective | Stable | Less equitable | | Stable | Medium |
| 6 | Indonesia | Community forestry | Users, frequently | Good | Very equitable | Negligible | High | High |
| | | Collaborative forest governance | In coordination | Degraded | Less equitable | Not given | Poor | Low |
| | | Hutan Kemasyarakatan/HKm | State | Not given | Less equitable | Not given | Poor | Low |
| | | Hutan Tanaman Rakyat/HTR | State | Not given | Less equitable | Not given | Poor | Low |
| | | Hutan Desa/HD | State | Not given | Less equitable | Not given | Poor | Low |

| # | Country | Management | In coordination | Condition | Equity | | | |
|---|---|---|---|---|---|---|---|---|
|  |  | Co-management between state and Nagari system | In coordination | Stable | Less equitable | Not given | Medium | Medium |
| 7 | Korea | Self-governed Nagari system | Users | Good | Less equitable | Medium | Medium | Medium |
|  |  | Self-governed village woods | Users, rarely | Good | Equitable | Negligible | High | High |
| 8 | Thailand | Self-initiated community forestry | Rarely | Good | Less equitable | Negligible | Medium | Medium |
|  |  | RFD supported/coordinated community forestry | Rarely | Good | Not equitable | Negligible | Poor | Low |
| 9 | China | Community-based co-management | Locally and state | Improved | Equitable | Negligible | High | Low |
| 10 | Laos | Village in a protected area | Locally, effective | Stable | Not equitable | More | High | Poor |
| 11 | Japan | Property Wards | Local | Stable | Equitable | Negligible | High | High |
|  |  | School Wards | Local | Stable | Equitable | Negligible | Medium | High |
| 12 | Malaysia | Forest management initiated by DFO with little community input | State | Stable | Equitable | Negligible | Medium | Medium |
|  |  | Forest management actively by local community | Locally | Stable | Equitable | Negligible | Medium | High |
| 13 | Vietnam | Community-based forest management (FLA) | Locally | Good | Equitable | Negligible | High | High |
| 14 | Philippines | State forest (before FLA) | State-not effective | Degraded | Less equitable | More | Low | Poor |
|  |  | Community-based forest management | In coordination with state | Good | Less equitable | Negligible | High | High |

*Source:* Authors.

The issue of resource dependence and autonomy vis-à-vis the role of external environment are interrelated. If the provision of technology and external support, including local and state support are easily available, the forest community will be more specialised on the harvest and benefit sharing from forest product. Almost all the co-management projects initiated by outsiders fall under this category; for example, the community-based co-management project in China, property ward and school ward forests in Japan, government-initiated forest management programmes in Malaysia, village protected forests in Lao, leasehold forestry programme in Nepal and community forestry programmes in Philippines and Thailand (Table 19.3). These cases imply that external support is contingent upon fulfilling conservation objectives only during the project period with induced external resources only for a limited time.

Table 19.4 presents the challenges of forestry as a resource system in Asian setting. It should be kept in mind that there is enormity of diverse challenges but several of these systems, which were initially managed by the local community were nationalised by the state and over time as the condition of the resources deteriorated, a multitude of governance arrangements surfaced including collaborative governance (Indonesia, Malaysia), co-management (China, Lao and Vietnam), joint management (India), self-management (Korea and Japan) and majority managed by the community (Bhutan, Bangladesh, Thailand, Philippines). As there is variation in governance arrangement, there is also variation in the challenges of these systems. While most of the community-managed forest systems are smaller in size and lower number of users, most of the co-managed systems are of large size and often users from multiple communities of heterogeneous endowments. This brings the challenge of managing large vs. small size forests. The challenge gets further complicated with high dependence of local community on forestry and poverty level of the users. The forest dwellers of Bangladesh, Indonesia, Lao and Vietnam are poor and depend on forest resources for their livelihood due to which the productivity of the resources are decreasing over time in these countries. On the other hand, although forestry dependence of poor in Nepal, Bhutan, Sri Lanka and Thailand (up to certain extent) is high but they also have higher level of autonomy in making rules for equitable distribution of benefits and management responsibilities. Thus, the conditions of the forest over time are improving in these systems.

The final set of evaluative criteria for multi-level governance and management outcomes include three important prototype design guidelines; namely graduated membership, commitment principle and fair benefit sharing as discussed earlier (Table 19.5). The analysis points out

**Table 19.3:**
*External environment, technology support and local acceptability*

| S. no. | Country | Case | Technology | Level of articulation | External support | State support | Local acceptance |
|---|---|---|---|---|---|---|---|
| 1 | Bangladesh | Upland Settlement Project | Medium | Medium | Medium | High | Medium |
| | | Community-initiated village common forests | Low | Low | Low | Low | High |
| 2 | India | Joint Forest Management | High | Medium | Medium | Medium | Low |
| 3 | Nepal | Centrally controlled national forestry | Not given | High | No | State owned | Low |
| | | Community forestry | Not given | High | Limited | Medium | High |
| | | community-based pro-poor leasehold forestry regime | Not given | High | High | High | High |
| 4 | Sri Lanka | Small Grant Programme for operations to promote tropical forests | Medium | High | Medium | Medium | Low |
| | | Farmer's wood lot | Medium | High | Medium | Medium | Medium |
| 5 | Bhutan | Jigne Singye Wangchuck National Park | Low | Medium | Low | High | Low |
| | | Community forestry | Low | High | Low | Medium | High |
| 6 | Indonesia | collaborative forest governance | Low | Low | Low | Low | Low |
| | | Hutan Kemasyarakatan/HKm | Low | Low | Low | Low | Medium |
| | | Hutan Tanaman Rakyat/HTR | Medium | Low | Low | Low | Medium |
| | | Hutan Desa/HD | Low | Low | Low | Low | Low |
| | | Co-management between State and Nagari system | Low | Low | Low | Medium | Medium |
| | | Self-Governed Nagari system | Low | High | Low | Low | High |

*(Table 19.3 Continued)*

*(Table 19.3 Continued)*

| S. no. | Country | Case | Technology | Level of articulation | External support | State support | Local acceptance |
|---|---|---|---|---|---|---|---|
| 7 | Korea | Self-governed village woods | High | High | Low | Medium | High |
| 8 | Thailand | Self-initiated community forestry | Medium | High | Medium | Medium | High |
| | | RFD supported/coordinated community forestry | Low | Low | Medium | Medium | Medium |
| 9 | China | Community-based co-management | High | High | High | High | High |
| 10 | Laos | Village in a protected area | Low | High | Low | Low | High |
| 11 | Japan | Property wards | Medium | High | High | Medium | High |
| | | School wards | Medium | High | Medium | Medium | High |
| 12 | Malaysia | Forest management initiated by DFO with little community input | Low | High | Medium | High | High |
| | | Forest management actively by local community | Low | High | Medium | Medium | High |
| 13 | Vietnam | Community-based forest management (FLA) | Low | High | High | Low | High |
| | | State forest (before FLA) | Poor | Low | Low | Medium | Low |
| 14 | Philippines | Community-based forest management | Medium | High | High | Medium | High |

*Source:* Authors.

**Table 19.4:**
*Challenges of the Asian forestry resources system*

| S. no. | Country | Cases | Prior history | Size of resource system | Size of user group | Well-defined boundaries | Heterogeneity of user group | Productivity of the system | Poverty level of user | Importance of resource |
|---|---|---|---|---|---|---|---|---|---|---|
| 1 | Bangladesh | Upland Settlement Project | Short term | Large | Large | Strong | Heterogeneous | Decreasing | Poor, landless | Settlement for landless |
| | | Community-initiated village common forests | Long term | Small | Small | Weak | Homogenous | Increasing | Above poor | Livelihood dependent |
| 2 | India | Joint Forest Management | Short term | Medium | Medium | Strong | Heterogeneous | Increasing | Poor, landless | Livelihood dependent |
| 3 | Nepal | Centrally controlled national forestry | Long term | Large | Large | Strong | Heterogeneous | Decreasing | Medium | Conservation, Livelihood |
| | | Community forestry | Medium term | Medium | Medium | Strong | Heterogeneous | Increasing | Poor, Medium | Livelihood dependent |
| | | Community-based pro-poor leasehold forestry regime | Medium term | small | small | Strong | Homogenous | Increasing | Poor, landless | Livelihood dependent |
| 4 | Sri Lanka | Small Grant Programme for operations to promote tropical forests | Very short term | Medium | Medium | Strong | Homogenous | Slightly increasing | Poor | Protection, poverty reduction |
| | | Farmer's wood lot | Short term | small | small | Strong | Homogenous | Slightly increasing | poor, shifting cultivators | Wood supply and livelihood |

*(Table 19.4 Continued)*

*(Table 19.4 Continued)*

| S. no. | Country | Cases | Prior history | Size of resource system | Size of user group | Well-defined boundaries | Heterogeneity of user group | Productivity of the system | Poverty level of user | Importance of resource |
|---|---|---|---|---|---|---|---|---|---|---|
| 5 | Bhutan | Jigne Singye Wangchuck National Park | Long term | Large | Large | Strong | Heterogeneous | Dlightly Increasing | Poor | Livelihood dependent |
| | | Community forestry | Very short term | Medium | Medium | Strong | Heterogeneous | Increasing | Poor | Fuel wood, food, NTFP |
| 6 | Indonesia | Collaborative forest governance | Long term | Large | Large | Weak | Heterogeneous | Decreasing | Poor | Livelihood dependent |
| | | Hutan Kemasyarakatan/ HKm | Long term | Large | Large | Strong | Heterogeneous | Decreasing | poor | Livelihood dependent |
| | | Hutan Tanaman Rakyat/ HTR | Short term | Large | Large | Strong | Heterogeneous | Decreasing | Medium poor | Economic benefit |
| | | Hutan Desa/HD | Short term | Large | Large | Weak | Heterogeneous | Decreasing | Medium poor | Livelihood dependent |
| | | Co-management between State and Nagari system | Medium term | Large | Large | Weak | Heterogeneous | Stable | Poor | Livelihood dependent |
| | | Self-governed Nagari system | Long term | Large | Large | Strong | Heterogeneous | Increasing | Poor | Livelihood dependent |
| 7 | Korea | Self-governed village woods | Long | Not given | Not given | Strong | Heterogeneous | Increasing | Not poor | Physical, cultural and recreational |

| # | Country | | Term | | | | | | | |
|---|---|---|---|---|---|---|---|---|---|---|
| 8 | Thailand | Self-initiated community forestry | Short term | Small | Small | Strong | Homogenous | Increasing | Medium | Livelihood, economic benefit |
| | | RFD supported/coordinated community forestry | Short term | Small | Small | Weak | Homogenous | Increasing | Medium | Livelihood, economic benefit |
| 9 | China | Community-based Co-management | Short term | Large | Large | strong | Heterogeneous | Increasing | Not poor | Income sources |
| 10 | Laos | Village in a Protected Area | Long term | Medium | Medium | Strong | Heterogeneous | Increasing | Mixed, 42% poor | Livelihood dependency |
| 11 | Japan | Property Wards | Long term | Medium | Small, Medium | Strong | Homogenous | Increasing | Not poor | Income |
| | | School Wards | Long term | Small | Medium | Strong | Homogenous | stable | Not poor | School fund |
| 12 | Malaysia | Forest management initiated by DFO with little community input | Long term | Large | Large | Strong | Not given | Increasing | Medium | Livelihood, timber |
| | | Forest management actively by local community | Long term | Small | Small | Strong | Not given | Increasing | Poor | Livelihood, timber |
| 13 | Vietnam | Community-based Forest Management (FLA) | Short term | Small | Small | Strong | Heterogeneous | Increasing | Poor | Livelihood dependency |
| | | State Forest (Before FLA) | Medium term | Medium | Large | Strong | Heterogeneous | Decreasing | Poor | Livelihood dependency |
| 14 | Philippines | Community-based Forest Management | Medium term | Medium to Large | Medium to Large | Strong | Heterogeneous | Increasing forest cover | Medium | Income and livelihood dependent |

*Source:* Authors.

**Table 19.5:**
*Prototype design guidelines and multi-level management outcomes of forestry in Asia*

| S. no. | Country | Case | *Graduated membership* | *Commitment principle* | *Fair benefit sharing* |
|---|---|---|---|---|---|
| 1 | Bangladesh | Upland Settlement Project | Low | Low | Low |
| | | Community initiated village common forests (VCF) | Medium | High | High |
| 2 | India | Joint Forest Management | Medium | Low | Low |
| 3 | Nepal | Centrally controlled national forestry | Low | Low | Low |
| | | Community forestry | High | High | High |
| | | Community-based pro-poor leasehold forestry regime | Medium | High | Medium |
| 4 | Sri Lanka | Small Grant Programme for operations to promote tropical forests | Low | Medium | Low |
| | | Farmer's wood lot | Low | Low | Low |
| 5 | Bhutan | Jigne Singye Wangchuck National Park | Low | Low | Low |
| | | Community forestry | High | High | High |
| 6 | Indonesia | Collaborative forest governance | Low | Low | Low |
| | | Hutan Kemasyarakatan/ HKm | Low | Low | Low |
| | | Hutan Tanaman Rakyat/ HTR | Low | Low | Low |
| | | Hutan Desa/HD | Low | Low | Low |
| | | Co-management between State and nagari system | High | Low | Low |
| | | Self-governed Nagari system | High | Low | Low |
| 7 | Korea | Self-governed village woods | Low | High | High |
| 8 | Thailand | Self-initiated community forestry | High | High | Low |
| | | RFD supported/ coordinated community forestry | High | High | Low |

*(Table 19.5 Continued)*

*(Table 19.5 Continued)*

| S. no. | Country | Case | Graduated membership | Commitment principle | Fair benefit sharing |
|---|---|---|---|---|---|
| 9 | China | Community-based co-management | High | High | High |
| 10 | Laos | Village in a protected area | High | Low | Low |
| 11 | Japan | Property wards | Low | Low | High |
| | | School wards | Low | Low | High |
| 12 | Malaysia | Forest management initiated by DFO with little community input | High | Low | Low |
| | | Forest management actively by local community | High | High | High |
| 13 | Vietnam | Community-based forest management (FLA) | High | Low | High |
| | | State forest(before FLA) | Low | Low | Low |
| 14 | Philippines | Community-based forest management | High | Low | Medium |

*Source:* Authors.

to the positive relationship between graduated membership and commitment principle, which results in fair benefit sharing among multiple levels of users. This relationship is more pronounced in the self-initiated community forestry systems across the countries in Asia. Many external initiated forest management regimes are poor on graduated membership criteria, commitment principle ultimately resulting in lower level of equity. Therefore, the higher level of autonomy coupled with higher dependency necessitates the application of design guidelines in managing a forest.

## Towards the Classification of Forests Based on Level of Dependency and Autonomy

We have categorised the cases from different countries with respect to their degrees of autonomy and dependency of local people on forest resources. Level of autonomy has been conceptualised as degree of freedom on decision making or degree of rule-making coupled with sense of responsibility. Property rights situation significantly affects the decision-making process, which categorises the kind and right of stakeholders to be involved in use, manage and control of the resources. There is a complex social relationship embedded in property rights, which provides a High degree of people's participation in forest management and capacity to delineate their constitution and harvesting rules. In addition, involvement of customary law, respect to legal pluralism and collective management of forest resources are the important parameters for defining autonomy. Hence, property right situation, power to constitutive and harvesting rules formation and manipulation is considered to be indicators for degree of autonomy.

Likewise, those people with the means of Livelihood security as forest resources are considered to be highly forest dependent. Forest also serves as an important asset to reduce vulnerability and increase adaptive capacity towards changing socio-ecological systems. In some instances, forest can be additional source for generating profitable income, while sometimes important for biodiversity and habitat conservation. For whatever the reasons forests are being used for, it will be useful to delineate forest dependency from highly forest dependable population like shifting cultivators to Low Livelihood dependable commercial plantations.

We have divided these 29 cases from 14 countries on the criteria of dependency and autonomy into four categories (Table 19.6). However, this criterion also does not provide the complete information. For instance, community forestry in Nepal, Bhutan and forest land allocation programme in Vietnam all falls into the same category 'I'. Nevertheless, in reality with relevance to property right situation, ability to design self-defined rules and freedom to manage the forest is higher in Nepal than the other counterparts. Similarly, people living within the forest and forest-dependent people who have no other means of living apart from the products they receive from the forest is considered to be highly forest dependable people. In the above-mentioned table, though both the Philippines and Vietnam are in the same category, the dependency of forest is much higher in Vietnam. For, the people living in Philippines forest is the means of livelihood and additional income and, for people of Vietnam, they will have to face problems of food scarcity to severely no food situation without a forest. To highlight these complexities, we have further presented the criteria of dependency and autonomy in Figure 19.1.

**Table 19.6:**
*Level of dependency and autonomy of community forestry in Asia*

| I (high dependency, high autonomy) | II (high dependency, low autonomy) |
| --- | --- |
| Community-initiated village common forests—Bangladesh | Upland settlement project—Bangladesh |
| Community forestry—Nepal | Joint Forest Management—India |
| Self-governed Nagari system—Indonesia | Centrally controlled national forest—Nepal |
| Community forestry—Bhutan | Community-based pro-poor leasehold forestry regime—Nepal |
| Forest land allocation—Vietnam | Self-initiated community forestry—Thailand |
| | RFD supported community forestry—Thailand |
| | Jigne Singye Wangchuck National Park—Bhutan |
| | Collaborative forest governance—Indonesia |
| | Hutan Kemasyarakatan/HKm—Indonesia |
| | Hutan Tanaman Rakyat/HTR—Indonesia |
| | Hutan Desa/HD—Indonesia |
| | Co-management between state and Nagari system—Indonesia |
| | Village in a protected Area—Laos |
| | State forest—Vietnam |
| | Community-based forest management—Philippines |

| III (low dependency, low autonomy) | IV (low dependency, high autonomy) |
| --- | --- |
| | Self-governed wood lots—Korea |
| Endogenous development of a mountain village—Japan | Property wards—Japan |
| Forest management initiated by DFO with little community input—Malaysia | School wards—Japan |
| Small grant programme for operations to promote tropical forests—Sri Lanka | |
| Farmer's wood lot—Sri Lanka | |
| Community-based co-management—China | |
| Forest management actively by local community—Malaysia | |

*Source:* Authors.

**Figure 19.1:**
*Level of dependency and autonomy of community forestry in Asia*

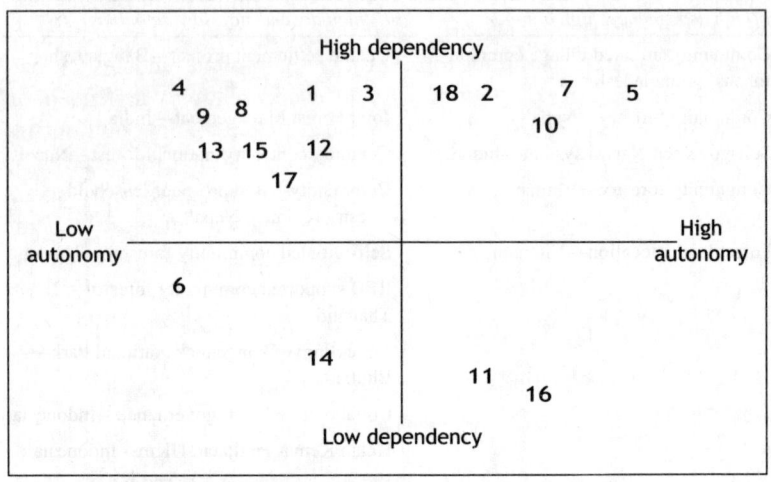

1 Upland Settlement Project—Bangladesh; Community-based Pro-poor
  Leasehold Forestry Regime—Nepal
2 Community-initiated Village Common Forests (VCF)—Bangladesh
3 Joint Forest Management—India
4 Centrally Controlled National Forestry—Nepal; Jigne Singye Wangchuck
  National Park—Bhutan; State Forest (also before FLA)—Vietnam
5 Community Forestry—Nepal
6 Small Grant Program for Operations to Promote Tropical Forests—
  Sri Lanka; Farmer's Wood Lot—Sri Lanka
7 Community Forestry—Bhutan
8 Collaborative Forest Governance—Indonesia; Co-management between
  State and *Nagari* System—Indonesia; Forest Management Initiated by DFO
  with Little Community Input—Malaysia
9 Hutan Kemasyarakatan/Hkm—Indonesia; Hutan Tanaman Rakyat/
  Htr—Indonesia; Hutan Desa/Hd—Indonesia
10 Community-based Forest Management (FLA)—Vietnam
11 Self-governed Village Woods—Korea
12 Self-initiated Community Forestry—Thailand; Community-based Forest
   Management—Philippines
13 RD-supported/coordinated Community Forestry—Thailand
14 Community Based Co-management—China
15 Village in a Protected Area—Laos
16 Property Wards—Japan; School Wards—Japan
17 Forest Management Activity by Local Community—Malaysia
18 Self-governed *Nagari* System—Indonesia

*Source:* Authors.

Based on these analyses, the forests that have reflected good performance are classified in the category of high dependency with high autonomy. The community-initiated and managed forests in Bangladesh, Nepal, West Sumatran Nagari systems in Indonesia and government-allocated long-term contract forests in Vietnam come under such categories. But majority of the forests where users are highly dependent on forest for their livelihood have experienced low autonomy and outsiders have major say on deciding rules for managing the forest as well as stringent harvesting and benefit sharing criteria. If we examine the social ecological condition of these forest systems, majority of these systems have been situated in the critical buffer regions either between the national parks and the local communities or forests allocated to the communities under forest improvement programmes initiated in collaborative arrangements in degraded lands. The forest-dependent communities in such situation are highly dependent on forest for their livelihood and they are the most vulnerable community both in terms of food and long-term land security. The analysis in the respective chapters clearly point out towards the policy reform on these cases both through the improvement in the structural and governance arrangement. Owing to the nature of fragile ecosystem and poverty alleviation, the long-term planning and management issue of these forests must be reoriented towards the effective collaborative multi-level governance by changing the lower level of autonomy of local people to the higher one.

The policy agenda for the third quadrant forest systems of Japan, Malaysia and Sri Lanka with lower dependency of local community and lower autonomy should include the larger agenda of environmental management under the context of climate change. However, the policy initiatives must be collaborative and of multi-scale in nature. Local participation can be enhanced through collaborative arrangements and polycentric approach can assure integrated management in a coordinated but with independent identity of multi-stakeholders. Perhaps these countries could even move towards higher level of autonomy following Korean model of self-governed wood lots and interact with unfettered outsiders to demolish vertical barriers keeping in mind the global nature of the forest resources as explained in Chapter 1 of this volume.

On the issue of autonomy, it is important that the policy domain makes a provision for the degree of freedom on decision making on the harvesting rules, rights and responsibilities of both forest user's executive committee and users. If the assembly of user's group is permitted by policy to make group's constitution and define rights to forest management decisions, benefit sharing and forest user's group fund mobilisation, the executive committee and the service providers are likely to be accountable to the users and in turn a high level of autonomy will be realised. This could be taken as the reason for high level of autonomy and dependency in community-initiated VCF in Bangladesh, community forest in Nepal and Bhutan. Similarly, a clear policy provision for autonomy, recognising local Nagari system in Western Sumatra for making decision on ownership of forest resources and forest land allocation with red certificates of land ownership in Central Vietnam are good examples of policy provision for local autonomy. But in some of the forest users groups, the service providers and the executive members make decisions on behalf of users and committee members, which means that they are less transparent and less accountable to the users. A mechanism should be made in order to avoid dominance of external service providers and executive committee members towards general users in decision making. Especially with Upland Settlement Project in Bangladesh, Joint Forest Management in India, state forest in Nepal, Forests of Indonesia and Village in protected area in Laos without autonomy, forest users and government officials tend to have poor accountability of duty bearers towards right bearers that have resulted in poor governance. In these forest groups, major decisions and economic transactions are carried out either by the executive committee or the local government. In this regard, the facilitation and co-ordination of government and related agencies without autonomy in order to address the diverse issues are not enough in relationship to fulfil increasing demand of forest dependent communities.

# References

Agrawal, A. & Ribot, J.C. (1999). Accountability in decentralization: A framework with South Asian and African cases. *Journal of Developing Areas*, 33(4), 473–502.
Ghate, R. & Nagendra, H. (2005). Role of monitoring in institutional performance: Forest management in Maharashtra, India. *Conservation and Society*, 3(2), 509–532.
Inoue, M. (2009). Design guidelines for collaborative governance (*kyouchi*) of natural resources. In T. Murota (ed.), *Local Commons in Globalized Era* (pp. 3–25). Tokyo: Minerva-shobou (in Japanese).
Mwangi, E. & Wardell, A. (2012). Multi-level governance of forest resources. *International Journal of the Commons*, 6 (2), 79–103.

Poteete, A. (2012). Levels, scales, linkages, and other 'multiples' affecting natural resources. *International Journal of the Commons*, 6(2), 134–150.

Potete, A. & Ostrom, E. (2004). Heterogeneity, group size and collective action: The role of institutions in forest management. *Development and Change*, 35(3), 435–461.

Presha, L., Agrawal, L., & Chhatre, A. (2011). Social and ecological synergy: Local rule-making, forest livelihoods, and biodiversity conservation. *Science*, 331, 1606–1608.

Nagendra, H. (2007). Drivers of reforestation in human-dominated forests. *PNAS*, 104(39), 15218–15223.

Ostrom, E. (2009). A general framework for analyzing sustainability of social ecological systems. *Science*, 325, 419–422.

Shivakoti, G. & Ostrom E. (2008). Facilitating decentralized policies for sustainable governance and management of forest resources in Asia. In E. Webb, & G.P. Shivakoti (eds), *Decentralization, Forests and Rural Communities: Policy Outcomes in South and Southeast Asia* (pp. 292–310). New Delhi/Thousand Oaks/London/Singapore: SAGE Publications.

# About the Editors and Contributors

## Editors

**Makoto Inoue** is a Professor at the Graduate School of Agricultural and Life Sciences, the University of Tokyo. He specialises in forest sociology and governance. He is the corresponding author of Chapter 1 and co-author of Chapters 2, 4, 5, 18 and 19.

**Ganesh P. Shivakoti** is a Professor at Asian Institute of Technology, Thailand. He specialises in Agricultural Development and Policy Analysis; Resource Development; Farming Systems; Natural Resources Management; Natural Resources Economics; Common Property Resources; NRM Policy Analysis; and Watershed Management. He is the corresponding author of Chapter 19 and co-author of Chapters 1, 4, 9, 13 and 15.

## Contributors

**Madhusudan Bandi** is an Assistant Professor at Gujarat Institute of Development Research, Ahmedabad, India. He is a political scientist with specialisation in Development Studies. His research interests include local/decentralised governance, participatory forest management, development concerns of the underprivileged sections in India and current affairs. He is a co-author of Chapter 3.

**Haiyun Chen** is a Lecturer in School of Economics and Management at Tongji University, Shanghai, People's Republic of China. He specialises in economics of environment and nature resources, institutional economics, sustainable development, community-based nature resource management, common pool resources governance, etc. He is the corresponding author of Chapter 15.

**Kweon Deogkyu** has graduated from Seoul National University with MSc major in forest environmental science and is a Kyomu, priest of Won-Buddhism in Korea. He is a co-author of Chapter 16.

**G. Simon Devung** is a Senior Researcher at the Center for Social Forestry, Mulawarman University, Samarinda, Indonesia. He specialises in anthropology, sociology of forestry, socio-cultural studies and social education. He is the corresponding author of Chapter 8.

**Mangala De Zoysa** is a Professor at Faculty of Agriculture, University of Ruhuna, Sri Lanka. He specialises in forest policy, agricultural policy and agricultural economics. He is the corresponding author of Chapter 5.

**Demis Galli** has an MSc from the Asian Institute of Technology and currently works as a freelance forestry consultant in Chiang Mai, Thailand. He is a co-author of Chapter 12.

**Ambika P. Gautam** is a natural resources management professional and academician with 28 years of experience in the field. Currently, he is serving as a professor and outreach coordinator at the Kathmandu Forestry College, Nepal. He is a co-author of Chapter 4.

**Kimihiko Hyakumura** is an Associate Professor at the Institute of Tropical Agriculture, Kyushu University, Fukuoka, Japan. He specialises in forest policy and area studies in Southeast Asia. He is the corresponding author of Chapter 14.

**Yukio Ikemoto** is a Vice Director of the Network for Education and Research on Asia and a Professor at the Institute for Advanced Studies on Asia, the University of Tokyo. He specialises in poverty issues in Southeast Asia and is the author of the Foreword.

**Ndan Imang** is an Associate Professor of Socio-Economic of Agriculture/senior researcher at the Center for Social Forestry (CSF) University of Mulawarman, Samarinda, Indonesia. He obtained his PhD from the Graduate School of Agriculture and Life Sciences, the University of

Tokyo. He specialises in rural sociology and shifting cultivation. He is a co-author of Chapter 7.

**Koo Ja-Choon** is a research associate at the Department of Forest Sciences, Seoul National University, where he obtained his PhD in forest environmental science. He is an ecological economist specialised in valuation of ecosystems services and is the leading author of Chapter 16.

**Mohammed Jashimuddin** is a Professor at the Institute of Forestry and Environmental Sciences, University of Chittagong, Bangladesh. He specialises in land and forest policy, forest and environmental economics, community-based forest management and climate change. He is a co-author of Chapter 2.

**Birendra K. Karna** is a natural resources management professional with 15 years of experience in research and development of common property institutions especially in area of community-based forest management. Currently, he is serving as an institutional development specialist at the Forest Action Nepal. He is a co-author of Chapter 4.

**Om N. Katel** is a Lecturer at the Department of Forestry, College of Natural Resources, Royal University of Bhutan, and teaches Applied Conservation Science, Ecotourism, Integrated Watershed Management and GIS. He has completed his PhD from Asian Institute of Technology, Thailand. He is the corresponding author of Chapter 6.

**Tapan K. Nath** is an Associate Professor at the School of Biosciences, University of Nottingham Malaysia Campus, Malaysia. He specialises in collaborative natural resources management, rural livelihood analysis, and project monitoring and evaluation. He is the corresponding author of Chapter 2 and co-author of Chapter 19.

**Mahdi** is a Chair of Integrated Natural Resources Management Field of Study at the Graduate Program of Andalas University in Padang, Indonesia. He obtained his PhD in Natural Resources Management from Asian Institute of Technology (AIT), Thailand. His research interest includes natural resource and economics of natural and environmental resources. He is a co-author of Chapter 9.

**Tri Martial** is a Lecturer at the Department of Agribusiness, Islamic University of North Sumatra, Medan, Indonesia. He is also in charge of quality control at the university. He obtained his PhD in Agricultural Development from Andalas University in Padang, Indonesia. He is a co-author of Chapter 9.

**A. Ainuddin Nuruddin** is Deputy Director at the Institute of Tropical Forestry and Forest Products and Associate Professor at the Faculty of Forestry, Universiti Putra Malaysia. He specialises in forest micro-climatology and fire management. He is the corresponding author of Chapter 10.

**Hemant R. Ojha** is a research fellow at the School of Social Sciences at the University of New South Wales (UNSW). He is also a senior fellow at the University of Melbourne, and Chair of South Asia Institute of Advanced Studies (SIAS), Kathmandu, Nepal. He specialises in development studies, climate change and public policy. He is a co-author of Chapter 1.

**Hironori Okuda** is an external cooperation coordinator at Kansai Research Center, Forestry and Forest Products Research Institute. He specialises in forest policy and the development policy of the mountain village. He is the corresponding author of Chapter 18.

**Rose Jane J. Peras** is an Assistant Professor at the Department of Social Forestry and Forest Governance, College of Forestry and Natural Resources, University of the Philippines Los Baños. She specialises in forest and natural resource governance and climate change studies. She is a co-author of Chapter 11.

**Juan M. Pulhin** is a Professor and Dean of the College of Forestry and Natural Resources, University of the Philippines Los Baños. He specialises in social forestry, natural resource governance and the human dimension of climate change. He is the corresponding author of Chapter 11 and co-author of Chapter 19.

**Haruo Saito** is an Assistant Professor at the University of Tokyo Forests, Graduate School for Agricultural and Life Sciences, the University of Tokyo. He specialises in forest policy and human–forest relationships. He is the corresponding author of Chapter 17.

**Mustofa Agung Sardjono** is a Professor at the Faculty of Forestry, Mulawarman University, Indonesia. He specialises in social forestry, forest sociology and forest politics. He is the corresponding author of Chapter 7.

**W.D. Lakmi Saubhagya** is an Assistant Director (Administrative Service) in the Ministry of Child Development & Women's Affairs, Sri Lanka. She conducts research on social harmony, gender issues and child protection. She is a co-author of Chapter 5.

**Dietrich Schmidt-Vogt** is a Professor at the Kunming Institute of Botany, Chinese Academy of Sciences, Kunming, China, and specialises in natural resources management and land use change. He is a co-author of Chapter 6.

**Shubhechchha Sharma** is pursuing a master's degree in Natural Resources Management in Asian Institute of Technology, Thailand. She is doing her thesis on 'Influence of REDD+ Projects on Robustness of Community Forestry of Nepal'. She is a co-author of Chapter 19.

**Taro Takemoto** is an Assistant Professor at the Graduate School for Agricultural and Life Sciences, the University of Tokyo, specialises in forest policy and forest history, and is a co-author of Chapter 17.

**Maricel A. Tapia** is a PhD student in Environmental Science and Assistant Professor at the College of Forestry and Natural Resources, both at University of the Philippines Los Baños. She works on community-based natural resources management and vulnerability and adaptation to climate change. She is a co-author of Chapter 11.

**Tran Nam Thang** completed his PhD from Asian Institute of Technology, Thailand. He is now a Lecturer in the Forestry Faculty, Hue University of Agriculture and Forestry, Vietnam. He is the corresponding author of Chapter 13.

**P.K. Viswanathan** is an Associate Professor at Gujarat Institute of Development Research, Ahmedabad, India. His research interests include institutional and governance aspects of NRM; collective action and sustainable livelihood outcomes in smallholder agriculture; ecological economics of mangrove restoration and community benefits; industrialisation and its impacts on natural resources and ecosystems, etc. He is the corresponding author of Chapter 3.

**Edward L. Webb** is an Associate Professor in the Department of Biological Sciences at the National University of Singapore, and corresponding author of Chapter 12 and co-author of Chapter 19.

**Huilan Wei** is a Professor in the School of Economics at Lanzhou University, Lanzhou, P.R. China, and co-author of Chapter 15.

**Yolamalinda** is a Lecturer at Padang Teacher Training Institute, field of economic teaching. She obtained her MS degree in Integrated Natural Resources Management from Graduate Program, Andalas University. She also serves as an expert for provincial government on competitive commodity development of West Sumatra Province. She is a co-author of Chapter 9.

**Yonariza** is a Professor in Forest Resources Management at Andalas University, Padang, Indonesia. He obtained his PhD in Natural Resources Management from Asian Institute of Technology, Thailand and specialises his work on forest governance in Southeast Asia. He is the corresponding author of Chapter 9.

**Youn Yeo-Chang** is a Professor of Ecological Economics and Forest Policy at Seoul National University. His research interests are governance of common pool resources and forest policy analysis. He is the corresponding author of Chapter 16.

**Ting Zhu** is a Lecturer in the School of Economics at Shanghai University, Shanghai, P.R. China. She specialises in institutional economics and mechanism design, economics of population, resource and environment, common pool resources governance, etc. She is a co-author of Chapter 15.

# Index